MATH.
STAT.
LIBRARY

This book presents a study of various problems related to arrangements of lines, segments, or curves in the plane. It starts with a discussion of Davenport-Schinzel sequences, which have been applied to obtain optimal or almost-optimal bounds for numerous combinatorial as well as algorithmic problems involving arrangements. The main result here establishes almost tight bounds on the maximal length of (n,s)–Davenport-Schinzel sequences.

The second problem studied is: given a collection of "red" Jordan arcs and another of "blue" arcs, determine whether any red arc intersects any blue arc. Some fast deterministic algorithms are presented, along with applications to many other problems, including collision detection.

Next, a partitioning algorithm is presented that improves the time complexity of a variety of problems involving lines or segments in the plane. Several applications are discussed, most importantly a fast deterministic algorithm to construct a family of spanning trees with low stabbing number. This is shown to be a versatile tool that provides efficient algorithms for a variety of query-type problems, including ray shooting in an arrangement of segments, polygon containment and implicit hidden surface removal.

Researchers in computational and combinatorial geometry should find much to interest them in this book.

Intersection and Decomposition Algorithms for Planar Arrangements

Intersection and Decomposition Algorithms for Planar Arrangements

PANKAJ K. AGARWAL

Department of Computer Science,
Duke University

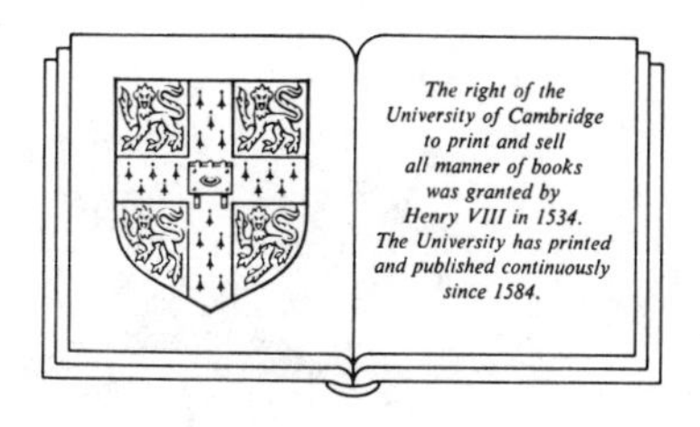

CAMBRIDGE UNIVERSITY PRESS

Cambridge

New York   Port Chester   Melbourne   Sydney

Published by the Press Syndicate of the University of Cambridge
The Pitt Building, Trumpington Street, Cambridge CB2 1RP
40 West 20th Street, New York, NY 10011, USA
10 Stamford Road, Oakleigh, Melbourne 3166, Australia

First published 1991

Printed in the United States of America

Library of Congress Cataloging in Publication applied for.

ISBN 0-521-40446-0 hardback

PREFACE

Computational geometry emerged as a scientific discipline about fifteen years back. It has developed very rapidly during the last few years because of its applications in areas such as robotics, computer graphics, VLSI, CAD, solid modeling, etc. Although it primarily concerns designing efficient algorithms for geometric problems, the algorithmic and combinatorial questions are so interwined in many problems that it is almost impossible to separate one from another. One such area in computational geometry is *arrangements* of lines or curves in the plane, or of hyperplanes and surfaces in higher dimensions. Recently arrangements have received a considerable amount of attention because several fundamental geometric problems can be formulated in terms of arrangements.

This book, based on the author's PhD thesis, studies several algorithmic as well as combinatorial problems involving arrangements of arcs in the plane. The first problem studied here concerns obtaining sharp bounds on the maximum length of *Davenport-Schinzel sequences.* These sequences have been applied to numerous combinatorial as well as algorithmic problems involving arrangements of arcs.

In Chapter 3 we study the *red-blue intersection* problem and its applications to other problems, including collision detection.

Chapter 4 and 5 discuss the following decomposition problem: Given a set of n lines in the plane and a parameter $r \leq n$, decompose the plane into $O(r^2)$ triangles, each of which intersects at most n/r original lines. Chapter 4 presents an efficient algorithm for computing such a partitioning. The application of this algorithm to various other problems is the topic of Chapter 5. The significance of this partitioning algorithm is evident from the problems discussed in Chapter 5.

Finally, in Chapter 6 we show that a spanning tree of a set of points in the plane with the property that each line intersects only few edges of the tree is a versatile tool and that it provides fast procedures for a variety of query-type problems including ray shooting and implicit hidden surface removal.

The open problems mentioned at the end of each chapter suggest further research problems in these areas.

I would like to express my deep sense of gratitude to my thesis advisor

Micha Sharir. It is difficult to imagine how I would have completed this work without his guidance, encouragement and unwavering support. I would like to thank him for reading various versions of the manuscript. Thanks are also due to Boris Aronov, Bernard Chazelle, Jiří Matoušek, Ricky Pollack, and two anonymous referees for several useful comments, and to Lauren Cowles for a careful reading of the manuscript.

Pankaj K. Agarwal

Contents

List of Figures

Chapter 1

Introduction

Computational geometry emerged as a scientific discipline about fifteen years ago. In its initial stages, it evolved around planar searching and sorting problems. Typical problems were convex hull, point location, closest point, Voronoi diagram, and orthogonal range searching, all of which can be easily cast into generalized two-dimensional searching and sorting. Gradually, the field expanded to include problems whose successful solution requires the application of sophisticated mathematical tools drawn from algebra, probability, topology, differential geometry, extremal graph theory and many other fields. Now a days the difference between combinatorial and computational geometry is becoming obscure. In many problems the combinatorial and algorithmic questions are so interwined that it is nearly impossible to separate one from the other, and progress in one of them leads to a satisfactory solution of the other. One such area in computational geometry, where algorithmic questions rely heavily on various mathematical properties, is the study of *arrangements* of lines or curves in the plane, or of hyperplanes or more general surfaces in higher dimensions. Many fundamental geometric problems can be formulated in terms of arrangements.

In this book, we study several problems involving arrangements of lines, segments, or curves in the plane. The book has two aims: to answer some combinatorial as well as algorithmic questions related to arrangements and to obtain fast algorithms for several other problems by formulating them in terms of arrangements.

We begin by defining arrangements and discussing some of the known results on arrangements. The remainder of the introduction gives a brief overview of the results described in the later chapters and of the tools used to obtain these results — Davenport-Schinzel sequences, random sampling and deterministic partitioning, and spanning trees with low stabbing number.

1.1 Arrangements

Let $\mathcal{L}$ be a finite set of lines in the Euclidean plane $\mathbb{R}^2$. The *arrangement* $\mathcal{A}(\mathcal{L})$ of $\mathcal{L}$ is the partition of the plane formed by $\mathcal{L}$, whose *vertices* are intersection points of lines in $\mathcal{L}$, *edges* are maximal connected (open) portions of lines in $\mathcal{L}$ not containing any vertex, and *faces* are maximal connected (open) portions of the plane not meeting any vertex or edge (see figure 1.1). The notion of arrangement can be easily generalized to curves in the plane, or to hyperplanes or hypersurfaces in higher dimensions. In $\mathbb{R}^d$, the arrangement $\mathcal{A}(\mathcal{H})$ of a finite family $\mathcal{H}$ of $(d-1)$-dimensional hyperplanes consists of open convex d-dimensional polyhedra (also referred to as *cells*) and various relatively open convex k-dimensional polyhedral faces bounding them, for $0 \le k \le d-1$.

The *combinatorial complexity* of an arrangement in the plane is the number of its vertices, edges and faces, and in higher dimensions it is the total number of faces of all dimensions. It is easy to see that an arrangement of n lines in $\mathbb{R}^2$ lying in general position has $\Theta(n^2)$ vertices, edges and faces. By generalizing this result to higher dimensions, it is shown that the combinatorial complexity of an arrangement of n hyperplanes in $\mathbb{R}^d$ is $\Theta(n^d)$. If Γ is a set of n hypersurfaces in $\mathbb{R}^d$, such that every d of them intersect in at most $O(1)$ points, e.g. graphs of $(d-1)$–variate polynomials of bounded degree, then the combinatorial complexity of $\mathcal{A}(\Gamma)$ is also $\Theta(n^d)$.

Arrangements of lines and "pseudo-lines" have been studied from various mathematical points of view for a very long time (see Steiner [123], von Staudt [122], Sylvester [126], Levi [85], Melchior [97] for some early works). The first systematic exposition of arrangements is due to Grünbaum [69], [71]. Arrangements in higher dimensions have received considerably less attention, and not much is known about them except for some basic results.

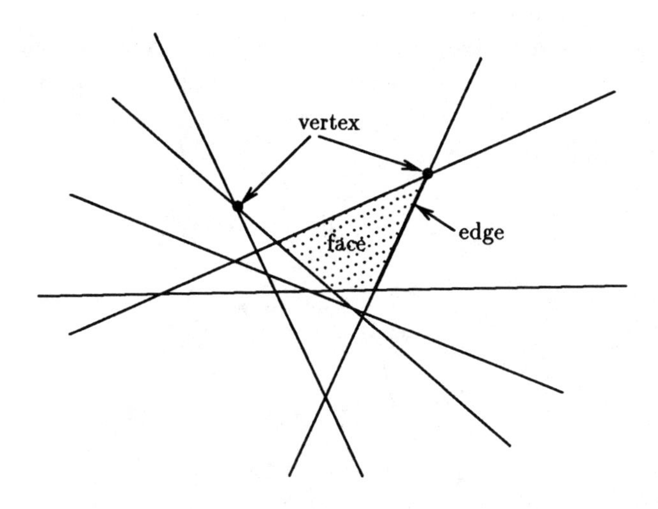

Figure 1.1: An arrangement of lines

Interested readers can find details on d-dimensional arrangements in [69], [70]. The recent book of Edelsbrunner [52] also provides an extensive survey of known results in this area.

The algorithmic study of arrangements, however, has started only recently, primarily because of their significance in many fundamental geometric problems. The arrangement of a set of n lines in the plane can be computed in time $O(n^2 \log n)$ using a line sweep approach (see [115]); this has been improved to $O(n^2)$ by Edelsbrunner et al. [62] (see also [30]). In fact the algorithm of [62] computes, in time $O(n^d)$, the arrangement of a given set of n hyperplanes in $\mathbb{R}^d$. Recently, Edelsbrunner and Guibas [54] presented another $O(n^2)$ algorithm to construct the arrangement of a given set of n lines using "topological sweep", which requires only $O(n)$ working storage. Constructing arrangements of curves is much more difficult; Edelsbrunner et al. [56] have generalized the approach of [62] to construct arrangements of arcs of some simple shape in the plane (in particular, it is required that any pair of these arcs intersect in at most some constant number of points). Under the assumption that certain basic operations on arcs, such as computing intersection points of a pair of arcs and points of vertical tangency of a given arc, can be performed in $O(1)$ time, the running time of their algorithm is almost (slightly superlinear–) quadratic. But unlike the algorithm of [62], this approach does not generalize to constructing arrangements of surfaces in higher dimensions.

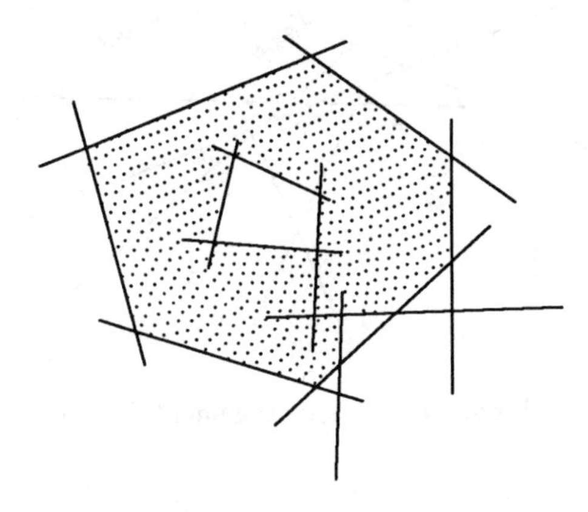

Figure 1.2: A non-convex face in an arrangement of segments

In several applications it suffices to compute a single face or few faces in an arrangement. For arrangements of lines it is obvious that a single face can have at most n edges, because each face is convex, and a single face can be easily computed in $O(n \log n)$ time using a divide and conquer algorithm [114]. But faces in arrangements of segments or of general arcs are not necessarily convex (see figure 1.1), and in fact the result of Wiernik and Sharir [133] shows that a single face in an arrangement of n segments can have $\Omega(n\alpha(n))$ edges in the worst case (see also [121]), where $\alpha(n)$ is a functional inverse of Ackermann's function (see Chapter 2 for details on Ackermann's function). A matching upper bound has been obtained in Pollack et al. [105] using Davenport-Schinzel sequences, which has been extended to general arcs by Guibas et al. [75]. Guibas et al. have also given an efficient algorithm to compute a single face in arrangements of arcs under an appropriate model of computation.

In some other applications, we are required to compute $m > 1$ faces; one such application is described in Chapter 3. If $K(m,n)$ denotes the maximum number of edges in m distinct faces of arrangements of n lines, then obviously $K(m,n) = O(mn)$. But a somewhat surprising result of Canham [16] shows that $K(m,n) = O(m^2 + n)$, implying that $K(\sqrt{n}, n)$ is not larger than $K(1,m)$ by more than a constant factor. Since then a great deal of effort has been made to obtain tight bounds on $K(m,n)$ (see [52]). Edelsbrunner and Welzl [66] proved that $K(m,n) = \Omega(m^{2/3}n^{2/3} + n)$, and very recently Clarkson et al. [38] have shown that indeed $K(m,n) = \Theta(m^{2/3}n^{2/3} + n)$. Similar, but not fully tight, bounds on the complexity of many faces in arrangements of segments have been obtained in [59], [4]. Edelsbrunner et al. [59] have given randomized algorithms to compute m distinct faces in arrangements of lines or segments (see also [55]).

Chapter 5 describes an $O(m^{2/3}n^{2/3} \log^{5/3} n \log^{\omega/3} \frac{m}{\sqrt{n}} + (m+n) \log n)$ deterministic algorithm for computing the faces in arrangements of n lines in the plane, containing m given points (cf. Section 5.3), where ω is a constant < 3.33. We also present an $O(m^{2/3}n^{2/3} \log n \log^{\omega/3+1} \frac{n}{\sqrt{m}} + n \log^3 n + m \log n)$ algorithm to compute the faces in arrangements of n segments in the plane, containing m given points (cf. Section 5.4). These algorithms are deterministic, faster than previously known (randomized) algorithms, and optimal up to a polylog factor.

These results use Davenport-Schinzel sequences, random sampling and other techniques, which we will discuss in Section 1.2.

A very simple example of a geometric problem that can be formulated in terms of arrangements is reporting or counting intersections in a given collection $\mathcal{G}$ of segments, because this problem is equivalent to asking for reporting/counting the vertices of $\mathcal{A}(\mathcal{G})$. A more interesting application area is *motion planning*, in which, given a set of obstacles in the plane, or in space, and a moving object B, we wish to plan a collision-free path for B from a given initial placement to a specified final placement. A well known technique for solving this problem is to represent the placements of B at which it makes contact with obstacles as a collection Γ of surfaces in an appropriate parametric "configuration space", so that B can be moved from the initial placement to the final placement without colliding with the obstacles if and only if the points of the configuration space representing the initial and final placements of B lie in the same cell of $\mathcal{A}(\Gamma)$. Thus the problem can be reduced to computing a single cell in $\mathcal{A}(\Gamma)$, and locating a point in it [105], [75]. Using this observation and efficient algorithms to compute a single cell in arrangements of arcs, a number of fast algorithms have been developed for various motion planning problems (see e.g. [75], [5]).

Many problems in computational geometry can be reduced to computing the *lower envelope* of the arrangement of a given collection Γ of surfaces or surface-patches in $\mathbb{R}^{d+1}$. We can view each surface $\gamma_i \in \Gamma$ as the graph of a partially defined function $x_{d+1} = f_i(x_1, \ldots, x_d)$, whose domain is the projection of γ_i on the hyperplane $x_{d+1} = 0$. The lower envelope $\mathcal{M}_\Gamma(x_1, \ldots, x_d)$ of $\mathcal{A}(\Gamma)$ is defined as the pointwise minimum of these functions, that is

$$\mathcal{M}_\Gamma(x_1, \ldots, x_d) = \min_{1 \le i \le n} \{f_i(x_1, \ldots, x_n)\}.$$

The combinatorial structure of $\mathcal{M}_\Gamma$ is a cell complex in $\mathbb{R}^d$ such that within each cell the lower envelope is attained by a single function f_i (see figure 1.1). This is basically the cell decomposition of the orthogonal projection of $\mathcal{M}_\Gamma$ on $x_{d+1} = 0$. Similarly, we can define the *upper envelope* of the arrangement of a collection of surfaces.

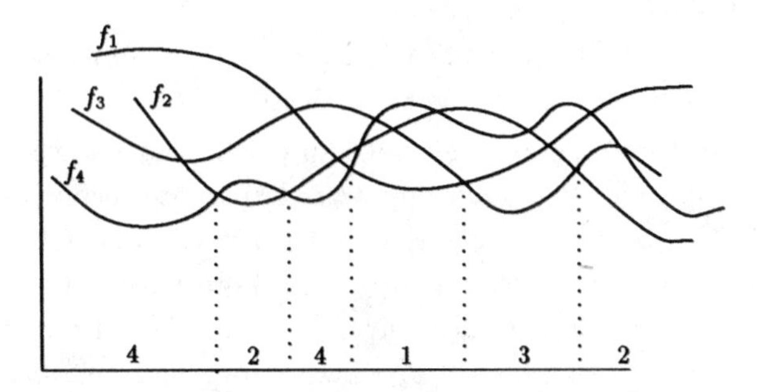

Figure 1.3: Lower envelope of a set of curves

It has been shown in [63] that Voronoi diagrams and all of its generalizations in $\mathbb{R}^d$ can be viewed as the lower envelope of arrangements of a collection of surfaces in $\mathbb{R}^{d+1}$, where the surfaces depend on the distance function used to define the Voronoi diagram, and on the shape of sites for which we want to compute the diagram. Other problems that can be reduced to computing the lower envelope of arrangements include transversals [64], [79], visibility [44], motion planning [83], and various questions in dynamic computational geometry [6]. Expressing problems in terms of arrangements also lead to efficient algorithms for computing stabbing lines for a set of segments in the plane [61], computing minimum volume simplices for a set of points in $\mathbb{R}^d$ [62], half-plane range searching, and many others.

Another application studied in this thesis is *red-blue intersection detection* problem, which is defined as follows: Given a collection Γ of n "red" Jordan arcs and another collection Γ' of m "blue" Jordan arcs, determine whether any arc of Γ intersects any arc of Γ'. Chapter 3 describes several efficient algorithms for detecting a red-blue intersection, by formulating it in terms of arrangements of arcs. For the general case, we reduce the problem to that of computing up to $2(m+n)$ faces of $\mathcal{A}(\Gamma)$ and $\mathcal{A}(\Gamma')$. In some special cases the running time can be significantly improved, e.g. if the arcs in Γ' form the boundary of a simply connected region, then the problem can be reduced to detecting an intersection between Γ' and a single cell of $\mathcal{A}(\Gamma)$. These algorithms rely heavily on Davenport-Schinzel sequences, which we describe and analyze in the next section.

Recently several researchers have considered the following decomposition problem involving arrangements of hyperplanes:

> *Given a set $\mathcal{H}$ of n hyperplanes in $\mathbb{R}^d$, and a parameter $1 < r < n$, partition the space into $O(r^d)$ simplices, each of which intersects $O(\frac{n}{r})$ hyperplanes of $\mathcal{H}$.*

This decomposition algorithm has turned out to be useful in a variety of geometric problems involving arrangements of lines in the plane, or hyperplanes in higher dimensions. The existence of a decomposition with somewhat weaker properties was originally proved in [78], [36], using probabilistic arguments. Later Chazelle and Friedman [26] gave a deterministic algorithm that runs in polynomial time. For higher dimensions, this is still the best known algorithm, but in the case of two-dimensions, faster algorithms have been developed by exploiting several properties of arrangements of lines in the plane. One such algorithm with a time complexity of $O(nr^2 \log^2 r)$ has been developed by Matoušek [89].

In Chapter 4 of this book we present a deterministic algorithm which runs in roughly nr time, and thus is almost an order of magnitude faster than that of Matoušek in terms of r. This improvement is quite crucial in many geometric problems (see Chapter 4 for details). We demonstrate the versatility and significance of our partitioning algorithm by describing fast solutions for numerous geometric problems, based on it. These applications include computing many faces in arrangements of lines or segments, counting intersections in a set of segments, computing incidences between points and lines, and computing spanning trees of low stabbing number.

In Chapter 5 we present fast algorithms for a variety of problems involving arrangements of segments, using a data structure called *spanning tree with low stabbing number*. The most interesting problem considered there is *ray shooting in arrangements of segments*: given an arrangement of segments in the plane, preprocess it so that the first intersection of a query ray with one of the given segments can be quickly reported. Ray shooting has applications in many other problems, including hidden surface removal [102], and computation a single cell in arrangements of triangles in $\mathbb{R}^3$ [5]. We will describe ray shooting and other related problems in greater detail in Section 1.4.

1.2 Davenport-Schinzel Sequences

Davenport-Schinzel sequences have a strong combinatorial structure that arises in many geometric problems. They were originally introduced by Davenport and Schinzel [48] in connection with solutions to linear differential equations. A sequence, $U = (u_1, u_2, \ldots, u_m)$, composed of n distinct symbols is a *Davenport-Schinzel* sequence of order s, for $s > 0$ (an (n,s)–DS sequence for short), if it satisfies the following two conditions:

(i) $\forall i < m,\ u_i \neq u_{i+1}$.

(ii) There do not exist $s+2$ indices $1 \leq i_1 < i_2 < \cdots < i_{s+2} \leq m$ such that

$$\begin{aligned} u_{i_1} &= u_{i_3} &= \cdots = a \\ u_{i_2} &= u_{i_4} &= \cdots = b \end{aligned}$$

and $a \neq b$.

These sequences characterize the combinatorial structure of the lower envelope of arrangements of n x-monotone curves in the plane (or n univariate functions), in the sense that the "lower-envelope sequence" of indices of curves as they appear along the envelope is a (n,s)–DS-sequence, where s is the maximum number of intersections between any pair of curves. Moreover, Atallah [6] proved that, for any given (n,s)–DS sequence, there exists a collection of n x-monotone curves, whose lower envelope sequence is the same as the given Davenport-Schinzel sequence.

Let $\lambda_s(n)$ denote the maximum length of a (n,s)–DS sequence. It is easy to see that $\lambda_1(n) = n$ and $\lambda_2(n) = 2n-1$. Hart and Sharir [77] proved that $\lambda_3(n) = \Theta(n\alpha(n))$, which implies that, for $s \geq 3$, $\lambda_s(n)$ is indeed non-linear. Our results in Chapter 2 prove that $\lambda_4(n) = \Theta(n2^{\alpha(n)})$, and

$$\begin{aligned} \lambda_{2s+2}(n) &= O\left(n \cdot 2^{(\alpha(n))^{s(1+o(1))}}\right) \\ \lambda_{2s+3}(n) &= O\left(n \cdot 2^{(\alpha(n))^{s(1+o(1))} \cdot \log \alpha(n)}\right) \\ \lambda_{2s+2}(n) &= \Omega\left(n \cdot 2^{K_s(\alpha(n))^s}\right) \end{aligned}$$

where, $K_s = \dfrac{1}{(s-1)!}$. Thus our lower and upper bounds are almost tight. For even values of s they are almost identical except for the constant K_s and lower order terms, but for odd values of s, the gap is slightly larger. Our upper bounds imply that $\lambda_s(n)$, for any fixed value of s, is almost linear, because $\alpha(n) \leq 5$ for all practical purposes (see Chapter 2). The proofs presented in this chapter are fairly complex, and exploit several properties of Ackermann's function and related functions.

This close relationship between Davenport-Schinzel sequences and lower envelopes of arrangements has made them a tool of of central significance in computational and combinatorial geometry, and numerous applications of

DS–sequences have been obtained, in diverse areas including arrangements [75], [103], [53], [57], [95], motion planning [101], [83], [81], shortest path [9], [35], visibility [44], transversals [64], [57], Voronoi diagrams [57] and many more. Guibas et al. [75] showed that the maximum number of edges in a single cell of an arrangement of n arcs is $O(\lambda_{s+2}(n))$, where s is the maximum number of intersections between any pair of arcs; if all arcs are closed curves or unbounded arcs, the complexity reduces to $O(\lambda_s(n))$. Using Davenport-Schinzel sequences, Edelsbrunner et al. [56] were able to prove that the m distinct faces in arrangements of n arcs have at most $O(m\sqrt{\lambda_{s+2}(n)})$ edges. Guibas et al. have described an $O(\lambda_{s+2}(n)\log^2 n)$ algorithm to compute a single cell in arrangements of n arcs, where s is as above; the algorithm can be generalized for computing more than one face (see [59]). Based on the result of Hart and Sharir [77], Pach and Sharir [103] have shown that the maximum number of facets in the the lower envelope of arrangements of n d-simplices in $\mathbb{R}^{d+1}$ is $\Theta(n^d\alpha(n))$ (see also [53]). An optimal algorithm for computing the lower envelope of arrangements of n triangles in $\mathbb{R}^3$ is described in [57].

Applications of DS-sequences in motion planning include an algorithm to plan a collision free path (allowing rotation) for a convex k-gon in presence of polygonal obstacles (cf. [81]), an efficient algorithm to separate a simple m-gon from another simple n-gon (cf. [105]), and a fast procedure to compute a shortest path in $\mathbb{R}^3$ in the presence of two convex polyhedral obstacles.

In Chapter 3, we describe efficient algorithms, using Davenport-Schinzel sequences, for the red-blue intersection detection problem defined in the previous section. We start with an almost linear algorithm for a special case, where the arcs in Γ' form the boundary of a simply connected region. Then we show that such an algorithm provides a fast procedure for the collision detection problem: given a set O of obstacles, a moving object B and a path Π, determine whether B collides with any obstacle of O while moving along the path Π. Some of the motion planning problems can be reduced to the collision detection problem by showing that there exists a canonical motion such that B can be moved from its initial position to a given final position, if and only if the canonical motion is collision free. One such example is given in Maddila and Yap [87], who considered the problem of moving a simple n-gon in an L-shaped corridor. They gave an $O(n^2)$ algorithm for this problem, which can be improved to $O(\lambda_s(n) \log^2 n)$, for some small value of s, using the above red-blue intersection detection algorithm.

Next, we consider red-blue intersection detection for general arcs, and describe a deterministic algorithm with $O((m\sqrt{\lambda_{s+2}(n)}+n\sqrt{\lambda_{s+2}(m)}) \log^{3/2}(m+n))$ running time. Finally, we obtain faster algorithms for some special cases by computing up to $2(m+n)$ faces of $\mathcal{A}(\Gamma)$ and $\mathcal{A}(\Gamma')$ and searching for an intersection only in these faces.

Interested readers can refer to [120], [77] and [119] for other applications, and for a survey of known results on Davenport-Schinzel sequences.

1.3 Random Sampling and Deterministic Partitioning

In the last few years several Las-Vegas algorithms have been developed for problems like Voronoi diagrams, segment intersections, motion planning, triangulation, k-sets, simplex range searching, point location, and many more ([5], [34], [36], [37], [39], [40], [59], [73], [74], [98], [99], [108], [109]). Most of these algorithms are based on a divide and conquer approach, and use ϵ-nets [78] or the random sampling technique of [36] (see also [109]).

Originally, random sampling was applied to specific problems like range searching [78] or k-sets [36]. Later it emerged as a much more general paradigm, which could be applied to obtain fast divide and conquer algorithms for a variety of problems. Random sampling can also be used to produce partition trees for a given set of points in the plane or in higher dimensions, and therefore provides fast procedures for various query type problems, including range searching [78], point location [37], and ray shooting [73].

The ϵ-net theory shows that, for a given set X of n objects, a set $\mathcal{R} \subseteq 2^X$ of *ranges* with finite *Vapnik-Chervonenkis dimension* (that is, the largest integer d such that there exists a subset A of X of cardinality d with the property that $\{A \cap r | r \in \mathcal{R}\} = 2^A$; notice that it does not depend on n, and it can be shown that many geometric cases have finite Vapnik-Chervonenkis dimension, see [78] for details) and an integer $1 \leq r \leq n$, there exists a subset $N \subset X$ of size r (called *ϵ-net*), such that if $\tau \in \mathcal{R}$ and $\tau \cap N =$

$\emptyset$, then $|\tau| = O(\frac{n}{r}\log r)$. Moreover, a random sample N of size r will be an ϵ-net with high probability. A similar result was independently obtained by Clarkson [36]. Roughly, it states that if a random sample N of size r is drawn from our set X and the underlying geometric space is partitioned into cells with the property that (i) each cell can be defined in terms of only a "constant" number of elements of N, and (ii) each cell does not meet any element of N, then, with high probability, no cell meets more than $O(\frac{n}{r}\log r)$ elements of X. Random sampling techniques allow us to split an original problem involving the set X into subproblems of small size, each of which can be solved recursively. For example, if the objects are "lines in the plane" and an element of $\mathcal{R}$ is the "set of lines intersecting a given triangle", then these results imply that, for a given set $\mathcal{L}$ of n lines in the plane, there exists a subset $N \subset \mathcal{L}$ of size r such that any triangle that misses all lines in N intersects at most $O(\frac{n}{r}\log r)$ lines of $\mathcal{L}$; moreover, a random sample $N \subset \mathcal{L}$ of size r will have this property with high probability. A problem involving the set $\mathcal{L}$ of lines can then be split into subproblems by computing the arrangement of the lines in N and by triangulating each of its faces; each resulting triangle Δ induces a subproblem involving only those $O(\frac{n}{r}\log r)$ lines of $\mathcal{L}$ intersecting Δ. As an example of a problem that benefits from this set-up, consider *Hopcroft's problem* (see [45]): Given m points and n lines in the plane, does any point lie on any line? Applying the above divide and conquer strategy to the given lines, and partitioning the points among the resulting triangles, each subproblem involves only the points that fall within some triangle and the lines cutting it. The same approach works for other geometric objects in the plane as well as in higher dimensions.

These techniques raise two obvious questions. The first question is whether we can determinize the random sampling based algorithms without an enormous increase in their time complexity. A second question is whether the factor $\log r$ can be eliminated from the above bound. (Intuitively, we can think of our drawing as picking every $O(\frac{n}{r})$-th element of X, so if a range τ misses all the sample objects, we do not expect it to meet many more than $\frac{n}{r}$ objects of X.) Recently Chazelle and Friedman [26] gave a deterministic construction for obtaining a "good" sample. In certain cases, such as the case of a set H of n hyperplanes in $\mathbb{R}^d$, they were able to get rid of the $\log r$ factor. Although their algorithm runs in polynomial time, it is quite inefficient for most of the applications. In the case of hyperplanes just mentioned, it requires $O(rn^{d(d+3)/2+1})$ time to partition $\mathbb{R}^d$ into $O(r^d)$ simplices, so that no simplex meets more than $O(\frac{n}{r})$ of the given hyperplanes. This has motivated researchers to seek fast deterministic procedures for special cases. Significant progress in this direction was made by Matoušek [89], who showed that, given a set of n lines in the plane and a parameter r, the plane can be partitioned, in time $O(nr \log^2 r)$, into $O(r^2)$ triangles, none of which meets more than $O(\frac{n}{r})$ lines. For constant values of r, this algorithm is optimal and can be used to determinize several randomized algorithms involving lines or segments. But this algorithm is still inefficient for larger values of r, and in many applications it is useful to choose large value of r. In such cases, it turns out that, to obtain improved (and sometimes close-to-optimal) performance, we require a partitioning algorithm with roughly nr time complexity (see Chapter 5 for details).

In Chapter 4 we obtain such an improvement by showing that, given a set $\mathcal{L}$ of n lines in the plane and a parameter r, we can partition the plane, in $O(nr \log n \log^{\omega} r)$ time, into $O(r^2)$ triangles so that no triangle meets more than $O(\frac{n}{r})$ lines of $\mathcal{L}$, where ω is a constant < 3.33. This algorithm consists of two phases. The first phase produces $O(r^2 \log^{\omega} r)$ triangles, each meeting at most $\frac{n}{r}$ lines of $\mathcal{L}$, and the second phase reduces the number of triangles to $O(r^2)$, still maintaining the property that no triangle meets more than $O(\frac{n}{r})$ lines of $\mathcal{L}$. The first phase of the algorithm is based on a recursive partitioning, which at every level of recursion partitions a convex quadrilateral into subquadrilaterals, each meeting few lines of $\mathcal{L}$ and containing few intersection points of $\mathcal{L}$. Although the algorithm is quite simple, its analysis is rather involved.

We demonstrate the significance of our partitioning algorithm in Chapter 5 by determinizing and improving the running time of the previously best known algorithms for several geometric problems. These applications include $O(m^{2/3}n^{2/3} \log^{2/3} n \log^{\omega/3} \frac{m}{\sqrt{n}} + (m+n) \log n)$ algorithm to compute incidences between m points and n lines in the plane, an $O(m^{2/3}n^{2/3} \cdot \log^{5/3} n \log^{\omega/3} \frac{m}{\sqrt{n}} + (m+n) \log n)$ algorithm to compute m distinct faces in arrangements of n lines, and an $O(n^{4/3} \log^{(\omega+2)/3} n)$ algorithm to count the number of intersections between n segments. See Chapter 5 for a longer list of all our applications. A significant application of our algorithm is to compute a family $\mathbf{T}$ of $O(\log n)$ spanning trees of a given set of n points with the property that, for any given line ℓ, there is a tree $\mathcal{T} \in \mathbf{T}$, such that ℓ intersects at most $O(\sqrt{n})$ edges of $\mathcal{T}$. These trees and their significance are discussed in Chapter 6. A brief introduction is given in the following subsection.

1.4 Spanning Trees with Low Stabbing Number

The *stabbing number* of a tree $\mathcal{T}$ in $\mathbb{R}^2$ is the maximum number of its edges that a line can cross, and the stabbing number of a family of trees $\mathbf{T}$ is s if for each line ℓ there is a tree $\mathcal{T} \in \mathbf{T}$ such that ℓ intersects at most s edges of $\mathcal{T}$. It has been shown that for a set S of n points in $\mathbb{R}^2$, there exists a spanning tree of S with stabbing number $O(\sqrt{n})$ [32]. A spanning tree $\mathcal{T}$ of S, with stabbing number s, can be easily converted into a spanning path of S with stabbing number $\leq 2s$ (see [32]), therefore S has a spanning path as well with low stabbing number. Spanning trees with low stabbing number were originally introduced by Welzl [132] to obtain an algorithm for simplex range searching that answers a query in almost optimal time using $O(n)$ space. Specifically, the algorithm preprocesses a given S in the plane into a data structure of linear size so that all K points of S lying in a query triangle can be reported (resp. counted) in time $O(\sqrt{n}\log n+K)$ (resp. $O(\sqrt{n}\log n)$). The basic idea underlying his approach is as follows. For the sake of exposition, assume that we want to count the number of points lying in a query half-plane; the number points lying in a triangle can be computed in a similar way. Let Π be a spanning path of S with stabbing number s. For a query half-plane h, let $q_1, \ldots, q_k$ be the intersection points of h and Π. Since the portion of Π between two consecutive intersection points lies either completely above or completely below h, either all points of S lying on that portion should be counted, or none of them should. Hence, after computing the intersection points $q_1, \ldots, q_k$, the number of points lying above h can be counted by summing up the sizes of $O(s)$ subpaths of Π, which can be done in time $O(s)$. Soon after Welzl's paper appeared, Edelsbrunner et al. [55] applied the spanning tree structure to obtain a fast algorithm for preprocessing a set $\mathcal{L}$ of lines so that, for a query point p, the face of $\mathcal{A}(\mathcal{L})$

containing p can be computed quickly. The main challenge in both of these papers is to use only roughly linear space (i.e. $O(n \log^{O(1)} n)$ space), because if we allow quadratic space, then a query can be easily answered in $O(\log n)$ time (see [52], [62], [54]).

In Chapter 5 of this book, as one of the applications of our partitioning algorithm, we present an efficient algorithm to compute a family $\mathbf{T}$ of $O(\log n)$ spanning trees of a given set of n points with stabbing number $O(\sqrt{n})$. We also show that spanning trees with low stabbing number are a versatile tool that provides fast algorithms for a number of diverse problems. Chapter 6 is devoted to the applications of spanning trees with low stabbing number. The most significant application described there is the "ray shooting" in arrangements of segments. In the special case, when given segments form the boundary of a simply connected region, ray shooting queries can be answered in $O(\log n)$ time using $O(n)$ space and $O(n \log n)$ preprocessing [27] (see also [72]). For the general case, however, the ray shooting and other visibility problems are much harder even for non-intersecting segments. For example, a result of Suri and O'Rourke [125] shows that the portion of a non-simple polygon visible from a fixed edge can have $\Omega(n^4)$ edges on its boundary, while for simple polygons such a region is bounded by only $O(n)$ edges. Here, we present an algorithm that preprocesses a given set $\mathcal{G}$ of n (possibly intersecting) segments, in time $O(n^{3/2} \log^{\omega+1} n)$, into a data structure of size $O(n\alpha(n) \log^4 n)$ so that given a query ray ρ, the first intersection of ρ with $\mathcal{G}$ can be computed in $O(\sqrt{n\alpha(n)} \log^2 n)$ time. If the segments are non-intersecting, the query time reduces to $O(\sqrt{n} \log^2 n)$ and the space required is $O(n \log^3 n)$. We also establish a trade-off between the storage requirement and query time for the ray shooting problem. In particular, we show that a query can be answered in $O(\frac{n\alpha(n)}{\sqrt{m}} \log^2 \frac{n\alpha(n)}{\sqrt{m}} + \log n)$ time using $O(m)$ space. Ray shooting algorithms have been applied to obtain efficient procedures for contour tracing, visibility, computing a single cell in arrangements of segments in the plane [105] or of triangles in $\mathbb{R}^3$ [5].

Another problem that benefits from the spanning tree structure is *implicit point location* , which is a generalization of the well studied planar point location problem [82], [60], [111]. In the planar point location problem one is to preprocess a given planar subdivision must be preprocessed so that, for a query point, the face containing p can be computed efficiently.

The above algorithms preprocess the planar map with n vertices, in time $O(n \log n)$, into a data structure of linear size so that a query point can be located in $O(\log n)$ time. In the implicit point location, the map is defined as the arrangement (that is, overlay) of objects of simple shape, and we wish to compute some information related to their arrangement. For example, given n triangles in the plane, we wish to preprocess them so that, given a query point x we can quickly determine whether x lies in the union of the triangles. This problem was first studied by Guibas et al. [73]; we improve the running time of their algorithm by using the spanning tree structure, obtaining algorithms with preprocessing time similar to that for the ray shooting problem. Once again the query time can be reduced by allowing more space. Some other applications of the spanning tree structure are also described in Chapter 6.

Chapter 2

Davenport-Schinzel Sequences

2.1 Introduction

In this chapter we obtain optimal bounds for the maximal length $\lambda_4(n)$ of an $(n, 4)$ *Davenport-Schinzel Sequence* (a $DS(n, 4)$ sequence in short), and then extend them to improve and almost tighten the lower and upper bounds for $\lambda_s(n)$, $s > 4$. A $DS(n, s)$ sequence, $U = (u_1, \ldots, u_m)$ is a sequence composed of n distinct symbols which satisfies the following two conditions:

1. $\forall i < m$, $u_i \neq u_{i+1}$.

2. There do not exist $s + 2$ indices $1 \leq i_1 < i_2 \cdots < i_{s+2} \leq m$ such that

$$\begin{aligned} u_{i_1} &= u_{i_3} = u_{i_5} = \cdots = a, \\ u_{i_2} &= u_{i_4} = u_{i_6} = \cdots = b \end{aligned}$$

   and $a \neq b$.

We refer to s as the *order* of the sequence U. We write $|U| = m$ for the *length* of the sequence U; thus

$$\lambda_s(n) = \max\{|U| : U \text{ is a } DS(n, s) \text{ sequence}\}.$$

Davenport-Schinzel sequences have turned out to be of central significance in computational and combinatorial geometry and related areas, and have many applications in diverse areas including motion planning, shortest path, visibility, transversals, Voronoi diagrams, arrangements and many more; see [6], [9], [35], [44], [64], [59], [75], [77], [81], [83], [101], [103], [105], [113], [133]. Atallah has shown [6] that $DS(n, s)$ sequences provide a combinatorial characterization of the lower envelope of n continuous univariate functions, each pair of which intersect in at most s points. Thus $\lambda_s(n)$ is the maximum number of connected portions of the graph of n such functions that constitute their lower envelope. Since minimization of functions is a central operation in many geometric and other combinatorial problems, sharp estimates of $\lambda_s(n)$ yield sharp and often near-optimal bounds for the complexity of these problems. This, combined with the highly non-trivial and surprising form of the bounds on $\lambda_s(n)$, as given below, makes Davenport-Schinzel sequences very powerful and versatile tools.

The problem of estimating $\lambda_s(n)$ has been studied by several authors [48], [47], [110], [127], [6], [77], [117], [118]. It is easy to show that $\lambda_1(n) = n$ and $\lambda_2(n) = 2n - 1$. For $\lambda_3(n)$ Davenport and Schinzel [48] proved that $\lambda_3(n) = O(n \log n)$, which was subsequently improved to $O(n \frac{\log n}{\log \log n})$ by Davenport. For higher order DS-sequences, the first non-trivial upper bound was $\lambda_s(n) = O(n 2^{A_s \sqrt{\log n}})$ (cf. [48]), where A_s is a constant depending on s; it was later improved to $O(n \log^* n)$ by Szemerédi [127]. Recently Hart and Sharir [77] proved that $\lambda_3(n) = \Theta(n\alpha(n))$, thus showing that DS-sequences are non-linear. Here is a functional inverse of Ackermann's function and is very slowly growing. Their proof proceeds by first showing an equivalence between Davenport-Schinzel sequences and generalized path compressions on an arbitrary rooted tree with a certain post-order restriction on the compressions, and then proving that this version of path compression has complexity $\Theta(n\alpha(n))$. Soon after this paper Sharir [117] extended the techniques to prove that

$$\lambda_s(n) = O(n\alpha(n)^{O(\alpha(n)^{s-3})}) \qquad \text{for } s \geq 4.$$

The previously best known lower bounds for higher order sequences are also due to [118], who proved that

$$\lambda_{2s+1}(n) = \Omega(n(\alpha(n))^s) \qquad \text{for } s \geq 2.$$

Thus for $s \geq 4$ there is still a gap between the lower and upper bounds for $\lambda_s(n)$. In what follows, we first establish tight upper and lower bounds for $\lambda_4(n)$ and then obtain sharp, and almost tight, upper and lower bounds on $\lambda_s(n)$ for higher values of s, by generalizing the techniques used in the case of $\lambda_4(n)$.

The main results of this chapter are as follows:

(i) The maximal length of a $DS(n, 4)$ sequence is

$$\lambda_4(n) = \Theta(n \cdot 2^{\alpha(n)}).$$

(ii) An upper bound on the maximal length of a $DS(n, s)$ sequence is

$$\lambda_s(n) \leq \begin{cases} n \cdot 2^{(\alpha(n))^{\frac{s-2}{2}} + C_s(n)} & \text{if } s \text{ is even,} \\ n \cdot 2^{(\alpha(n))^{\frac{s-3}{2}} \log(\alpha(n)) + C_s(n)} & \text{if } s \text{ is odd,} \end{cases}$$

where $C_s(n)$ is a function of $\alpha(n)$ and s. For a fixed value of s, $C_s(n)$ is asymptotically smaller than the first term of the exponent and therefore for sufficiently (and extremely) large values of n the first term of the exponent dominates.

(iii) A lower bound on the maximal length of a $DS(n, s)$ sequence of an even order is

$$\lambda_s(n) = \Omega\left(n \cdot 2^{K_s(\alpha(n))^{\frac{s-2}{2}} + Q_s(n)},\right)$$

where $K_s = \frac{1}{\frac{(s-2)}{2}!}$ and $Q_s(n)$ is a polynomial in $\alpha(n)$ of degree at most $\frac{s-4}{2}$.

Thus our lower and upper bounds are much closer than the previous bounds although they are still not tight. For even s they are almost identical except for the constant K_s and the lower order additive terms $C_s(n)$, $Q_s(n)$, appearing in the exponents. For odd s the gap is more "substantial".

The proofs are fairly complicated and involve a lot of technical details. For the sake of exposition, we first present the derivation of the tight bounds for $\lambda_4(n)$, which gives the general flavor of the techniques used in establishing the bounds, but is relatively much simpler. Then we generalize these techniques for higher values of s. Another reason for considering $\lambda_4(n)$ separately is that we solve the recurrence relation that gives an upper bound for $\lambda_4(n)$ in a slightly more "efficient" way, which enables us to get tight bounds. While for general values of s, where no such refinement could be obtained, the proofs are slightly different.

In Section 2.2 we give the upper bounds for $\lambda_4(n)$; in Section 2.3 we construct a class of $DS(n, 4)$ sequences and prove that their length is $\Omega(n \cdot 2^{\alpha(n)})$; in Section 2.4 we prove the upper bounds for general values of s and Section 2.5 establishs our lower bounds for higher values of s. The proofs introduce and exploit several variants of Ackermann's functions. A large technical part of the proofs involves derivation of various proprties of these functions.

2.2 The Upper Bound for $\lambda_4(n)$

The best previously known upper bound for $\lambda_4(n)$ was $O(n \cdot \alpha(n)^{O(\alpha(n))})$, as follows from [117]. In this section we improve his bound by showing that $\lambda_4(n) = O(n \cdot 2^{\alpha(n)})$.

2.2.1 Decomposition of DS-sequences into chains

We begin by reviewing some definitions and facts from [117].

Definition 2.1: Let U be a $DS(n, s)$ sequence, and let $1 \leq t < s$. A *t-chain* c is a contiguous subsequence of U which is a Davenport-Schinzel sequence of order t.

Given n, s, t and U as above, we partition U into disjoint t-chains, proceeding from left to right in the following inductive manner. Suppose that the initial portion $(u_1, \ldots, u_j)$ of U has already been decomposed into t-chains. The next t-chain in our partitioning is then the largest subsequence of U of the form $(u_{j+1}, \ldots, u_k)$ which is still a Davenport-Schinzel sequence of order t. We refer to this partitioning as the *canonical decomposition* of U into t-chains, and let $m = m_t(U)$ denote the number of t-chains in this decomposition.

The problem of obtaining good upper bounds for the quantities

$$\mu_{s,t}(n) = \max\{\, m_t(U) \,:\, U \text{ is a } DS(n,s) \text{ sequence }\}$$

seems quite hard for general s and t. Sharir [117] has proved the following result:

Lemma 2.2 (Hart and Sharir [77]) $\mu_{s,s-1}(n) \leq n$ *and* $\mu_{s,s-2}(n) \leq 2n-1$.

□

The above result shows, in particular, that a $DS(n,4)$ sequence can be decomposed into at most $2n-1$ 2-chains.

Lemma 2.3 *Given a $DS(n,4)$ sequence U composed of m 2-chains, we can construct another $DS(n,4)$ sequence U' composed of m 1-chains such that* $|U'| \geq \frac{1}{2}(|U| - m)$.

Proof: Replace each 2-chain c by a 1-chain c' composed of the same symbols of c in the order of their leftmost appearances in c. Since $\lambda_2(n) = 2n-1$, we have $|c'| \geq \frac{1}{2}|c| + \frac{1}{2}$. Take the concatenation of all these 1-chains, erasing each first element of a chain that is equal to its preceding element. The resulting sequence U' is clearly a $DS(n,4)$ sequence composed of at most m 1-chains, whose length is $|U'| \geq \frac{1}{2}|U| + \frac{m}{2} - m = \frac{1}{2}(|U| - m)$.

□

Definition 2.4: Let n, m and s be positive integers. We denote by $\Psi_s^t(m,n)$ the maximum length of a $DS(n,s)$ sequence composed of at most m t-chains. For $t = 1$, we denote it as $\Psi_s(m, n)$ also.

Corollary 2.5 $\lambda_4(n) \leq 2\Psi_4(2n-1,\, n) + 2n - 1.$

Proof: The proof directly follows from Lemma 2.2 and 2.3.

□

The main result of this section is to prove that $\Psi_4(m,\, n) = O((m+n)\cdot 2^{\alpha(m)})$. This upper bound for $\Psi_4(m,\, n)$ in conjunction with Corollary 2.5 gives the desired upper bound for $\lambda_4(n)$.

2.2.2 Properties of Ackermann's and related functions

Before proving the main result, we will prove certain properties of Ackermann's function and some auxiliary functions which we need in establishing the desired upper bound. For a more basic review of Ackermann's function see [1], [77].

We first review the definition of Ackermann's function. Let $\mathcal{N}$ be the set of positive integers $1, 2, \ldots$. Given a function g from a set into itself, denote by $g^{(s)}$ the composition $g \circ g \circ \cdots \circ g$ of g with itself s times, for $s \in \mathcal{N}$. Define inductively a sequence $\{A_k\}_{k=1}^{\infty}$ of functions from $\mathcal{N}$ into itself as follows:

$$\begin{aligned} A_1(n) &= 2n, \\ A_k(n) &= A_{k-1}^{(n)}(1), \qquad k \geq 2 \end{aligned}$$

for all $n \in \mathcal{N}$. Note that for all $k \geq 2$, the function A_k satisfies

$$\begin{aligned} A_k(1) &= 2, \\ A_k(n) &= A_{k-1}(A_k(n-1)), \qquad n \geq 2. \end{aligned}$$

In particular $A_2(n) = 2^n$ and $A_3(n) = 2^{2^{\cdot^{\cdot^{2}}}}$ with n 2's in the exponential tower. Finally, Ackermann's function is defined as:

$$A(n) = A_n(n).$$

For any (weakly) monotonic function $g : \mathcal{N} \to \mathcal{N}$ its functional inverse $\gamma(n)$ is defined as

$$\gamma(n) = \min\{j : g(j) \geq n\}.$$

Let α_k and α denote the functional inverses of A_k and A, respectively. Then, for all $n \in \mathcal{N}$, the functions $\alpha_k(n)$ are given by the following recursive formula:

$$\alpha_k(n) = \min\{s \geq 1 : \alpha_{k-1}^{(s)}(n) = 1\};$$

that is, $\alpha_k(n)$ is the number of iterations of α_{k-1} needed to go from n to 1.

All the functions $\alpha_k(n)$ are non-decreasing, and converge to infinity with their argument. The same holds for $\alpha(n)$, which grows more slowly than any of the $\alpha_k(n)$.

The following property, which follows immediately from the above definitions, will be used in the sequel

$$\alpha(n)_{\alpha}(n)(n) \leq \alpha(n). \tag{2.1}$$

In the following lemmas we prove some more properties of $A_k(n)$ and other auxiliary functions $\beta_k(n)$ defined below.

Lemma 2.6 *For all $k \geq 1$, $A_k(2) = 4$ and $A_k(3) \geq 2k$.*

Proof: First consider $A_k(2)$. For $k = 1$, $A_1(2) = 2 \times 2 = 4$. By induction,

$$\begin{aligned} A_{k+1}(2) &= A_k(A_{k+1}(1)) \\ &= A_k(2) = 4. \end{aligned}$$

As to $A_k(3)$, For $k = 1$, $A_1(3) = 6 > 2$.
For $k = 2$, $A_2(3) = 8 > 2 \times 2$.
For $k > 2$, assume the inequality is true for all $k' < k$, then

$$\begin{aligned} A_k(3) &= A_{k-1}(A_k(2)) \\ &= A_{k-1}(4) \\ &= A_{k-2}(A_{k-1}(3)) \\ &> A_{k-2}(2(k-1)) \\ &\geq 4(k-1) \\ &> 2k. \end{aligned}$$

□

The above lemma implies that $\alpha_k(4) = 2$ and $\alpha_k(k) \leq \alpha_k(2k) \leq 3$ for all $k \geq 1$. We use these results in the next lemma.

Lemma 2.7 *For all $n \geq 1$, $\alpha_{\alpha(n)+1}(n) \leq 4$.*

Proof: By the definition of $\alpha_k(n)$,

$$\begin{aligned} \alpha_{\alpha(n)+1}(n) &= \min\{\, s \geq 1 \,:\, \alpha^{(s)}_{\alpha(n)}(n) = 1\} \\ &= \min\{\, s \geq 1 \,:\, \alpha^{(s)}_{\alpha(n)}(\alpha(n)) = 1\} + 1. \end{aligned}$$

By Lemma 2.6, $\alpha_{\alpha(n)}(\alpha(n)) \leq 3$, therefore after applying $\alpha_{\alpha(n)}$ once more:

$$\alpha_{\alpha(n)+1}(n) \leq \min\{\, s \geq 1 \,:\, \alpha^{(s)}_{\alpha(n)}(4) = 1\} + 2.$$

But, by Lemma 2.6, $\alpha_k(4) = 2$. Therefore $\alpha_{\alpha(n)+1}(n) \leq 4$.

□

Lemma 2.8 *For all $k \geq 4$ and $s \geq 3$,*

$$2^{A_k(s)} \leq A_{k-1}(\log(A_k(s))).$$

Proof:

$$\begin{aligned} A_{k-1}(\log(A_k(s))) &= A_{k-2}(A_{k-1}(\log(A_k(s)) - 1)) \\ &= A_{k-2}(A_{k-2}(A_{k-1}(\log(A_k(s)) - 2))) \\ &\geq A_2(A_2(2^{\log(A_k(s))-2})) \\ &= A_2\left(2^{\frac{A_k(s)}{4}}\right). \end{aligned}$$

For $x \geq 16$, $2^{\frac{x}{4}} \geq x$. For $k \geq 3$ and $s \geq 3$, $A_k(s) \geq 2^{2^2} = 16$. Therefore

$$A_{k-1}(\log(A_k(s))) \geq A_2(A_k(s)) = 2^{A_k(s)}.$$

□

Lemma 2.9 *Let $\xi_k(n)$ be $2^{\alpha_k(n)}$. Then for $k \geq 3$, $n \geq A_{k+1}(4)$,*

$$\min\{\, s' \geq 1 \,:\, \xi_k^{(s')}(n) \leq A_{k+1}(4)\} \leq 2 \cdot \alpha_{k+1}(n) - 2.$$

Proof: We first prove the lemma in the case, where n has the form $n = A_{k+1}(q)$ by induction on q. It is obvious that it holds for $n = A_{k+1}(4)$ since the left hand side is 1. Let us assume it is true for all $q' \leq q$. Now consider $n = A_{k+1}(q+1)$.

$$\begin{aligned} &\min\{\, s' \geq 1 \,:\, \xi_k^{(s')}(A_{k+1}(q+1)) \leq A_{k+1}(4)\} \\ =\; &\min\{\, s' \geq 1 \,:\, \xi_k^{(s')}(A_k(A_{k+1}(q))) \leq A_{k+1}(4)\} \end{aligned}$$

$$
\begin{aligned}
&= \min\{\, s' \geq 1 : \xi_k^{(s')}(2^{A_{k+1}(q)}) \leq A_{k+1}(4)\} + 1 \\
&\leq \min\{\, s' \geq 1 : \xi_k^{(s')}(A_k(\log(A_{k+1}(q)))) \leq A_{k+1}(4)\} + 1 \\
&\quad \text{(Using Lemma 2.8)} \\
&= \min\{\, s' \geq 1 : \xi_k^{(s')}(A_{k+1}(q)) \leq A_{k+1}(4)\} + 2 \\
&\leq 2 \cdot \alpha_{k+1}(A_{k+1}(q)) - 2 + 2 \\
&\quad \text{(By inductive hypothesis)} \\
&= 2q - 2 + 2 = 2 \cdot (q+1) - 2 \\
&= 2 \cdot \alpha_{k+1}(A_{k+1}(q+1)) - 2.
\end{aligned}
$$

For general values of n,

$$A_{k+1}(\alpha_{k+1}(n) - 1) < n \leq A_{k+1}(\alpha_{k+1}(n))$$

and also $\alpha_{k+1}(n) = \alpha_{k+1}(A_{k+1}(\alpha_{k+1}(n)))$. Therefore

$$
\begin{aligned}
& \min\{\, s' \geq 1 : \xi_k^{(s')}(n) \leq A_{k+1}(4)\} \\
\leq\ & \min\{\, s' \geq 1 : \xi_k^{(s')}(A_{k+1}(\alpha_{k+1}(n))) \leq A_{k+1}(4)\} \\
=\ & 2 \cdot \alpha_{k+1}(A_{k+1}(\alpha_{k+1}(n))) - 2 \\
=\ & 2 \cdot \alpha_{k+1}(n) - 2.
\end{aligned}
$$

□

We define a sequence of functions $\beta_k(n)$ which are related to the inverse Ackerman functions as follows:

$$
\begin{aligned}
\beta_1(n) &= \alpha_1(n), \\
\beta_2(n) &= \alpha_2(n), \\
\beta_k(n) &= \min\{\, s \geq 1 : (\alpha_{k-1} \cdot \beta_{k-1})^{(s)}(n) \leq 64\}.
\end{aligned}
$$

The functions $\beta_k(n)$ are non-decreasing, and converge to infinity with their argument. Note that

$$\beta_3(n) = \min\{\, s \geq 1 : \left(\lceil \log \rceil^2\right)^{(s)}(n) \leq 64\}.$$

In the next lemma we give an upper bound on $\beta_k(n)$ which shows that they grow at the same rate as $\alpha_k(n)$.

Lemma 2.10 *For all $k \geq 1$, $n \geq 2$, $\beta_k(n) \leq 2\alpha_k(n)$.*

Proof: For $k \leq 2$, the claim follows directly from the definition of $\beta_k(n)$. For $k = 3$

$$\beta_3(n) = \min\{ s' \geq 1 : (\alpha_2 \cdot \alpha_2)^{(s')}(n) \leq 64\}.$$

We first prove this for n of the form $A_3(q)$. For $n = A_3(2) = 4$, and $n = A_3(3) = 16$ it is true since $\beta_3(n)$ is simply 1. Assume that it is true for some $q \geq 3$; then

$$\begin{aligned}
\beta_3(A_3(q+1)) &= \min\{ s' \geq 1 : (\log^2)^{(s')}(A_3(q+1)) \leq 64\} \\
&= \min\{ s' \geq 1 : (\log^2)^{(s')}(A_2(A_3(q))) \leq 64\} \\
&= \min\{ s' \geq 1 : (\log^2)^{(s')}(A_3(q) \cdot A_3(q)) \leq 64\} + 1 \\
&= \min\{ s' \geq 1 : (\log^2)^{(s')}(4 \log^2 A_3(q)) \leq 64\} + 2.
\end{aligned}$$

For $q = 3$, $A_3(q) = 16$ and therefore $4 \log^2 A_3(q) = 64$, which implies

$$\beta_3(A_3(q+1)) = 3 \leq 2 \cdot \alpha_3(A_3(q)).$$

For $q > 3$, $\log A_3(q) \geq 16$ and for $x \geq 16$, $4x^2 \leq 2^x$; therefore

$$\begin{aligned}
\beta_3(A_3(q+1)) &\leq \min\{ s' \geq 1 : (\log^2)^{(s')}(A_3(q)) \leq 64\} + 2 \\
&= \beta_3(A_3(q)) + 2 \\
&\leq 2\alpha_3(A_3(q)) + 2 \\
&= 2\alpha_3(A_3(q+1)).
\end{aligned}$$

For general values of n,

$$A_3(\alpha_3(n) - 1) < n \leq A_3(\alpha_3(n))$$

and $\alpha_3(n) = \alpha_3(A_3(\alpha_3(n)))$. Using the same argument as in the previous lemma we can show that $\beta_3(n) \leq 2\alpha_3(n)$.

For $k > 3$, $n \leq A_k(4) = A_{k-1}(A_k(3))$, we have $\alpha_{k-1}(n) \leq A_k(3)$ and by induction hypothesis $\beta_{k-1}(n) \leq 2A_k(3)$. Hence

$$\alpha_{k-1}(n) \cdot \beta_{k-1}(n) \leq 2A_k^2(3).$$

But for $k > 3$, $A_k(3) \geq 8$ and for $x \geq 8$, $2x^2 \leq 2^x$, hence

$$\begin{aligned} q &\equiv \alpha_{k-1}(n) \cdot \beta_{k-1}(n) \\ &\leq 2^{A_k(3)} = 2^{A_{k-1}(A_k(2))} \\ &= 2^{A_{k-1}(4)} \text{ (Using Lemma 2.6)} \\ &\leq A_{k-2}(A_{k-1}(4)) = A_{k-1}(5) \end{aligned}$$

and therefore

$$\alpha_{k-1}(q) \cdot \beta_{k-1}(q) \leq 5 \times 10 < 64.$$

Thus for $n \leq A_k(4)$, $\beta_k(n) \leq 2$, which clearly implies the assertion. For $n > A_k(4) = A_{k-1}(A_k(3))$, we have

$$\begin{aligned} \beta_k(n) &= \min\{ s' \geq 1 : (\alpha_{k-1} \cdot \beta_{k-1})^{(s')}(n) \leq 64\} \\ &= \min\{ s' \geq 1 : q \equiv (\alpha_{k-1} \cdot \beta_{k-1})^{(s')}(n) \leq A_k(4)\} + \\ &\quad \min\{t \geq 1 : (\alpha_{k-1} \cdot \beta_{k-1})^{(t)}(q) \leq 64\}. \end{aligned}$$

By induction hypothesis,

$$\alpha_{k-1}(n') \cdot \beta_{k-1}(n') \leq \alpha_{k-1}(n') \cdot 2\alpha_{k-1}(n').$$

But as long as $n' > A_k(4)$, we have $\alpha_{k-1}(n') \geq A_k(3) \geq 8$, so that

$$\begin{aligned} \alpha_{k-1}(n') \cdot \beta_{k-1}(n') &\leq 2\alpha_{k-1}^2(n') \\ &\leq 2^{\alpha_{k-1}(n')} = \xi_{k-1}(n'). \end{aligned}$$

Therefore

$$\begin{aligned} \beta_k(n) &\leq \min\{ s' \geq 1 : q' \equiv \xi_{k-1}^{(s')}(n) \leq A_k(4)\} + \\ &\quad \min\{t \geq 1 : (\alpha_{k-1} \cdot \beta_{k-1})^{(t)}(q') \leq 64\} \\ &\leq 2 \cdot \alpha_k(n) - 2 + \beta_k(A_k(4)) \\ &\leq 2 \cdot \alpha_k(n) - 2 + 2 \\ &= 2\alpha_k(n). \end{aligned}$$

□

2.2.3 Upper bound for $\Psi_4(m, n)$

In this subsection we establish an upper bound on the maximal length $\Psi_4(m, n)$ of an $(n, 4)$ Davenport Schinzel sequence composed of at most m 1-chains. The following lemma is a refinement of proposition 4.1 of [117].

Lemma 2.11 *Let $m, n \geq 1$, and $1 < b < m$ be integers. Then for any partitioning $m = \sum_{i=1}^{b} m_i$ with $m_1, \dots, m_b \geq 1$ there exist integers $n^\star, n_1, n_2, \dots, n_b \geq 0$ such that*

$$n^\star + \sum_{i=1}^{b} n_i = n,$$

and

$$\Psi_4(m, n) \leq \sum_{i=1}^{b} \Psi_4(m_i, n_i) + 2\Psi_4(b, n^\star) + \Psi_3(m, 2n^\star) + 3m. \qquad (2.2)$$

Proof: Let U be a $DS(n, 4)$-sequence consisting of at most m 1-chains $c_1, \dots, c_m$ such that $|U| = \Psi_4(m, n)$, and let $m = \sum_{i=1}^{b} m_i$ be a partitioning of m as above. Partition the sequence U into b *layers* (i.e. contiguous subsequences) $L_1, \dots, L_b$ so that layer L_i consists of m_i 1-chains. Call a symbol a "internal" to layer L_i if all the occurrences of a in U are within L_i. A symbol will be called "external" if it is not internal to any layer. Suppose that there are n_i internal symbols in layer L_i and $n^\star$ external symbols (thus $n^\star + \sum_{i=1}^{b} n_i = n$).

To estimate the total number of occurrences in U of symbols that are internal to L_i, erase all external symbols from L_i. Next scan L_i from left to right and erase each element that has become equal to the element immediately preceding it. This leaves us with a sequence $L_i^\star$ which is clearly a $DS(n_i, 4)$ sequence consisting of at most m_i 1-chains, and thus its length is at most $\Psi_4(m_i, n_i)$. Moreover, if two equal internal elements in L_i have become adjacent after erasing the external symbols, then these two elements must have belonged to two distinct 1-chains, thus the total number of deletions of internal symbols is at most $m_i - 1$.

Hence, summing over all layers, we conclude that the total contribution of internal symbols to $|U|$ is at most

$$m - b + \sum_{i=1}^{b} \Psi_4(m_i, n_i).$$

We estimate the total number of occurrences of external symbols in two parts. For each layer L_i, we call an external symbol a *middle symbol* if it neither starts in L_i nor ends in L_i. If an external symbol is not a middle symbol, we call it an *end symbol.* An external symbol appears as an end symbol exactly in two layers. First we estimate the contribution of middle symbols. For each layer L_i we erase all internal symbols and end symbols and if necessary, also erase each occurrence of a middle symbol which has become equal to the element immediately preceding it. The above process deletes at most $m_i - 1$ middle symbols. Let us denote the resultant sequence by $L_i^\star$.

We claim that the $L_i^\star$ is a $DS(p_i, 2)$ sequence, where p_i is the number of distinct symbols in $L_i^\star$. Suppose the contrary; then $L_i^\star$ has a subsequence of the form

$$a \ldots b \ldots a \ldots b$$

where a and b are two distinct symbols of $L_i^\star$. But they are middle symbols, i.e. each appears in a layer before $L_i^\star$ as well as in a layer after $L_i^\star$. This implies that U has a subsequence of the form

$$(b \ldots a) \ldots a \ldots b \ldots a \ldots b \ldots (b \ldots a)$$

in which each of the first and last pairs may appear in reverse order. But this alternation of length ≥ 6 contradicts the fact that U is a $DS(n, 4)$ sequence. Therefore, $L_i^\star$ is a $DS(p_i, 2)$ sequence. Thus, the concatenation of all sequences $L_i^\star$, with the additional possible deletions of any first element of $L_i^\star$ which happens to be equal to the last element of $L_{i-1}^\star$, is a $DS(n^\star, 4)$ sequence V composed of b 2-chains, and it follows from (2.3) that we can replace this sequence by another $DS(n^\star, 4)$ sequence $V^\star$ composed of b 1-chains so that $|V| \leq 2|V^\star| + b$. Hence, the contribution of middle symbols to $|U|$ is at most

$$2\Psi_4(b, n^\star) + m + b.$$

Now, we consider the contribution of end symbols. For each layer L_i, we erase all internal symbols and middle symbols and if necessary also erase each occurrence of an end symbol if it is equal to the element immediately preceding it. We erase at most $m_i - 1$ end symbols. Let us denote the resultant sequence by $L_i^{\#}$. Let q_i be the number of distinct symbols in $L_i^{\#}$. We claim that $L_i^{\#}$ is a $DS(q_i, 3)$ sequence. Indeed, if this were not the case, $L_i^{\#}$ would have contained an alternating subsequence of the form

$$a \ldots b \ldots a \ldots b \ldots a.$$

Since b is an external symbol, it also appears in a sequence $L_j^{\#}$ other than $L_i^{\#}$. But then U has an alternation of length six which is impossible. Hence, $L_i^{\#}$ is a $DS(q_i, 3)$ sequence consisting of m_i 1-chains, so its length is at most $\Psi_3(m_i, q_i)$. Summing over all the layers, the contribution of the end symbols is at most

$$m + \sum_{i=1}^{b} \Psi_3(m_i, q_i) \leq m + \Psi_3(m, \sum_{i=1}^{b} q_i).$$

But an external symbol appears as an end symbol only in two layers, therefore $\sum_{i=1}^{b} q_i = 2n^\star$. Hence, the total contribution of external symbols is at most

$$2m + b + 2\Psi_4(b, n^\star) + \Psi_3(m, 2n^\star).$$

Thus, we obtain the asserted inequality:

$$\Psi_4(m, n) \leq \sum_{i=1}^{b} \Psi_4(m_i, n_i) + 3m + \Psi_3(m, 2n^\star) + 2\Psi_4(b, n^\star).$$

□

Lemma 2.12 *Let $n, m \geq 1$, $k \geq 2$. Then*

$$\Psi_4(m, n) \leq (\frac{15}{2} \cdot 2^k - 4k - 11) m \cdot \alpha_k(m) \cdot \beta_k(m) + (\frac{21}{2} \cdot 2^k - 4k - 8) \cdot n. \quad (2.3)$$

Proof: We use equation 2.2 repeatedly to obtain the above upper bounds for $k = 2, 3, \ldots$. At each step we choose b appropriately and estimate $\Psi_4(b, n^\star)$ using a technique similar to that in [77] and [117]. At the kth step we refine the bound of $\Psi_3(m, 2n^\star)$ using the inequality

$$\Psi_3(m, n) \leq 4km \cdot \alpha_k(m) + 2kn$$

obtained in [77].

We proceed by double induction on k and m. Initially $k = 2$, $m \geq 1$, and $\beta_k(m) = \alpha_k(m) = \lceil \log m \rceil$. Choose $b = 2$, $m_1 = \lfloor \frac{m}{2} \rfloor$, $m_2 = \lceil \frac{m}{2} \rceil$ in the equation (2.2):

$$\Psi_4(b, n^\star) = \Psi_4(2, n^\star) = 2n^\star$$

for all $n^\star$, and $\Psi_3(m, 2n^\star)$ is

$$\leq 8m \lceil \log m \rceil + 8n^\star$$

so the equation (2.2) becomes:

$$\begin{aligned} \Psi_4(m, n) &= \Psi_4\left(\left\lfloor \frac{m}{2} \right\rfloor, n_1\right) + \Psi_4\left(\left\lceil \frac{m}{2} \right\rceil, n_2\right) + 12n^\star + \\ &\quad 3m + 8m \lceil \log m \rceil \\ &\leq \Psi_4\left(\left\lfloor \frac{m}{2} \right\rfloor, n_1\right) + \Psi_4\left(\left\lceil \frac{m}{2} \right\rceil, n_2\right) + 12n^\star + 11m \lceil \log m \rceil \end{aligned}$$

where $n = n_1 + n_2 + n^\star$. The solution of the above equation is easily seen to be

$$\begin{aligned} \Psi_4(m, n) &\leq 11m(\lceil \log m \rceil)^2 + 12n \\ &= 11m \cdot \alpha_2(m) \cdot \beta_2(m) + 12n \\ &\leq (\frac{15}{2} \cdot 2^k - 4k - 11)m\alpha_k(m)\beta_k(m) + (\frac{21}{2} \cdot 2^k - 4k - 8)n. \end{aligned}$$

For $k > 2$ and $m \leq 64$ the inequality holds because $\Psi_4(m, n) \leq 64n$, which is less than the right hand side of the inequality.

For $k > 2$ and $m > 64$, assume that the inductive hypothesis is true for all $k' < k$ and $m' \geq 1$ and for $k' = k$ and $m' < m$; choose

$$t = \left\lceil \frac{\alpha_{k-1}(m) \cdot \beta_{k-1}(m)}{\alpha_k(m)} \right\rceil, \text{ and } b = \left\lfloor \frac{m}{t} \right\rfloor.$$

For $m > 64$ and $k > 2$, $\alpha_k(m) > 2$ and $\alpha_{k-1}(m) \cdot \beta_{k-1}(m) \leq \lceil \log m \rceil^2$, thus

$$t \leq \left\lceil \frac{1}{2} \lceil \log m \rceil^2 \right\rceil < m - 1.$$

Suppose $m = b \cdot t + r$, then for the first r layers $L_1, \ldots, L_r$ choose $m_i = t+1$ and for the remaining layers choose $m_i = t$; therefore $m_i \leq t + 1 < m$ for all i.

By induction hypothesis (for $k-1$ and b) we have

$$\begin{aligned}\Psi_4(b, n^\star) &\leq (\frac{15}{2} \cdot 2^{(k-1)} - 4(k-1) - 11)b \cdot \alpha_{k-1}(b) \cdot \beta_{k-1}(b) \\ &+ (\frac{21}{2} \cdot 2^{(k-1)} - 4(k-1) - 8)n^\star.\end{aligned}$$

But

$$\begin{aligned}b &\leq \frac{m}{t} \\ &= \frac{m}{\left\lceil \frac{\alpha_{k-1}(m) \cdot \beta_{k-1}(m)}{\alpha_k(m)} \right\rceil} \\ &\leq \frac{m \cdot \alpha_k(m)}{\alpha_{k-1}(m) \cdot \beta_{k-1}(m)}.\end{aligned}$$

Since clearly $b \leq m$,

$$\Psi_4(b, n^\star) \leq (\frac{15}{2} \cdot 2^{(k-1)} - 4(k-1) - 11) \cdot m\alpha_k(m) + (\frac{21}{2} \cdot 2^{(k-1)} - 4(k-1) - 8) \cdot n^\star.$$

Since each $m_i < m$, by inductive hypothesis (for $k-1$ and m_i) equation (2.3) becomes

$$\begin{aligned}\Psi_4(m, n) \leq\ & 2(\frac{15}{2} \cdot 2^{(k-1)} - 4(k-1) - 11) \cdot m\alpha_k(m) + \\ & 2(\frac{21}{2} \cdot 2^{(k-1)} - 4(k-1) - 8)n^\star + 4km\alpha_k(m) + 4kn^\star + \\ & 3m + \sum_{i=1}^{b} (\frac{15}{2} \cdot 2^k - 4k - 11)m_i\alpha_k(m_i) \cdot \beta_k(m_i) + \\ & \sum_{i=1}^{b} \cdot (\frac{21}{2} \cdot 2^k - 4k - 8)n_i.\end{aligned}$$

The value of $\beta_k(m_i)$ can be estimated as follows:

$$\begin{aligned}\beta_k(m_i) &\leq \beta_k(t+1) \\ &= \beta_k\left(\left\lceil \frac{\alpha_{k-1}(m) \cdot \beta_{k-1}(m)}{\alpha_k(m)} \right\rceil + 1\right) \\ &\leq \beta_k\left(\frac{\alpha_{k-1}(m) \cdot \beta_{k-1}(m)}{\alpha_k(m)} + 2\right).\end{aligned}$$

But for all $m > 4$, $\alpha_k(m) \geq 3$ (and $\alpha_{k-1}(m) \cdot \beta_{k-1}(m) \geq 3$ too) and for $x \geq 3, y \geq 3$, $\frac{x}{y} + 2 \leq x$, therefore

$$\begin{aligned} \beta_k(m_i) &\leq \beta_k(\alpha_{k-1}(m) \cdot \beta_{k-1}(m)) \\ &= \beta_k(m) - 1, \end{aligned}$$

which implies

$$\begin{aligned} \Psi_4(m, n) &\leq (\frac{15}{2} \cdot 2^k - 4k - 11) \cdot \alpha_k(m) \cdot (\beta_k(m) - 1) \cdot \sum_{i=1}^{b} m_i + \\ & (\frac{15}{2} \cdot 2^k - 4k - 11) \cdot m \cdot \alpha_k(m) + \\ & (\frac{21}{2} \cdot 2^k - 4k - 8) \cdot (\sum_{i=1}^{b} n_i + n^*) \\ &= (\frac{15}{2} \cdot 2^k - 4k - 11) m \cdot \alpha_k(m) \beta_k(m) + (\frac{21}{2} \cdot 2^k - 4k - 8) n \end{aligned}$$

because $\sum_{i=1}^{b} m_i = m$ and $n^* + \sum_{i=1}^{b} n_i = n$.

□

Theorem 2.13 $\Psi_4(m, n) = O((m + n) \cdot 2^{\alpha(m)})$.

Proof: By Lemma 2.10, $\beta_k(m) \leq 2\alpha_k(m)$, therefore

$$\Psi_4(m, n) \leq 2 \cdot (\frac{15}{2} \cdot 2^k - 4k - 11) m \cdot (\alpha_k(m))^2 + (\frac{21}{2} \cdot 2^k - 4k - 8) \cdot n.$$

Choose $k = \alpha(m) + 1$. By Lemma 2.7, $\alpha_{\alpha(m)+1}(m) \leq 4$. Substituting this value of k in the above inequality we get

$$\begin{aligned} \Psi_4(m, n) &\leq 2 \cdot \frac{15}{2} \cdot 2^{\alpha(m)+1} \cdot m \cdot 16 + \frac{21}{2} \cdot 2^{\alpha(m)+1} \cdot n \\ &= (480m + 21n) \cdot 2^{\alpha(m)}. \end{aligned} \quad (2.4)$$

Therefore,

$$\Psi_4(m, n) = O((m + n) \cdot 2^{\alpha(m)}).$$

□

Corollary 2.5 therefore yields:

Theorem 2.14 $\lambda_4(n) = O(n \cdot 2^{\alpha(n)})$.

□

2.3 The Lower Bound for $\lambda_4(n)$

In this section we establish the matching lower bound for $\lambda_4(n) = \Omega(n \cdot 2^{\alpha(n)})$ which improves the previous bounds given by [118].

Our construction is based on a doubly inductive process which some what resembles that of [133]. In this construction we use a sequence of functions $F_k(m)$ which grow faster than $A_k(m)$ but nevertheless asymptotically at the same rate.

2.3.1 The functions $\mathbf{F_k(m)}$ and their properties

Define inductively a sequence $\{F_k\}_{k=1}^{\infty}$ of functions from the set $\mathcal{N}$ to itself as follows:

$$\begin{array}{rcll} F_1(m) & = & 1, & m \geq 1, \\ F_k(1) & = & (2^k - 1) \cdot F_{k-1}(2^{k-1}), & k \geq 2, \\ F_k(m) & = & 2F_k(m-1) \cdot F_{k-1}(F_k(m-1)), & k \geq 2, m > 1. \end{array}$$

Here are some properties of $F_k(m)$.

Lemma 2.15 $F_2(m) = 3 \cdot 2^{m-1} \geq A_2(m)$.

Lemma 2.16 *Each function $F_k(m)$ is strictly increasing in m for all $k \geq 2$. Thus*

$$F_k(m) \geq m + 1 \quad \textit{and} \quad \rho \cdot F_k(m) \leq F_k(\rho \cdot m) \ \forall k \geq 2.$$

Lemma 2.17 $\{F_k(m)\}_{k \geq 1}^{\infty}$ *is strictly increasing for a fixed $m \geq 1$.*

Proof: For $k = 2$, $F_2(m) = 3 \cdot 2^{m-1} \geq 1 = F_1(m)$. For $k > 2$ and $m = 1$

$$\begin{array}{rcl} F_k(1) & = & (2^k - 1) \cdot F_{k-1}(2^{k-1}) \\ & > & F_{k-1}(2^{k-1}) \\ & > & F_{k-1}(1). \end{array}$$

Now for $k > 2$, $m \geq 1$, assume the assertion is true for all $k' < k$ and for $k' = k$ and $m' < m$. Then we have

$$F_k(m) \quad = \quad 2F_k(m-1) \cdot F_{k-1}(F_k(m-1))$$

$$\begin{aligned} &\geq 2F_k(m-1)\cdot F_{k-1}(m) \\ &> F_{k-1}(m). \end{aligned}$$

□

Lemma 2.18 $F_k(m) \geq A_k(m)$, *for* $k \geq 2$, $m \geq 1$.

Proof: For $k = 2$, the assertion is true by Lemma 2.15, for $k > 2$, $m = 1$,

$$\begin{aligned} F_k(1) &= (2^k - 1)\cdot F_{k-1}(2^{k-1}) \geq 2 \\ &= A_k(1). \end{aligned}$$

For $k \geq 2$, $m \geq 1$, assume it is true for all $k' < k$, and all $k' = k$ and $m' < m$. We have

$$\begin{aligned} F_k(m) &= 2F_k(m-1)\cdot F_{k-1}(F_k(m-1)) \\ &\geq F_{k-1}(F_k(m-1)) \\ &\geq A_{k-1}(A_k(m-1)) \\ &\geq A_k(m). \end{aligned}$$

□

Lemma 2.19 $2^{F_k(m)} \leq A_k(m+4)$, *for* $k \geq 3$, $m \geq 1$.

Proof: For $k = 3$ we prove the stronger inequality $F_3(m) \leq \frac{1}{2}A_3(m+3)$, which implies

$$2^{F_3(m)} \leq 2^{A_3(m+3)} = A_3(m+4).$$

Indeed,

$$\begin{aligned} F_3(1) &= (2^3 - 1)\cdot F_2(2^2) \\ &\leq 2^3 \times 3 \times 2^4 \\ &< 2^{3+2+4} < \frac{1}{2}\cdot 2^{16} \\ &= \frac{1}{2}A_3(4). \end{aligned}$$

For $k = 3$ and $m > 1$, assume the assertion is true for all $m' < m$, then

$$\begin{aligned} F_3(m) &= 2F_3(m-1)\cdot F_2(F_3(m-1)) \\ &= 3F_3(m-1)\cdot 2^{F_3(m-1)} \end{aligned}$$

$$
\begin{aligned}
&\le \frac{3}{2} A_3(m+2) \cdot 2^{A_3(m+2)/2} \\
&\le \frac{1}{2} \cdot 2^{A_3(m+2)} \\
&= \frac{1}{2} A_3(m+3).
\end{aligned}
$$

The last inequality follows from the fact that $3x \le 2^{x/2}$ for $x = A_3(m+2) \ge 10$.

Now for $k > 3$ and $m = 1$,

$$
\begin{aligned}
F_k(1) &= (2^k - 1) \cdot F_{k-1}(2^{k-1}) \le 2^k \cdot F_{k-1}(2^{k-1}) \\
&\le F_{k-1}(2^k \cdot 2^k) \le F_{k-1}(2^{2k}).
\end{aligned}
$$

Therefore,

$$
2^{F_k(1)} \le 2^{F_{k-1}(2^{2k})} \le A_{k-1}(2^{2k} + 4).
$$

But for $k \ge 4$,

$$
\begin{aligned}
A_k(4) &\ge A_3(A_k(3)) \ge A_3(2k) \qquad \text{(By Lemma 2.6)} \\
&= A_2(A_3(2k-1)) \ge A_2(2(2k-1)).
\end{aligned}
$$

For $k \ge 4$, $2(2k-1) > 2k+1$, so

$$
\begin{aligned}
A_k(4) &\ge A_2(2k+1) = 2^{2k+1} \\
&\ge 2^{2k} + 4
\end{aligned}
$$

and thus

$$
2^{F_k(1)} \le A_{k-1}(A_k(4)) \le A_k(5).
$$

Finally for $k > 3$ and $m > 1$, assume the assertion is true for all $k' < k$ and $m \ge 1$ and for $k' = k$ and $m' < m$, then

$$
\begin{aligned}
F_k(m) &= 2F_k(m-1) \cdot F_{k-1}(F_k(m-1)) \\
&\le F_{k-1}(2F_k(m-1) \cdot F_k(m-1)).
\end{aligned}
$$

Thus $\qquad 2^{F_k(m)} \le 2^{F_{k-1}(2F_k(m-1) \cdot F_k(m-1))}$

$$\begin{aligned}
&\leq A_{k-1}(2F_k(m-1)\cdot F_k(m-1)+4)\\
&\leq A_{k-1}(2^{F_k(m-1)})\\
&\quad\text{(because } F_k(m-1)\geq F_4(1)\geq 15)\\
&\leq A_{k-1}(A_k(m+3))\\
&\leq A_k(m+4).
\end{aligned}$$

□

Lemma 2.20 *Hence,* $A_k(m) \leq F_k(m) \leq A_k(m+4)\ \forall k \geq 3.$

We will also use an auxiliary sequence $\{N_k\}_{k\geq 1}^{\infty}$ of functions defined on the integers as follows

$$\begin{aligned}
N_1(m) &= m, && m\geq 1,\\
N_k(1) &= N_{k-1}(2^{k-1}), && k\geq 2,\\
N_k(m) &= 2N_k(m-1)\cdot F_{k-1}(F_k(m-1))+ &&\\
&\quad N_{k-1}(F_k(m-1)) && k\geq 2, m>1.
\end{aligned}$$

2.3.2 The sequence $S_k(m)$

In this section we construct an order 4 *DS*-sequence consisting of n symbols whose length is $\Omega(n2^{\alpha}(n))$. Our construction is similar to that of Wiernik and Sharir [133] for $\lambda_3(n)$. For each pair of integers $k, m \geq 1$ we define a sequence $S_k(m)$ so that

(i) $S_k(m)$ is composed of $N_k(m)$ symbols;

(ii) $S_k(m)$ is the concatenation of $F_k(m)$ *fans*, where each fan is composed of m distinct symbols $a_1, \ldots, a_m$ and has the form

$$(a_1\, a_2\ \cdots\ a_{m-1}\, a_m a_{m-1}\ \cdots\ a_2\, a_1)$$

so its length is $2m-1$ (we call m the *fan size*);

(iii) $S_k(m)$ is a Davenport Schinzel sequence of order 4.

The doubly inductive definition of $S_k(m)$ proceeds as follows.

I. $S_1(m) = (1\ 2\ \cdots\ m-1\ m\ m-1\ \cdots\ 2\ 1)$ for each $m\geq 1$.

II. $S_k(1)$ is the sequence $S_{k-1}(2^{k-1})$; each fan of length $2^k - 1$ in $S_{k-1}(2^{k-1})$ is regarded in $S_k(1)$ as $2^k - 1$ fans of size (and length) 1.

III. To obtain $S_k(m)$ for $k > 1$, $m > 1$, we proceed as follows.

(a) Construct $S' = S_k(m-1)$. S' has $F_k(m-1)$ fans, each of size $m-1$.

(b) Create $2F_{k-1}(F_k(m-1))$ distinct copies of S' (with pairwise disjoint sets of symbols). These copies have $2F_k(m-1) \cdot F_{k-1}(F_k(m-1)) = F_k(m)$ fans altogether.

(c) Construct $S^\star = S_{k-1}(F_k(m-1))$. $S^\star$ has $F_{k-1}(F_k(m-1))$ fans, each of size $F_k(m-1)$. Duplicate the middle element of each fan of $S^\star$. The total length of the modified $S^\star$ is $2F_k(m-1) \cdot F_{k-1}(F_k(m-1)) = F_k(m)$.

(d) For each $\beta \leq F_{k-1}(F_k(m-1))$, merge the βth expanded fan of the modified $S^\star$ with the $(2\beta-1)^{th}$ and the $(2\beta)^{th}$ copies of S', by inserting the αth element of the first half (resp. the second half) of the fan into the middle place of the αth fan of the $(2\beta-1)^{th}$ (resp. the $(2\beta)^{th}$) copy of S', for each $\alpha \leq F_k(m-1)$, thereby duplicating the formerly middle element of each of these fans.

(e) $S_k(m)$ is just the concatenation of all these modified copies of S'.

Theorem 2.21 *$S_k(m)$ satisfies conditions (i)-(iii) stated above.*

Proof: The proof proceeds by double induction on k and m. Clearly $S_1(m)$ satisfies the conditions for each $m \geq 1$. For arbitrary k, m, condition (i) is a direct consequence of the inductive construction and definition of $N_k(m)$.

As for condition (ii), the inductive construction and definition of $F_k(m)$ imply that $S_k(m)$ is the concatenation of $F_k(m)$ fans. That each fan consists of m distinct symbols and has the required form also follows from the inductive construction of the sequences.

As for condition (iii), we first observe that no pair of adjacent elements of $S_k(m)$ can be identical. Indeed, by the induction hypothesis this is the case for each copy of S' and for $S^\star$. The only duplications of adjacent elements caused by our construction is of the middle elements of all the fans of the copies of S' and of $S^\star$. However, in $S_k(m)$, an element of $S^\star$ is inserted between the two duplicated appearances of the middle element of each fan of any copy of S', and the two duplicated appearances of the middle element of a fan of $S^\star$ are inserted into two different fans in two different copies of S'. Thus $S_k(m)$ contains no pair of adjacent equal elements.

We also claim that $S_k(m)$ does not contain an alternation of the form

$$a \cdots b \cdots a \cdots b \cdots a \cdots b$$

for any pair of distinct symbols a and b. Indeed, by the induction hypothesis this holds if both a and b belong to $S^\star$ or if both belong to the same copy of S'. If a and b belong to two different copies of S' then these two copies are not interspersed at all in $S_k(m)$. The only remaining cases are when a belongs to $S^\star$ and b to some copy S'_β of S' or vice versa. In the first case only a single appearance of a (in the first or second half of the corresponding fan of $S^\star$) is inserted into S'_β, so the largest possible alternation between a and b in $S_k(m)$ is $a \cdots b \cdots a \cdots b \cdots a$. This same observation rules out the latter possibility (a belongs to S'_β and b to $S^\star$). Thus it follows by induction that $S_k(m)$ is a Davenport Schinzel sequence of order 4.

□

It remains to estimate the length $|S_k(m)|$ of $S_k(m)$ as a function of its number of symbols $N_k(m)$. Clearly

$$\frac{|S_k(m)|}{N_k(m)} = \frac{(2m-1)F_k(m)}{N_k(m)}.$$

To bound this from below, we will obtain an upper bound on $\frac{N_k(m)}{F_k(m)}$, as follows.

Theorem 2.22 $\frac{N_k(m)}{F_k(m)} \le m \cdot D_k$, *where* $D_k = \prod_{j=1}^{k} c_j$ *for* $k \ge 1$ *and*

$$c_j = \frac{1}{2 - \frac{1}{2^{j-1}}} \qquad \textit{for } j \ge 1.$$

Proof: For $k = 1$ we have $D_1 = c_1 = 1$ and $\frac{N_1(m)}{F_1(m)} = m = m \cdot D_1$, as required.

For $k > 1$ and $m = 1$ we have

$$\begin{aligned}\frac{N_k(1)}{F_k(1)} &= \frac{N_{k-1}(2^{k-1})}{(2^k-1)F_{k-1}(2^{k-1})} \\ &\le \frac{2^{k-1}}{2^k-1} \cdot D_{k-1} \\ &= c_k \cdot D_{k-1} = 1 \cdot D_k\end{aligned}$$

as required.

For $k > 1$ and $m > 1$ we have

$$\begin{aligned}\frac{N_k(m)}{F_k(m)} &= \frac{N_k(m-1)}{F_k(m-1)} + \frac{1}{2F_k(m-1)} \cdot \frac{N_{k-1}(F_k(m-1))}{F_{k-1}(F_k(m-1))} \\ &\le (m-1) \cdot D_k + \frac{1}{2F_k(m-1)} \cdot F_k(m-1) \cdot D_{k-1} \\ &= (m-1)D_k + \frac{1}{2}D_{k-1} \\ &\le m \cdot D_k \qquad \text{(because } \tfrac{1}{2} < c_k\text{)}.\end{aligned}$$

□

Corollary 2.23

$$\frac{|S_k(m)|}{N_k(m)} \ge \frac{2m-1}{m} \cdot 2^k \cdot \prod_{j=1}^{\infty} (1 - \frac{1}{2^j}).$$

Proof: By Theorem 2.22,

$$\begin{aligned}\frac{|S_k(m)|}{N_k(m)} &\ge \frac{2m-1}{m \cdot D_k} = \frac{2m-1}{m} \cdot \prod_{j=1}^{k}(2(1 - \frac{1}{2^j})) \\ &\ge \frac{2m-1}{m} \cdot 2^k \cdot \prod_{j=1}^{\infty} (1 - \frac{1}{2^j}).\end{aligned}$$

(the limit of the last infinite product is easily seen to be positive).

□

Theorem 2.24 $\lambda_4(n) = \Omega(n \cdot 2^{\alpha(n)})$.

Proof: Put $\beta = \prod_{j=1}^{\infty}(1 - \frac{1}{2^j})$. Clearly $0 < \beta < 1$. Theorem 2.22 and property Lemma 2.20 imply $N_k(1) \leq F_k(1) \leq A_k(5)$ for all $k \geq 3$. Hence for each $k \geq 5$ we have

$$n_k \equiv N_k(1) \leq A_k(k) = A(k)$$

so that $\alpha(n_k) \leq k$. On the other hand, the sequence $\{n_k\}_{k\geq 1}$ is easily seen to converge to infinity. Thus, for any given n, we find k such that

$$n_k \leq n < n_{k+1}.$$

Assume with out loss of generality that $k \geq 4$. Put $t = \left\lfloor \frac{n}{n_k} \right\rfloor$ so that

$$t \cdot n_k \leq n < (t+1) \cdot n_k < 2t \cdot n_k.$$

Clearly,
$$\begin{aligned} \lambda_4(n) &\geq t\lambda_4(n_k) \geq t \cdot |S_k(1)| \\ &\geq \beta t \cdot n_k \cdot 2^k \qquad \text{(by Corollary 2.23)} \\ &> \frac{\beta}{2} n \cdot 2^k. \end{aligned}$$

But $\alpha(n) \leq \alpha(n_{k+1}) \leq k+1$ so that $k \geq \alpha(n) - 1$, and we thus have

$$\lambda_4(n) \geq \frac{\beta}{4} n \cdot 2^{\alpha(n)}.$$

for all $n \geq N_4(1)$. For smaller values of n we have $\alpha(n) \leq 5$, $\beta < \frac{3}{8}$, so we have to show that $\lambda_4(n) \geq 3n$, which is easily seen to hold for all $n \geq 3$. For $n = 1, 2$ the asserted inequality is trivial, thus we have for each $n \geq 1$

$$\lambda_4(n) \geq \frac{\beta}{4} n \cdot 2^{\alpha(n)} = \Omega(n \cdot 2^{\alpha(n)}).$$

□

Corollary 2.25 $\lambda_4(n) = \Theta(n \cdot 2^{\alpha(n)})$.

Proof: The above relation immediately follows from the results of Theorem 2.14 and 2.24.

□

2.4 The Upper Bounds for $\lambda_s(n)$

In this section we extend the approach of section 2 to obtain improved upper bounds for $\lambda_s(n)$. In particular, we show that

$$\lambda_s(n) \le \begin{cases} n \cdot 2^{(\alpha(n))^{\frac{s-2}{2}}+C_s(n)} & \text{if } s \text{ is even,} \\ n \cdot 2^{(\alpha(n))^{\frac{s-3}{2}}\log(\alpha(n))+C_s(n)} & \text{if } s \text{ is odd,} \end{cases} \tag{2.5}$$

where $C_s(n)$ satisfies the following bound

$$3+s \le C_s(n) = \begin{cases} 6 & \text{if } s = 3, \\ 11 & \text{if } s = 4, \\ O\left((\alpha(n))^{\frac{s-4}{2}} \cdot \log(\alpha(n))\right) & \text{if } s > 4 \text{ is even,} \\ O\left((\alpha(n))^{\frac{s-3}{2}}\right) & \text{if } s > 3 \text{ is odd.} \end{cases} \tag{2.6}$$

A more precise definition of $C_s(n)$ is given in (2.8).

In [48] , [6], it has been proved that $\lambda_s(n) \le \frac{n(n-1)}{2} \cdot s + 1$. For $n \le 4$ and $s \ge 3$ we can directly verify that

$$\lambda_s(n) \le n \cdot 2^{(\alpha(n))^{\frac{s-3}{2}}+C_s(n)}.$$

For $4 < n \le 16$ we have $\alpha(n) \ge 2$ and

$$\begin{aligned} \lambda_s(n) &\le 8s \cdot n \\ &= 2^{3+\log s} \cdot n \\ &\le n \cdot 2^{2^{\frac{s-3}{2}}+3+s} \\ &\le n \cdot 2^{(\alpha(n))^{\frac{s-3}{2}}+C_s(n)}. \end{aligned}$$

Thus for $n \le 16$, $\lambda_s(n)$ satisfies the desired inequality. Therefore, we restrict our attention to $n > 16$. It can be easily verified that the above inequality holds for $s = 3$ and $s = 4$. For $s = 3$, Hart and Sharir [77] proved that

$$\begin{aligned} \lambda_3(n) &\le 52 \cdot n\alpha(n) \\ &= n \cdot 2^{\log 52+\log(\alpha(n))} \\ &\le n \cdot 2^{6+\log(\alpha(n))} \\ &= n \cdot 2^{\log(\alpha(n))+C_3(n)}. \end{aligned}$$

For $s = 4$, the equation 2.4 actually gives, for $n > 16$

$$\begin{aligned} \lambda_4(n) &\leq 2^{11} \cdot n \cdot 2^{\alpha(n)} \\ &= n \cdot 2^{\alpha(n)+C_4(n)}. \end{aligned}$$

For $s > 4$, we prove the desired upper bound (2.5) for $\lambda_s(n)$ by induction on s.

In this section, apart from the Ackermann's function, we need some more functions defined in terms of $\alpha(n)$. Let $\{\Gamma_s\}_{s\geq 2}$ be a sequence of functions defined on $\mathcal{N}$ by:

$$\Gamma_s(n) = \begin{cases} (\alpha(n))^{\frac{s-2}{2}} & \text{if } s \text{ is even,} \\ (\alpha(n))^{\frac{s-3}{2}} \cdot \log(\alpha(n)) & \text{if } s \text{ is odd.} \end{cases} \tag{2.7}$$

Therefore, $\Gamma_2(n) = 1$, $\Gamma_3(n) = \log(\alpha(n))$ and for all $s \geq 4$, $\Gamma_s(n) = \Gamma_{s-2}(n) \cdot \alpha(n)$.

We define $\{C_s(n)\}_{s\geq 3}$ as follows:

$$C_s(n) = \sum_{i=2}^{s-1} a_i^s \cdot \Gamma_i(n) \tag{2.8}$$

where a_i^s is a constant depending on the value of i and s and is defined recursively as follows:

$$a_2^3 = 6 \qquad a_2^4 = 11 \qquad a_3^4 = 0$$

and for $s > 4$

$$a_i^s = \begin{cases} a_{s-3}^{s-2} + 1 & \text{if } i = s-1 \\ a_{i-2}^{s-2} + a_i^{s-1} & \text{if } 3 < i < s-1 \\ a_i^{s-1} & \text{if } i \leq 3 \end{cases} \tag{2.9}$$

and finally, let $\Pi_s(n)$ be

$$\Pi_s(n) = 2^{\Gamma_s(n)+C_s(n)} \qquad \text{for } s \geq 3. \tag{2.10}$$

Note that for each fixed n, $\{\Pi_s(n)\}_{s\geq 3}$ is increasing and for each fixed s, $\{\Pi_s(n)\}_{n\geq 1}$ is also increasing. From the definition of $\Pi_s(n)$ it follows that to prove the desired upper bound for $\lambda_s(n)$, we have to show that

$$\lambda_s(n) \leq n \cdot \Pi_s(n).$$

2.4.1 Upper bounds for $\Psi_s(m, n)$

In this subsection we establish an upper bound on the maximal length $\Psi_s(m, n)$ of an (n, s) of Davenport-Schinzel sequence composed of at most m 1-chains and having maximal length. The following lemma is a (somewhat modified) extension of Lemma 2.11.

Lemma 2.26 *Let $m, n \geq 1$ and $1 < b < m$ be integers. Then for any partitioning $m = \sum_{i=1}^{b} m_i$ with $m_1, \ldots, m_b \geq 1$, there exist integers $n^\star, n_1, n_2, \ldots, n_b \geq 0$ such that*

$$n^\star + \sum_{i=1}^{b} n_i = n$$

and

$$\Psi_s(m, n) \leq \Psi_s^{s-2}(b, n^\star) + 2 \cdot \Psi_{s-1}(m, n^\star) + 4m + \sum_{i=1}^{b} \Psi_s(m_i, n_i). \quad (2.11)$$

Proof: The proof given in Lemma 2.11 can be extended to handle the general case. Let U be a $DS(n, s)$-sequence consisting of at most m 1-chains $c_1, \ldots, c_m$ such that $|U| = \Psi_s(m, n)$. Partition the sequence into b layers (i.e. disjoint contiguous subsequences) $L_1, \ldots, L_b$ so that the layer L_i consists of m_i chains. Call a symbol a *internal* or *external* as in Lemma 2.11. Suppose there are n_i internal symbols in layer L_i, and $n^\star$ external symbols (thus $n^\star + \sum_{i=1}^{b} n_i = n$).

Using the same argument as in Lemma 2.11 we can show that the total contribution of internal symbols to $|U|$ is at most

$$m - b + \sum_{i=1}^{b} \Psi_s(m_i, n_i).$$

We bound the total number of occurrences of external symbols in three parts instead of two as in Lemma 2.11. For each layer L_i, we call an external symbol a a *starting symbol* if its first (i.e. leftmost) occurrence is in L_i, an *ending symbol* if its last (i.e. rightmost) occurrence is in L_i, and a *middle symbol* if it is neither a starting nor an ending symbol. An external symbol appears as a starting symbol or an ending symbol exactly in one layer. First we estimate the total number of occurrences of middle symbols. For each layer L_i we erase all internal symbols, starting symbols and ending symbols. We also erase each occurrence of a middle symbol which has become equal to the element immediately preceding it (there are at most $m_i - 1$ such erasures). Let us denote the resultant sequence by $L_i^\star$.

By generalizing the argument given in the proof of Lemma 2.11, it can be easily shown that $L_i^\star$ is a $DS(p_i, s-2)$ sequence. Thus the concatenation of all sequences $L_i^\star$, with the additional possible deletions of any first element of $L_i^\star$ which happens to be equal to the last element of $L_{i-1}^\star$, is a $DS(n^\star, s)$ sequence composed of b $(s-2)$-chains, and therefore the contribution of the middle symbols is at most

$$\Psi_s^{s-2}(b, n^\star) + m.$$

We now consider the contribution of the starting external symbols. For each layer $L_i^\star$, we erase all internal symbols, middle symbols and ending symbols and if necessary also erase each occurrence of a starting symbol if it is equal to the element immediately preceding it. This process deletes at most $m_i - 1$ starting symbols. Let us denote the resultant sequence by $L_i^\#$. Let q_i be the number of distinct symbols in $L_i^\#$. We claim that $L_i^\#$ is a $DS(q_i, s-1)$ sequence. Indeed, if this were not the case, $L_i^\#$ would have contained an alternating sequence of the form

$$\underbrace{a \quad b \quad a \ldots a \quad b}_{s+1}$$

if s is odd and

$$\underbrace{a \quad b \quad a \ldots b \quad a}_{s+1}$$

if s is even. Since b and a are external symbols and their first appearance is in $L_i^\#$, they also appear in some layers after $L_i^\#$. But then U contains an alternation of a and b having length $s+2$, which is impossible. Hence $L_i^\#$ is a $DS(q_i, s-1)$ sequence consisting of m_i 1-chains, so its length is at most $\Psi_{s-1}(m_i, q_i)$. Summing over all the layers, the contribution of starting symbols is at most

$$m + \sum_{i=1}^{b} \Psi_{s-1}(m_i, q_i) \le m + \Psi_{s-1}(m, \sum_{i=1}^{b} q_i).$$

But an external symbol appears as a starting symbol only in one layer, therefore $\sum_{i=1}^{b} q_i = n^\star$. Hence the total contribution of starting symbols is bounded by

$$m + \Psi_{s-1}(m, n^\star).$$

Since the ending symbols are symmetric to the starting symbols, the same bound holds for the number of appearances of ending symbols also. Therefore the total contribution of the external symbols is bounded by

$$3m + 2 \cdot \Psi_{s-1}(m, n^\star) + \Psi_s^{s-2}(b, n^\star).$$

Thus we obtain the desired inequality

$$\Psi_s(m, n) \leq \Psi_s^{s-2}(b, n^\star) + 2 \cdot \Psi_{s-1}(m, n^\star) + 4m + \sum_{i=1}^{b} \Psi_s(m_i, n_i).$$

□

Remark 2.27: Note that in the above proof, we estimate the contribution of external symbols in three parts instead of two as in Lemma 2.11. The reason is that while the treatment of starting and ending external symbols as a single case can be extended to *even* values of s, it fails for odd values, because the resulting sequence $L_i^\#$ might be of order s rather than $s-1$, e.g. if a is a starting symbol and b is an ending symbol, then it is possible that a and b have $s+1$ alternations in the layer L_i (starting with a and ending with b). That is why, in general, partitioning the external symbols into two parts is not enough. Also the extra overhead for even values of s is negligible.

The proof of our upper bound proceeds by induction on s. The base cases $s = 3$ and $s = 4$ have already been discussed above. Let $s > 4$ and suppose the upper bound holds for each $t < s$, i.e. $\lambda_t(n) \leq n \cdot \Pi_t(n)$. Before giving the solution of the equation 2.11, we bound $\Psi_s^t(m, n)$ in terms of $\Psi_s(m, n)$.

Lemma 2.28 *Let $m, n \geq 1$ and $3 \leq t < s$ be integers; then*

$$\Psi_s^t(m, n) \leq \Pi_t(n) \cdot \Psi_s(m, n) + (m-1) \cdot \Pi_t(n).$$

Proof: This lemma is basically a generalization of Lemma 2.3. Let U be a $DS(n, s)$ sequence composed of m t-chains and having maximal length. Replace each chain c_i of U by the 1-chain c'_i composed of the same symbols in the order of their leftmost appearance in c_i. Since by the inductive hypothesis $\lambda_t(n) \le n \cdot \Pi_t(n)$, we have $|c_i| \le |c'_i| \cdot \Pi_t(n)$. Construct another sequence U' by concatenating all the 1-chains c'_i and erasing each first symbol of c'_i which is equal to its immediately preceding element. It is clear that U' is a $DS(n, s)$ sequence composed of at most m 1-chains and its length is at least $\sum_{i=1}^{m} |c'_i| - (m-1)$. Therefore

$$\Psi_s(m, n) \ge \frac{1}{\Pi_t(n)} \cdot \sum_{i=1}^{m} |c_i| - (m-1).$$

But, $\Psi_s^t(m, n) = \sum_{i=1}^{m} |c_i|$. Thus

$$\Psi_s^t(m, n) \le \Pi_t(n) \cdot \Psi_s(m, n) + (m-1) \cdot \Pi_t(n).$$

□

Corollary 2.29 *For $s \ge 4$,*

$$\lambda_s(n) \le \Psi_s(2n-1, n) \cdot \Pi_{s-2}(n) + (2n-2) \cdot \Pi_{s-2}(n). \tag{2.12}$$

Proof: The proof directly follows from Lemma 2.2 and 2.28.

□

Lemma 2.30 *Let $m, n \ge 1$, and $k \ge 2$. Then*

$$\Psi_s(m, n) \le \mathcal{F}_k(n) \cdot m \cdot \alpha_k(m) + \mathcal{G}_k(n) \cdot n \tag{2.13}$$

where $\mathcal{F}_k(n)$ and $\mathcal{G}_k(n)$ are defined recursively as follows:

$$\begin{aligned}
\mathcal{F}_2(n) &= 4 \\
\mathcal{F}_k(n) &= 2\Pi_{s-2}(n) \cdot \mathcal{F}_{k-1}(n) + (\Pi_{s-2}(n) + 4) && (2.14) \\
\mathcal{G}_2(n) &= 5\Pi_{s-1}(n) \\
\mathcal{G}_k(n) &= \Pi_{s-2}(n) \cdot \mathcal{G}_{k-1}(n) + 2\Pi_{s-1}(n). && (2.15)
\end{aligned}$$

Proof: $\Psi_s(m, n) \leq \lambda_s(n)$, therefore $\Psi_{s-1}(m, n^\star) \leq n^\star \cdot \Pi_{s-1}(n^\star)$. If we replace $\Psi_{s-1}(m, n^\star)$ by this bound in the equation (2.11) and also replace $\Psi_s^{s-2}(b, n^\star)$ by the right hand side of the bound of lemma 2.28, we get:

$$\begin{aligned}\Psi_s(m, n) \quad \leq \quad & \Pi_{s-2}(n^\star) \cdot \Psi_s(b, n^\star) + (b-1) \cdot \Pi_{s-2}(n^\star) + 4m + \\ & 2n^\star \cdot \Pi_{s-1}(n^\star) + \sum_{i=1}^{b} \Psi_s(m_i, n_i).\end{aligned} \tag{2.16}$$

We use equation (2.16) repeatedly to obtain the desired bound for $k = 2, 3, \cdots$. At each step we choose b appropriately and estimate $\Psi_s(b, n^\star)$ using a technique similar to Lemma 2.12.

We proceed by double induction on k and m. Initially $k = 2$, and $\alpha_k(m) = \log m$ for $m \geq 1$. For $k = 2$ choose $b = 2$, $m_1 = \lfloor \frac{m}{2} \rfloor$ and $m_2 = \lceil \frac{m}{2} \rceil$ in equation (2.16); $\Psi_s(b, n^\star) = \Psi_s(2, n^\star) = 2n^\star$ for all $n^\star$ so (2.16) becomes

$$\begin{aligned}\Psi_s(m, n) \quad \leq \quad & \Psi_s(\lfloor \tfrac{m}{2} \rfloor, n_1) + \Psi_s(\lceil \tfrac{m}{2} \rceil, n_2) + \Pi_{s-2}(n^\star) + 4m + \\ & 2n^\star \cdot (\Pi_{s-1}(n^\star) + \Pi_{s-2}(n^\star)) \\ \leq \quad & \Psi_s(\lfloor \tfrac{m}{2} \rfloor, n_1) + \Psi_s(\lceil \tfrac{m}{2} \rceil, n_2) + 4m + \\ & n^\star \cdot (2\Pi_{s-1}(n^\star) + 3\Pi_{s-2}(n^\star))\end{aligned}$$

where $n = n_1 + n_2 + n^\star$. The solution of this recurrence relation is easily seen to be

$$\Psi_s(m, n) \leq 4m \cdot \lceil \log m \rceil + n(2\Pi_{s-1}(n) + 3\Pi_{s-2}(n)).$$

Since $\Pi_{s-1}(n) > \Pi_{s-2}(n)$, for $k = 2$ we have

$$\begin{aligned}\Psi_s(m, n) \quad & \leq \quad 4m \cdot \lceil \log m \rceil + 5n \cdot \Pi_{s-1}(n) \\ & = \quad m \cdot \mathcal{F}_2(n) \cdot \alpha_2(m) + \mathcal{G}_2(n) \cdot n\end{aligned}$$

as asserted. For $k > 2$ and $m \leq 16$ the inequality (2.13) obviously holds as $\Psi_s(m, n) \leq 16n$ and the right hand side of (2.13) is $\geq 16n$. Now suppose that $k > 2$ and $m > 16$ and that the inductive hypothesis holds for all $k' < k$ and $m' \geq 1$ and for $k' = k$ and for all $m' < m$. Choose $t = \left\lceil \frac{\alpha_{k-1}(m)}{2} \right\rceil$, and $b = \lfloor \frac{m}{t} \rfloor$. For $k > 2$, $\alpha_{k-1}(m) \leq \lceil \log m \rceil$; thus

$$t \leq \left\lceil \frac{\lceil \log m \rceil}{2} \right\rceil < m - 1.$$

Suppose $m = b \cdot t + r$, then any $DS(n, s)$ sequence U composed of m 1-chains can be decomposed into b layers, $L_1, \ldots, L_b$ containing $m_1, \ldots, m_b$ 1-chains, so that $m_i = t + 1$ for $i \leq r$ and $m_i = t$ for the remaining layers; therefore $m_i \leq t + 1 < m$ for all i. By induction hypothesis (for $k - 1$ and b) we have

$$\Psi_s(b, n^*) \leq \mathcal{F}_{k-1}(n^*) \cdot b \cdot \alpha_{k-1}(b) + \mathcal{G}_{k-1}(n^*) \cdot n^*.$$

But

$$\begin{aligned} b &\leq \frac{m}{t} = \frac{m}{\left\lceil \frac{\alpha_{k-1}(m)}{2} \right\rceil} \\ &\leq \frac{2m}{\alpha_{k-1}(m)}. \end{aligned}$$

Clearly $b \leq m$, therefore $\alpha_{k-1}(b) \leq \alpha_{k-1}(m)$ and we have

$$\Psi_s(b, n^*) \leq \mathcal{F}_{k-1}(n^*) \cdot 2m + \mathcal{G}_{k-1}(n^*) \cdot n^*.$$

Since each $m_i < m$, by inductive hypothesis:

$$\sum_{i=1}^{b} \Psi_s(m_i, n_i) \leq \sum_{i=1}^{b} \left(\mathcal{F}_k(n_i) \cdot m_i \cdot \alpha_k(m_i) + \mathcal{G}_k(n_i) \cdot n_i\right).$$

The value of $\alpha_k(m_i)$ can be estimated as follows:

$$\begin{aligned} \alpha_k(m_i) &\leq \alpha_k(t+1) \\ &= \alpha_k\left(\left\lceil \frac{\alpha_{k-1}(m)}{2} \right\rceil + 1\right). \end{aligned}$$

Now for $m \geq 16$, $\alpha_{k-1}(m) \geq 3$ and it is easy to check that

$$\left\lceil \frac{\alpha_{k-1}(m)}{2} \right\rceil + 1 \leq \alpha_{k-1}(m).$$

Thus

$$\begin{aligned} \alpha_k(m_i) &\leq \alpha_k(\alpha_{k-1}(m)) \\ &= \alpha_k(m) - 1. \end{aligned}$$

which implies

$$\begin{aligned}\sum_{i=1}^{b} \Psi_s(m_i, n_i) &\le \sum_{i=1}^{b} [\mathcal{F}_k(n_i) \cdot m_i \cdot (\alpha_k(m) - 1) + \mathcal{G}_k(n_i) \cdot n_i] \\ &\le m \cdot \mathcal{F}_k(n) \cdot (\alpha_k(m) - 1) + \mathcal{G}_k(n) \cdot \sum_{i=1}^{b} n_i.\end{aligned}$$

If we substitute these values of $\Psi_s(b, n^\star)$ and $\sum_i \Psi_s(m_i, n_i)$ in (2.16) and use the fact that $b \le m$, we get

$$\begin{aligned}\Psi_s(m, n) &\le \Pi_{s-2}(n) \cdot [\mathcal{F}_{k-1}(n) \cdot 2m + \mathcal{G}_{k-1}(n) \cdot n^\star] + \\ &\quad (\Pi_{s-2}(n) + 4) \cdot m + 2n^\star \cdot \Pi_{s-1}(n) + \\ &\quad \mathcal{F}_k(n) \cdot m(\alpha_k(m) - 1) + \mathcal{G}_k(n) \cdot \sum_{i=1}^{b} n_i \\ &\le [2\Pi_{s-2}(n) \cdot \mathcal{F}_{k-1}(n) + (\Pi_{s-2}(n) + 4)] \cdot m + \\ &\quad [\Pi_{s-2}(n) \cdot \mathcal{G}_{k-1}(n) + 2\Pi_{s-1}(n)] \cdot n^\star + \\ &\quad \mathcal{F}_k(n) \cdot m(\alpha_k(m) - 1) + \mathcal{G}_k(n) \cdot \sum_{i=1}^{b} n_i\end{aligned}$$

which by definition of $\mathcal{F}_k(n)$ and $\mathcal{G}_k(n)$ is

$$\begin{aligned}\Psi_s(m, n) &\le \mathcal{F}_k(n) \cdot m\alpha_k(m) + \mathcal{G}_k(n) \cdot (n^\star + \sum_{i=1}^{b} n_i) \\ &= \mathcal{F}_k(n) \cdot m\alpha_k(m) + \mathcal{G}_k(n) \cdot n.\end{aligned}$$

Hence, the lemma is true.

□

Lemma 2.31 *For $k \ge 2$ $\mathcal{F}_k(n)$ and $\mathcal{G}_k(n)$ satisfy the following inequalities:*

$$\mathcal{F}_k(n) \le 5 \cdot (2\Pi_{s-2}(n))^{k-2}, \tag{2.17}$$

$$\mathcal{G}_k(n) \le 6\Pi_{s-1}(n) \cdot (\Pi_{s-2}(n))^{k-2}. \tag{2.18}$$

Proof: It is not difficult to see that a recurrence relation of the form

$$\begin{aligned} T(2) &= c \\ T(k) &= aT(k-1)+b \end{aligned}$$

has the following solution

$$T(k) = c \cdot a^{k-2} + \frac{a^{k-2}-1}{a-1} \cdot b.$$

The recursive definition of $\mathcal{F}_k(n)$ given in (2.14) has the same form with $a = 2\Pi_{s-2}(n)$, $b = (\Pi_{s-2}(n) + 4)$ and $c = 4$. Therefore

$$\mathcal{F}_k(n) \leq 4 \cdot (2\Pi_{s-2}(n))^{k-2} + \frac{\left[(2\Pi_{s-2}(n))^{k-2} - 1\right]}{[2\Pi_{s-2}(n) - 1]} \cdot (\Pi_{s-2}(n) + 4).$$

But for $x > 5$ $\frac{x+4}{2x-1} < 1$. Since $\Pi_{s-2}(n) > 5$, we get

$$\mathcal{F}_k(n) \leq 5 \cdot (2\Pi_{s-2}(n))^{k-2}.$$

Similarly, the recursive definition of $\mathcal{G}_k(n)$ given in (2.15) also has the same form with $a = \Pi_{s-2}(n)$, $b = 2\Pi_{s-1}(n)$ and $c = 5\Pi_{s-1}(n)$. Hence

$$\mathcal{G}_k(n) \leq 5\Pi_{s-1}(n) \cdot (\Pi_{s-2}(n))^{k-2} + \frac{\left[(\Pi_{s-2}(n))^{k-2} - 1\right]}{[\Pi_{s-2}(n) - 1]} \cdot 2\Pi_{s-1}(n).$$

Since $\Pi_{s-2}(n) > 4$, $\frac{2}{[\Pi_{s-2}(n)-1]} < 1$, we get

$$\mathcal{G}_k(n) \leq 6\Pi_{s-1}(n) \cdot (\Pi_{s-2}(n))^{k-2}.$$

□

Theorem 2.32 *For $s \geq 2$, $n \geq 1$*

$$\lambda_s(n) \leq n \cdot \Pi_s(n).$$

Proof: If we substitute $k = \alpha(n)$ in (2.13) we get

$$\Psi_s(m, n) \leq \mathcal{F}_{\alpha(n)}(n) \cdot m\alpha_{\alpha(n)}(m) + \mathcal{G}_{\alpha(n)}(n) \cdot n.$$

Now we can use the corollary 2.29 to bound $\lambda_s(n)$. We substitute the above value of $\Psi_s(m, n)$ in the equation (2.12):

$$\begin{aligned}\lambda_s(n) &\leq \Pi_{s-2}(n) \cdot \mathcal{F}_{\alpha(n)}(n) \cdot (2n-1) \cdot \alpha_{\alpha(n)}(2n-1) + \\ &\quad \Pi_{s-2}(n) \cdot \mathcal{G}_{\alpha(n)}(n) \cdot n + (2n-2) \cdot \Pi_{s-2}(n).\end{aligned}$$

For $k \geq 2$, $\alpha_k(2n) \leq (\alpha_k(n) + 1)$ and $\alpha_{\alpha(n)}(n) \leq \alpha(n)$, so we have

$$\begin{aligned}\lambda_s(n) &\leq n \cdot \Pi_{s-2}(n) \cdot \left[2\mathcal{F}_{\alpha(n)}(n) \cdot (\alpha(n)+1) + \mathcal{G}_{\alpha(n)}(n) + 2\right] \\ &\leq n \cdot \Pi_{s-2}(n) \cdot \left[4\mathcal{F}_{\alpha(n)}(n) \cdot \alpha(n) + \mathcal{G}_{\alpha(n)}(n)\right].\end{aligned}$$

After substitution of the values of $\mathcal{F}_{\alpha(n)}(n)$ and $\mathcal{G}_{\alpha(n)}(n)$ from (2.17) and (2.18), the above inequality becomes

$$\begin{aligned}\lambda_s(n) &\leq n \cdot \Pi_{s-2}(n) \cdot \Big[4 \cdot 5(2\Pi_{s-2}(n))^{\alpha(n)-2} \cdot \alpha(n) + \\ &\qquad 6(\Pi_{s-2}(n))^{\alpha(n)-2} \cdot \Pi_{s-1}(n)\Big] \\ &\leq n \cdot (\Pi_{s-2}(n))^{\alpha(n)-1} \cdot \left[5\alpha(n) \cdot 2^{\alpha(n)} + 6\Pi_{s-1}(n)\right].\end{aligned}$$

Since for all $s > 4$,

$$\Pi_{s-1}(n) \geq \Pi_4(n) = 2^{\alpha(n)+8},$$

we get

$$\lambda_s(n) \leq n \cdot (\Pi_{s-2}(n))^{\alpha(n)} \cdot \Pi_{s-1}(n) \cdot \frac{\left[\frac{5\alpha(n)}{2^8} + 6\right]}{\Pi_{s-2}(n)}.$$

But for $n > 16$, $\Pi_{s-2}(n) \geq \Pi_3(n) = 64\alpha(n)$, therefore

$$\lambda_s(n) \leq n \cdot (\Pi_{s-2}(n))^{\alpha(n)} \cdot \Pi_{s-1}(n) \qquad = n \cdot 2^{\vartheta}.$$

Putting in the values of Π_{s-1} and Π_{s-2} from (2.10) we get

$$\begin{aligned}\vartheta &= \Gamma_{s-2}(n) \cdot \alpha(n) + \sum_{i=2}^{s-3} a_i^{s-2} \cdot \Gamma_i(n) \cdot \alpha(n) + \\ &\quad \Gamma_{s-1}(n) + \sum_{i=2}^{s-2} a_i^{s-1} \cdot \Gamma_i(n).\end{aligned}$$

But by definition, $\Gamma_i(n) \cdot \alpha(n) = \Gamma_{i+2}(n)$, therefore

$$\begin{aligned}
\vartheta &= \Gamma_s(n) + \sum_{i=2}^{s-3} a_i^{s-2} \cdot \Gamma_{i+2}(n) + \Gamma_{s-1}(n) + \sum_{i=2}^{s-2} a_i^{s-1} \cdot \Gamma_i(n) \\
&= \Gamma_s(n) + (1 + a_{s-3}^{s-2}) \cdot \Gamma_{s-1}(n) + \sum_{i=4}^{s-2} (a_{i-2}^{s-2} + a_i^{s-1}) \cdot \Gamma_i(n) + \\
&\quad a_3^{s-1} \cdot \Gamma_3(n) + a_2^{s-1} \cdot \Gamma_2(n) \\
&= \Gamma_s(n) + \sum_{i=2}^{s-1} a_i^s \cdot \Gamma_i(n) \\
&= \Gamma_s(n) + C_s(n).
\end{aligned}$$

Thus, we get the desired upper bound for $\lambda_s(n)$.

□

Corollary 2.33 *For $s \geq 2$ and for sufficiently large n*

$$\lambda_s(n) = n \cdot 2^{\Gamma_s(n) \cdot (1+o(1))}.$$

Proof: We have already shown that the above equality holds for $s \leq 4$. Therefore, we assume that $s > 4$. By the definition of Γ_s, for all $i < s$

$$\lim_{n \to \infty} \frac{\Gamma_i(n)}{\Gamma_s(n)} = 0.$$

Thus

$$\lim_{n \to \infty} \frac{C_s(n)}{\Gamma_s(n)} = 0$$

so that

$$\begin{aligned}
\lambda_s(n) &\leq n \cdot 2^{\Gamma_s(n) \cdot \left(1 + \frac{C_s(n)}{\Gamma_s(n)}\right)} \\
&= n \cdot 2^{\Gamma_s(n)(1+o(1))}.
\end{aligned}$$

□

Remark 2.34:

(i) The above proof of the upper bound is similar to the one given by Sharir in [117]. The main difference between the two proofs is that he estimated the contribution of the external symbols by dividing them into two parts: *starting symbols* and *non-starting symbols* while we divide them into three parts *starting*, *middle* and *ending* which allows us to write $\Psi_s(m, n)$ in terms of $\Psi_{s-2}(b, n^\star)$ instead of $\Psi_{s-1}(b, n^\star)$. Since we go two steps down at a time instead of one, we get a new $\alpha(n)$ term in the exponent only after increasing the value of s by 2 instead of every step. Moreover, in our bounds, we do not have any $\log(\alpha(n))$ term for even values of s, i.e. the base of the exponent in this case is 2, not $\alpha(n)$.

(ii) Note that in equation 2.16 we approximate $\Psi_{s-1}(m, n)$ to $\lambda_{s-1}(n)$ instead of substituting the bound achieved from (2.16) inductively as we do in the case of $s = 4$. If we substitute $\Psi_{s-1}(m, n)$ by (2.16) instead of approximating it to $\lambda_{s-1}(n)$, we can improve the upper bound for $\lambda_s(n)$ a little bit by optimizing the polynomial $C_s(n)$, however it does not affect the leading term and also as we will see in the next section, even then the bounds we obtain still do not match our lower bounds. Moreover, the proof also becomes much more complicated.

2.5 The Lower Bounds for $\lambda_s(n)$

In this section we establish the lower bounds for $\lambda_s(n)$. We show that for $n \geq A(7)$ and even $s \geq 6$,

$$\lambda_s(n) \geq n \cdot 2^{K_s \alpha(n)^{\frac{s-2}{2}} + Q_s(n)}$$

where $K_s = \frac{1}{\frac{s-2}{2}!}$ and Q_s is a polynomial in $\alpha(n)$ of degree at most $\frac{s-4}{2}$ defined later in this section. These bounds improve significantly the previous lower bounds given by Sharir [118] and almost match the upper bounds given in the previous section for even values of s.

The proof of this bound is quite similar to the proof of the lower bound for $s = 4$, only it is more complicated. Before we give the proof, we will need to define several functions which behave similarly to the Ackermann function, and prove certain properties about them. We will then define a collection of Davenport Schinzel sequences $S_k^s(m)$ of order s that realize our lower bounds.

2.5.1 The functions $\mathbf{F}^s_k(m)$, $\mathbf{N}^s_k(m)$, $\mathbf{F}^\omega_k(m)$ and their properties

For the lower bounds we will need two classes of functions that grow faster than Ackermann's functions though roughly at the same rate. These functions, $F^s_k(m)$ and $N^s_k(m)$, are defined for integral $k \geq 1$, integral $m \geq 1$ and even $s \geq 2$. $N^s_k(m)$ gives the number of symbols composing the sequence $S^s_k(m)$, and $F^s_k(m)$ gives the number of fans in $S^s_k(m)$ (see below for more details). These functions are defined inductively by the following equations.

$$
\begin{array}{lcll}
F^s_1(m) & = & 1 & m \geq 1,\ s \geq 2 \\
F^2_k(m) & = & 1 & m \geq 1,\ k \geq 1 \\
F^s_k(1) & = & (2^k - 1) \cdot F^{s-2}_{k-1}(2^{k-1}) \cdot F^s_{k-1}(N^{s-2}_{k-1}(2^{k-1})) & k \geq 2,\ s \geq 4 \\
F^s_k(m) & = & 2F^s_k(m-1) \cdot F^{s-2}_{k-1}(F^s_k(m-1)) \cdot & \\
 & & F^s_{k-1}(N^{s-2}_{k-1}(F^s_k(m-1))) & m \geq 2, k \geq 2, \\
 & & & s \geq 4
\end{array}
$$

$$
\begin{array}{lcll}
N^s_1(m) & = & m & m \geq 1, s \geq 2 \\
N^2_k(m) & = & m & m \geq 1, k \geq 1 \\
N^s_k(1) & = & N^s_{k-1}(N^{s-2}_{k-1}(2^{k-1})) & k \geq 2, s \geq 4 \\
N^s_k(m) & = & N^s_{k-1}(N^{s-2}_{k-1}(F^s_k(m-1))) + 2N^s_k(m-1) \cdot & \\
 & & F^{s-2}_{k-1}(F^s_k(m-1)) \cdot F^s_{k-1}(N^{s-2}_{k-1}(F^s_k(m-1))) & m \geq 2, k \geq 2, \\
 & & & s \geq 4
\end{array}
$$

For $s = 4$, these formulas define the functions $F_k(m)$ and $N_k(m)$ that we used in the lower bounds for $\lambda_4(n)$.

We will now state several facts about the functions $F^s_k(m)$ and $N^s_k(m)$. Their proofs are given in Appendix 3. Notice that it is clear from the definitions that these functions are always positive integers.

Lemma 2.35 *For* $m \geq 2$, $F^s_k(m) \geq F^s_k(m-1)$.

Lemma 2.36 *For* $m \geq 2$, $N^s_k(m) \geq N^s_k(m-1)$.

These facts are trivially true when $k = 1$ or $s = 2$. For $k > 1$ we see that $F_k^s(m) \geq 2F_k^s(m-1)$ and $N_k^s(m) \geq 2N_k^s(m-1)$.

Recall the product D_k we used in the lower bound for $s = 4$. We will also need it in this bound, as well as some of its properties. The definition of D_k was

$$D_k = \prod_{j=1}^{k} \frac{2^{j-1}}{2^j - 1}.$$

We need

Lemma 2.37 *For $k \geq 2$, $D_k \leq 2^{-(k-2)}$.*

Proof:

$$\begin{aligned}
\prod_{j=1}^{k} \frac{2^{j-1}}{2^j - 1} &\leq 1 \cdot \prod_{j=2}^{k} \frac{2^j - 1}{2^{j+1} - 4} \\
&= \frac{\prod_{j=2}^{k}(2^j - 1)}{\prod_{j=2}^{k}(2^{j+1} - 4)} \\
&= \frac{\prod_{j=2}^{k}(2^j - 1)}{4^{k-1} \cdot \prod_{j=1}^{k-1}(2^j - 1)} \\
&= \frac{2^k - 1}{4^{k-1}} \\
&\leq 2^{-k+2}.
\end{aligned}$$

□

Another function we need is $P(k, s)$, defined on positive integers k and even positive integers s as follows

$$P(k, s) = \sum_{i=1}^{\frac{s}{2}-1} \binom{k-2}{i}$$

where we define the binomial coefficient $\binom{a}{b}$ to be 0 if $a < b$.

Lemma 2.38 *For $k \geq 2$,*

$$P(k, s) = \sum_{i=1}^{k-1} P(i, s-2) + k - 2.$$

Proof: We prove this by induction. The assertion is clearly true when $k = 2$, since all terms in the summation are zero. Now, assume this is true for all $k' < k$. Then

$$\begin{aligned}
\sum_{i=1}^{k-1} P(i,\, s-2) + k - 2 &= 1 + P(k-1, s-2) + \sum_{i=1}^{k-2} P(i,\, s-2) + k - 3 \\
&= 1 + P(k-1,\, s-2) + P(k-1,\, s) \\
&= 1 + \sum_{i=1}^{\frac{s}{2}-2} \binom{k-3}{i} + \binom{k-3}{i} \\
&= \sum_{i=1}^{\frac{s}{2}-1} \left(\binom{k-3}{i-1} + \binom{k-3}{i} \right) \\
&= \sum_{i=1}^{\frac{s}{2}-1} \binom{k-2}{i} \\
&= P(k, s)
\end{aligned}$$

which is what we were trying to prove.

□

We now show

Lemma 2.39

$$\frac{N_k^s(m)}{F_k^s(m)} \le m \cdot 2^{-P(k,s)}. \tag{2.19}$$

Proof: We will actually show that

$$\frac{N_k^s(m)}{F_k^s(m)} \le m \cdot D_k \cdot 2^{-\sum_{i=1}^{k-1} P(i,s-2)}, \tag{2.20}$$

Lemmas 2.37 and 2.38 above show that this implies Lemma 2.39. The proof of this lemma will be by induction. During the induction, we use both this inequality and the one stated in the lemma for *smaller* values of s, k and m.

We prove that the inequality holds for m, k, and s, assuming that it holds for m', k' and s' whenever $s' < s$, or $k' < k$ and $s' = s$, or $m' < m$, $k' = k$ and $s' = s$.

Case 1: $s = 4$. In this case $P(i, s-2) = 0$ for $i \geq 1$, so we have to show

$$\frac{N_k^4(m)}{F_k^4(m)} \leq m \cdot D_k$$

which is what we have shown in Theorem 2.22.

Case 2: $k = 1$. In this case

$$\begin{aligned}\frac{N_1^s(m)}{F_1^s(m)} &= m = m \cdot 1 \\ &= m \cdot D_1 = m \cdot 2^{-P(1,s-2)}.\end{aligned}$$

Case 3: $m = 1$. We have

$$\begin{aligned}\frac{N_k^s(1)}{F_k^s(1)} &= \frac{N_{k-1}^s(N_{k-1}^{s-2}(2^{k-1}))}{(2^k-1)\cdot F_{k-1}^{s-2}(2^{k-1})\cdot F_{k-1}^s(N_{k-1}^{s-2}(2^{k-1}))} \\ &\leq \frac{N_{k-1}^{s-2}(2^{k-1})}{(2^k-1)\cdot F_{k-1}^{s-2}(2^{k-1})}\cdot D_{k-1}\cdot 2^{-\sum_{i=1}^{k-2}P(i,s-2)} \\ &\quad \text{(Using equation 2.20)} \\ &\leq \frac{2^{k-1}}{2^k-1}\cdot D_{k-1}\cdot 2^{-P(k-1,s-2)}\cdot 2^{-\sum_{i=1}^{k-2}P(i,s-2)} \\ &\quad \text{(Using equation 2.19)} \\ &= D_k \cdot 2^{-\sum_{i=1}^{k-1}P(i,s-2)}.\end{aligned}$$

Case 4: $m > 1$. We have

$$\begin{aligned}\frac{N_k^s(m)}{F_k^s(m)} &= \frac{N_{k-1}^s(N_{k-1}^{s-2}(F_k^s(m-1)))}{2F_k^s(m-1)\cdot F_{k-1}^{s-2}(F_k^s(m-1))\cdot F_{k-1}^s(N_{k-1}^{s-2}(F_k^s(m-1)))} + \\ &\quad \frac{N_k^s(m-1)}{F_k^s(m-1)} \\ &\leq \frac{N_{k-1}^{s-2}(F_k^s(m-1))}{2F_k^s(m-1)\cdot F_{k-1}^{s-2}(F_k^s(m-1))}\cdot D_{k-1}\cdot 2^{-\sum_{i=1}^{k-2}P(i,s-2)} + \\ &\quad (m-1)\cdot D_k\cdot 2^{-\sum_{i=1}^{k-1}P(i,s-2)} \\ &\quad \text{(Using equation 2.20)}\end{aligned}$$

$$\begin{aligned}
&\leq \frac{D_{k-1}}{2} \cdot 2^{-P(k-1,s-2)} \cdot 2^{-\sum_{i=1}^{k-2} P(i,s-2)} + \\
&\quad (m-1) \cdot D_k \cdot 2^{-\sum_{i=1}^{k-1} P(i,s-2)} \\
&\quad \text{(Using equation 2.19)} \\
&\leq m \cdot D_k \cdot 2^{-\sum_{i=1}^{k-1} P(i,s-2)}
\end{aligned}$$

since $\frac{D_{k-1}}{2} < D_k$.

□

We must still relate the functions F_k^s to the Ackermann's function. We will do this by using limit functions $F_k^\omega(m)$ such that $F_k^\omega(m) \geq F_k^s(m)$ for all s. We define the limit function $F_k^\omega(m)$ by

$$\begin{aligned}
F_1^\omega(m) &= 1 & m \geq 1 \\
F_k^\omega(1) &= (2^k - 1) \cdot F_{k-1}^\omega(2^{k-1}) \cdot F_{k-1}^\omega(2^{k-1} \cdot F_{k-1}^\omega(2^{k-1})) & k \geq 2 \\
F_k^\omega(m) &= 2 \cdot F_k^\omega(m-1) \cdot F_{k-1}^\omega(F_k^\omega(m-1)) \cdot \\
&\quad F_{k-1}^\omega(F_k^\omega(m-1) \cdot F_{k-1}^\omega(F_k^\omega(m-1))) & m \geq 2, k \geq 2.
\end{aligned}$$

We will now show

Lemma 2.40 *For all s, $F_k^\omega(m) \geq F_k^s(m)$.*

Proof: We proceed using induction, Lemmas 2.35 and 2.36, and the inequality $m \cdot F_k^s(m) \geq N_k^s(m)$, which follows from Lemma 2.39.
For $k = 1$ or $s = 2$, the theorem is trivial since $F_1^s(m) = F_k^2(m) = 1$.

We now assume that we have shown this for smaller s, for smaller k with the same s, and for smaller m with the same k and s. For $m = 1$, we have

$$\begin{aligned}
F_k^s(1) &= (2^k - 1) \cdot F_{k-1}^{s-2}(2^{k-1}) \cdot F_{k-1}^s(N_{k-1}^{s-2}(2^{k-1})) \\
&\leq (2^k - 1) \cdot F_{k-1}^\omega(2^{k-1}) \cdot F_{k-1}^s(2^{k-1} \cdot F_{k-1}^{s-2}(2^{k-1})) \\
&\leq (2^k - 1) \cdot F_{k-1}^\omega(2^{k-1}) \cdot F_{k-1}^s(2^{k-1} \cdot F_{k-1}^\omega(2^{k-1})) \\
&\leq (2^k - 1) \cdot F_{k-1}^\omega(2^{k-1}) \cdot F_{k-1}^\omega(2^{k-1} \cdot F_{k-1}^\omega(2^{k-1})) \\
&= F_k^\omega(1).
\end{aligned}$$

Similarly, for $k, m \geq 1$, we have

$$F_k^s(m) = 2F_k^s(m-1) \cdot F_{k-1}^{s-2}(F_k^s(m-1)) \cdot F_{k-1}^s(N_{k-1}^{s-2}(F_k^s(m-1)))$$

$$
\begin{aligned}
&\leq\ 2F_k^\omega(m-1)\cdot F_{k-1}^{s-2}(F_k^\omega(m-1))\cdot F_{k-1}^s(N_{k-1}^{s-2}(F_k^\omega(m-1)))\\
&\leq\ 2F_k^\omega(m-1)\cdot F_{k-1}^\omega(F_k^\omega(m-1))\cdot F_{k-1}^s(F_k^\omega(m-1)\cdot\\
&\qquad F_{k-1}^{s-2}(F_k^\omega(m-1)))\\
&\leq\ 2F_k^\omega(m-1)\cdot F_{k-1}^\omega(F_k^\omega(m-1))\cdot F_{k-1}^\omega(F_k^\omega(m-1)\cdot\\
&\qquad F_{k-1}^\omega(F_k^\omega(m-1)))\\
&=\ F_k^\omega(m).
\end{aligned}
$$

□

We will now need to prove various facts about the functions F^ω.

Lemma 2.41 $F_2^\omega(m) = 3\cdot 2^{m-1}$.

This follows from the definitions: substituting $F_1^\omega(m) = 1$ in the recursions, we get $F_2^\omega(m) = 2\cdot F_2^\omega(m-1)$ and $F_2^\omega(1) = 3$.

Lemma 2.42 *For $k \geq 2$, $2^a\cdot F_k^\omega(m) \leq F_k^\omega(m+a)$.*

This follows from $F_k^\omega(m) \geq 2\cdot F_k^\omega(m-1)$ which is immediate from the definition of F_k^ω. Since $2^a > a$ for $a \geq 1$, Lemma fact:l6 implies

Lemma 2.43 *For $k \geq 2$, $a\cdot F_k^\omega(m) \leq F_k^\omega(m+a)$.*

Finally, we show that

Lemma 2.44 *For $k \geq 2$, $A_k(m+1) \geq 2A_k(m)$.*

Proof: This is clear for $k = 2$. For $k \geq 3$, we assume it is true for smaller k. This gives

$$
\begin{aligned}
A_k(m+1) &= A_{k-1}(A_k(m))\\
&\geq 2A_{k-1}(A_k(m)-1)\\
&\geq 2A_{k-1}(A_k(m-1))\\
&= 2A_k(m).
\end{aligned}
$$

□

We are now ready to prove

Lemma 2.45 $F_k^\omega(m) \leq A_k(7m)$.

Proof: We first show several horrible inequalities on F^ω. Using Lemmas 2.41 through 2.43 extensively, we see that for $k \geq 3$,

$$\begin{aligned}
F_k^\omega(1) &= (2^k-1)\cdot F_{k-1}^\omega(2^{k-1})\cdot F_{k-1}^\omega(2^{k-1}\cdot F_{k-1}^\omega(2^{k-1}))\\
&\leq F_{k-1}^\omega(k+2^{k-1})\cdot F_{k-1}^\omega(F_{k-1}^\omega(k-1+2^{k-1}))\\
&\leq F_{k-1}^\omega(2\cdot F_{k-1}^\omega(k+2^{k-1}))\\
&\leq F_{k-1}^\omega(F_{k-1}^\omega(k+1+2^{k-1}))\\
&\leq F_{k-1}^\omega(F_{k-1}^\omega(2^k))\\
&\leq F_{k-1}^\omega(F_{k-1}^\omega(F_{k-1}^\omega(k)))\\
&\leq F_{k-1}^\omega(F_{k-1}^\omega(F_{k-1}^\omega(F_{k-1}^\omega(1)))).
\end{aligned}$$

The last step follows from

$$F_{k-1}^\omega(1) \geq 2^{k-1}-1 \geq k \text{ for } k \geq 3.$$

Similarly, for $k \geq 3$,

$$\begin{aligned}
F_k^\omega(m) &= 2\cdot F_k^\omega(m-1)\cdot F_{k-1}^\omega(F_k^\omega(m-1))\cdot F_{k-1}^\omega(F_k^\omega(m-1)\cdot\\
&\quad F_{k-1}^\omega(F_k^\omega(m-1)))\\
&\leq 2\cdot F_{k-1}^\omega(2\cdot F_k^\omega(m-1))\cdot F_{k-1}^\omega(F_{k-1}^\omega(2\cdot F_k^\omega(m-1)))\\
&\leq 2\cdot F_{k-1}^\omega(2\cdot F_{k-1}^\omega(2\cdot F_k^\omega(m-1)))\\
&\leq F_{k-1}^\omega(1+2\cdot F_{k-1}^\omega(2\cdot F_k^\omega(m-1)))\\
&\leq F_{k-1}^\omega(4\cdot F_{k-1}^\omega(2\cdot F_k^\omega(m-1)))\\
&\leq F_{k-1}^\omega(F_{k-1}^\omega(2+2\cdot F_k^\omega(m-1)))\\
&\leq F_{k-1}^\omega(F_{k-1}^\omega(4\cdot F_k^\omega(m-1)))\\
&\leq F_{k-1}^\omega(F_{k-1}^\omega(F_{k-1}^\omega(F_k^\omega(m-1)))).
\end{aligned}$$

This last step follows from the inequalities $F_{k-1}^\omega(x) \geq 2^x \geq 4x$ when $x \geq 4$, $k \geq 3$ and $F_k^\omega(x) \geq 2^k - 1 \geq 4$ when $k \geq 3$.

We will now show $F_k^\omega(m) \leq A_k(7m)$. This is easy for $k = 1$ and $k = 2$. For $k \geq 3$, we assume that it is true for k', m' when $k' < k$ and when $k' = k$ and $m' < m$. This gives

$$F_k^\omega(1) \leq F_{k-1}^\omega(F_{k-1}^\omega(F_{k-1}^\omega(F_{k-1}^\omega(1))))$$

$$
\begin{aligned}
&\leq A_{k-1}(7 \cdot A_{k-1}(7 \cdot A_{k-1}(7 \cdot A_{k-1}(7)))) \\
&\leq A_{k-1}(8 \cdot A_{k-1}(7 \cdot A_{k-1}(7 \cdot A_{k-1}(7)))) \\
&\leq A_{k-1}(A_{k-1}(3 + 7 \cdot A_{k-1}(7 \cdot A_{k-1}(7)))) \\
&\leq A_{k-1}(A_{k-1}(16 \cdot A_{k-1}(7 \cdot A_{k-1}(7)))) \\
&\leq A_{k-1}(A_{k-1}(A_{k-1}(4 + 7 \cdot A_{k-1}(7)))) \\
&\leq A_{k-1}(A_{k-1}(A_{k-1}(A_{k-1}(11)))) \\
&\leq A_{k-1}(A_{k-1}(A_{k-1}(A_{k-1}(A_{k-1}(A_{k-1}(2)))))) \\
&= A_k(7).
\end{aligned}
$$

Similarly, for $m > 1$, we get

$$
\begin{aligned}
F_k^\omega(m) &\leq F_{k-1}^\omega(F_{k-1}^\omega(F_{k-1}^\omega(F_k^\omega(m-1)))) \\
&\leq A_{k-1}(7 \cdot A_{k-1}(7 \cdot A_{k-1}(7 \cdot A_k(7m-7)))) \\
&\leq A_{k-1}(A_{k-1}(3 + 7 \cdot A_{k-1}(7 \cdot A_k(7m-7)))) \\
&\leq A_{k-1}(A_{k-1}(A_{k-1}(4 + 7 \cdot A_k(7m-7)))) \\
&\leq A_{k-1}(A_{k-1}(A_{k-1}(A_k(7m-3)))) \\
&= A_k(7m).
\end{aligned}
$$

□

2.5.2 The sequences $\mathbf{S_k^s(m)}$

We will now define the Davenport Schinzel sequences of order s that we will use to prove our lower bound. The sequences of order s will be indexed by two variables, k and m, and called $S_k^s(m)$. The sequence $S_k^s(m)$ will be composed of $N_k^s(m)$ symbols. As in the case $s = 4$, the sequence $S_k^s(m)$ will be a concatenation of $F_k^s(m)$ fans of size m, where a fan of size m is composed of m distinct symbols $a_1, a_2, \cdots, a_m$ and has the form $(a_1 a_2 \cdots a_{m-1} a_m a_{m-1} \cdots a_2 a_1)$, so its length is $2m - 1$. In our construction, we will be replacing fans in certain subsequences by Davenport Schinzel sequences of order $s - 2$. When we replace a fan by a sequence, the sequence will contain the same symbols as in the replaced fan, and the first appearance of these symbols in the sequence will be in the same order as it was in the fan.

We will define $S_k^s(m)$ for even $s \geq 2$, and integral $k \geq 1$, $m \geq 1$. The definition of $S_k^s(m)$ proceeds inductively as follows

I. $S_1^s(m) = (1\,2\,\cdots\,m-1\;m\;m-1\,\cdots\,2\,1)$ for $s \geq 2$, and $m \geq 1$.

II. $S_k^2(m) = (1\,2\,\cdots\,m-1\;m\;m-1\,\cdots\,2\,1)$ for $k, m \geq 1$.

III. To obtain $S_k^s(1)$ for $k > 1$, $s > 2$,

(a) Construct the sequence

$$S' = S_{k-1}^s(N_{k-1}^{s-2}(2^{k-1})).$$

S' has $F_{k-1}^s(N_{k-1}^{s-2}(2^{k-1}))$ fans, each of size $N_{k-1}^{s-2}(2^{k-1})$.

(b) Replace each fan of S' by the sequence $S_{k-1}^{s-2}(2^{k-1})$ using the same set of $N_{k-1}^{s-2}(2^{k-1})$ symbols as the replaced fan, with the first appearance of symbols in the same order.

(c) Regard each element of the resulting sequence as its own singleton fan.

IV. To obtain $S_k^s(m)$, for $k > 1$, $s > 2$, $m > 1$,

(a) First construct the sequence $S_k^s(m-1)$. It has $F_k^s(m-1)$ fans, each of size $m-1$.

(b) Create $2 \cdot F_{k-1}^{s-2}(F_k^s(m-1)) \cdot F_{k-1}^s(N_{k-1}^{s-2}(F_k^s(m-1)))$ distinct copies of it, having pairwise disjoint sets of symbols. Duplicate the middle element of each fan in these copies of $S_k^s(m-1)$ and concatenate all these copies into a long sequence. Call this sequence S'. These copies have

$$2 \cdot F_k^s(m-1) \cdot F_{k-1}^{s-2}(F_k^s(m-1)) \cdot F_{k-1}^s(N_{k-1}^{s-2}(F_k^s(m-1))) = F_k^s(m)$$

fans altogether.

(c) Now construct the sequence

$$S_{k-1}^s(N_{k-1}^{s-2}(F_k^s(m-1))).$$

It has $F_{k-1}^s(N_{k-1}^{s-2}(F_k^s(m-1)))$ fans, each of size $N_{k-1}^{s-2}(F_k^s(m-1))$.

(d) Replace each of its fan of it by the sequence $S_{k-1}^{s-2}(F_k^s(m-1))$ using the same set of $N_{k-1}^{s-2}(F_k^s(m-1))$ symbols as in the replaced fan, making their first appearance in the same order. Duplicate the middle element of each fan of this sequence; these fans come from the sequences $S_{k-1}^{s-2}(F_k^s(m-1))$ and thus have size $F_k^s(m-1)$. Call this sequence $S^\star$.

(e) Notice that the sequence $S^\star$ has

$$2 \cdot F_k^s(m-1) \cdot F_{k-1}^{s-2}(F_k^s(m-1)) \cdot F_{k-1}^s(N_{k-1}^{s-2}(F_k^s(m-1))) = F_k(m)$$

elements, which is the same as the number of fans in S'. To obtain $S_k^s(m)$, insert the sequence $S^\star$ into the sequence S', with each element of $S^\star$ going into the middle of a corresponding fan of S'. The fans of $S_k^s(m)$ are the fans of S', with the extra symbol from $S^\star$ added in the middle.

We will show that the sequence $S_k^s(m)$ has the following properties:

(i) $S_k^s(m)$ is composed of $N_k^s(m)$ symbols.

(ii) $S_k^s(m)$ is the concatenation of $F_k^s(m)$ fans of size m, where each fan is composed of m distinct symbols $a_1, a_2, \cdots, a_m$ and has the form

$$(a_1 a_2 \cdots a_{m-1} a_m a_{m-1} \cdots a_2 a_1),$$

so its length is $2m-1$.

(iii) $S_k^s(m)$ is a Davenport Schinzel sequence of order s.

(iv) If every fan in $S_k^s(m)$ is replaced by any Davenport Schinzel sequence of order $s-2$ on the same m symbols, with the first appearances of these symbols in the same order, the resulting sequence is still a Davenport Schinzel sequence of order s.

Theorem 2.46 $S_k^s(m)$ *satisfies conditions (i)-(iv).*

Proof: By induction on s, k and m. Assume that they hold for $S_{k'}^{s'}(m')$ with $s' < s$, with $s' = s$ and $k' < k$, or with $s' = s$, $k' = k$ and $m' < m$.

Checking properties (i) and (ii) is straightforward from the definition of the functions $N_k^s(m)$ and $F_k^s(m)$. Properties (iii) and (iv) obviously hold when $S_k^s(m) = (1\ 2\ \cdots\ m-1\ m\ m-1\ \cdots\ 2\ 1)$, which is the case when $S_k^s(m)$ is generated by method I or II in the definition. We must then prove that (iii) and (iv) hold when $S_k^s(m)$ is generated by method III or IV.

Consider first the case where $S_k^s(1)$ is obtained by method III. We must first show that $S_k^s(1)$ is a Davenport Schinzel sequence of order s. This is true because, by our induction hypothesis, property (iv) holds for the sequence $S_{k-1}^s(N_{k-1}^{s-2}(2^{k-1}))$. The sequence $S_k^s(1)$ is obtained by replacing every fan of this sequence by a sequence of order $s-2$. Property (iv) for $S_k^s(1)$ follows trivially form property (iii) because all fans have size 1.

Consider next the case of $S_k^s(m)$, when obtained by method IV. We first show that it is a Davenport Schinzel sequence of order s. It is easy to check that the same symbol cannot occur twice in a row, because the duplicated symbols in S' and $S^\star$ always have an element of the other sequence placed between them. It remains to check that $S_k^s(m)$ contains no subsequence

$$\underbrace{a\ \ldots\ b\ \ldots\ a\ \ldots\ b\ \ldots}_{s+2}$$

of length $s+2$, where $a \neq b$. If a and b are both from $S^\star$ there is no such alternating subsequence because, by property (iv) applied to the sequence $S_{k-1}^s(N_{k-1}^{s-2}(F_k^s(m-1)))$, $S^\star$ is a Davenport Schinzel sequence of order s with some elements duplicated. If a and b are both from S', there is no subsequence of length $s+2$ because S' is the concatenation of Davenport-Schinzel sequences of order s on pairwise disjoint sets of symbols, again with some elements duplicated. This leaves the case when a belongs to S' and b to $S^\star$ (or vice versa; the proof for both cases is the same). We are safe here too because into each copy of the sequence $S_{k-1}^s(m-1)$ contained in S', we have only inserted either the *ascending* or the *descending* half of a fan of $S^\star$, and all the symbols in half of a fan are distinct. Thus, between two a's from the sequence S', there can only be one occurrence of b. We can thus get at worst the alternating subsequence

$$b \quad a \quad b \quad a \quad b.$$

We must now show property (iv). We first show that no two adjacent elements are the same. For this, it suffices to show that the first element of every fan is not contained in the preceding fan. We show this by induction. It is clearly true for $S_1^s(m)$, $S_k^2(m)$, and $S_k^s(1)$. The first symbol in a fan in $S_k^s(m)$, $m > 1$, is the first symbol in the corresponding fan of the copy of $S_k^s(m-1)$ that it came from. The preceding fan either contains symbols from the previous copy of $S_k^s(m-1)$ or from the same copy of $S_k^s(m-1)$. In the first case, the two fans share no symbols from S'. In the second case, the first symbol of the fan is not in the preceding fan by our induction hypothesis (the preceding fan has been extended by an element of $S^\star$, not of S'). Thus, when every fan is replaced by a sequence of order $s-2$, two adjacent elements from different sequences of order $s-2$ cannot be the same. Two adjacent elements within a sequence cannot be the same by the definition of a Davenport Schinzel sequence. Thus, no two adjacent elements are the same.

We must now show that when the fans are replaced, no alternating sequence

$$a \quad b \ldots a \quad b$$

of length $s+2$ appears. Suppose that such a sequence appears among the elements of S'. It must appear within one copy of $S_k^s(m-1)$, because different copies contain distinct symbols and do not interleave. If we delete the symbols from $S^\star$ in this copy of $S_k^s(m-1)$ and combine equal elements that have become adjacent, we obtain a copy of $S_k^s(m-1)$ where all the fans have been replaced by Davenport Schinzel sequences of order $s-2$, with symbols making their first appearance in the proper order. Such a sequence cannot contain a subsequence of length $s+2$ because, by condition (iv) applied to $S_k^s(m-1)$, it is a Davenport Schinzel sequence of order s.

When all the fans are replaced by Davenport Schinzel sequences of order $s-2$, a subsequence

$$a \quad b \ldots a \quad b$$

of length $s+2$ cannot appear among the elements of $S^\star$ because each fan of $S_k^s(m)$ contains only one element of $S^\star$. Replacing each fan by a subsequence containing the same symbols thus cannot introduce any new alternations among the elements of $S^\star$.

We must still show that we cannot have a bad subsequence

$$a \quad b \ldots a \quad b$$

of length $s+2$ when a belongs to S' and b to $S^\star$ (the argument for the reverse situation is identical). Recall that for each copy of $S_k^s(m-1)$ in S', a different symbol of $S^\star$ is added to each fan. Since the only appearances of a are in a single copy of $S_k^s(m-1)$, we can restrict our attention to this copy and assume (in the worst case) that b's occur on both sides of it. The symbol b can appear in only one fan of this copy. After this fan is replaced by a Davenport Schinzel sequence of order $s-2$, this sequence will contain at worst a subsequence

$$\underbrace{a \ldots b \ldots a \ldots b \ldots a \ldots}_{s-1}$$

of length $s-1$ (a appears first because b was the middle element of the fan). There may be a's before and after this fan within the copy of $S_k^s(m-1)$, and b's before and after this copy of $S_k^s(m-1)$. We thus get, at worst, the alternating subsequence

$$b \quad a \quad (a \quad b \quad a \ldots b \quad a) \quad a \quad b,$$

of length $s+1$. Property (iv) thus holds for $S_k^s(m)$.

□

Remark 2.47: The above proof fails for odd values of s. In particular, the last argument depends crucially on s being even, so that the alternating sequence of length $s-1$ starts and ends with a.

Theorem 2.48 *When* $n \geq A(7)$,

$$\lambda_s(n) \geq n \cdot 2^{K_s \alpha(n)^{\frac{s-2}{2}} + Q_s(n)}$$

where $K_s = \dfrac{1}{\frac{s-2}{2}!}$ *and* $Q_s(n)$ *is a polynomial in* $\alpha(n)$ *of degree at most* $\frac{s-4}{2}$.

Proof: Let $n_k^s = N_k^s(1)$. Then, for $k \geq 7$, we have

$$\begin{aligned} n_k^s &= N_k^s(1) \leq F_k^s(1) \\ &\leq F_k^\omega(1) \leq A_k(7) \\ &\leq A(k). \end{aligned}$$

We first show that $N_k^s(1) > N_{k-1}^s(1)$, since

$$N_k^s(1) = N_{k-1}^s(N_{k-1}^{s-2}(2^{k-1})) > N_{k-1}^s(1).$$

Thus, for any n, we can find k such that

$$n_k^s \leq n < n_{k+1}^s.$$

Put $t = \left\lfloor \frac{n}{n_k^s} \right\rfloor$, so

$$t \cdot n_k^s \leq n < (t+1) \cdot n_k^s < 2t \cdot n_k^s.$$

Now, using Lemma 2.39

$$\begin{aligned} \lambda_s(n) &\geq t \cdot \lambda_s(n_k^s) \geq t \cdot |S_k^s(1)| \\ &= t \cdot F_k^s(1) \\ &\geq t \cdot N_k^s(1) \cdot 2^{P(k,s)} \\ &> n \cdot 2^{P(k,s)-1}. \end{aligned}$$

The definition of $P(k, s)$ gives

$$P(\alpha(n) - 1, s) - 1 = K_s \cdot \alpha(n)^{\frac{s-2}{2}} + Q_s(n),$$

where Q_s is a polynomial in $\alpha(n)$ of degree at most $\frac{s-4}{2}$ and $K_s = \frac{1}{\frac{s-2}{2}!}$.

If $n \geq A(7)$, then we have $\alpha(n) \leq \alpha(n_{k+1}^s) \leq k+1$. Since P is an increasing function of k, this gives

$$\begin{aligned} \lambda_s(n) &\geq n \cdot 2^{P(\alpha(n)-1,s)-1} \\ &= n \cdot 2^{K_s \alpha(n)^{\frac{s-2}{2}} + Q_s(n)}. \end{aligned}$$

□

Chapter 3

Red-blue Intersection Detection Algorithms

3.1 Introduction

Consider the following problem:

> *Let Γ be a collection of n (possibly intersecting) "red" Jordan arcs in the plane, and let Γ' be a similar collection of m "blue" arcs. Does any red arc intersect any blue arc?*

Suppose the arcs have relatively simple shape; for example, suppose they are line segments, or x-monotonic algebraic arcs of some small fixed degree, so that in particular each pair of them intersects in at most some fixed number of points. There is a simple way to detect a red-blue intersection, using a standard sweep-line algorithm, such as that of Bentley and Ottmann [11]. However, this algorithm runs in time $O((m + n + t) \log(m + n))$, where t is the number of red-red and blue-blue intersections that the algorithm has to sweep through before detecting a red-blue intersection. Since t can be quadratic (that is $\Omega(m^2 + n^2)$) in the worst case, the algorithm is not generally efficient, and in fact might probably be inferior to the naive $O(mn)$ algorithm that checks all possible pairs of a red arc and a blue arc for intersection.

In this chapter we present several efficient algorithms for detecting a red-blue intersection that avoid the overhead of having to examine many red-red or blue-blue intersections. Our first algorithm is for the case when the red arcs in Γ do not cross one another and form the boundary of a simply connected region, for example segments bounding a simple polygon. In this case we can detect a red-blue intersection in time

$$O\left(\lambda_{s+2}(m)\log^2 m + (\lambda_{s+2}(m) + n)\log(m+n)\right),$$

where s is the maximum number of intersection points between a pair of arcs of Γ'. Thus we obtain an almost linear solution to the intersection detection problem in this case. In the more special case, where each red arc and each blue arc is a straight line segment, we can use the ray-shooting technique of Chazelle and Guibas [27] to detect an intersection of each red segment with the blue polygon in $O(\log n)$ time, after $O(n \log\log n)$ preprocessing, thus obtaining an improved algorithm for intersection detection. However, for general arcs, no such efficient "arc-shooting" procedure is known.

Our algorithm uses a recent result of Guibas et al. [75] which shows that the combinatorial complexity of a single face in an arrangement of n arcs, each pair intersecting in at most s points, is $O(\lambda_{s+2}(n))$, and that such a face can be computed in time $O(\lambda_{s+2}(n)\log^2 n)$. We also extend our algorithm to arbitrary collections of intersecting arcs, Γ and Γ', provided the union of all red arcs is a connected set, as is the union of all blue arcs; in this extension the algorithm runs in time $O(\lambda_{s+2}(m+n)\log^2(m+n))$.

These results have several applications in motion planning and collision detection for a simple polygon. For example, Maddila and Yap [87] study the problem of moving (translating and rotating) an n-sided simple polygon $\mathcal{P}$ around a right-angle corner in a corridor. They show that the problem can be reduced to testing a single canonical motion of $\mathcal{P}$ for intersection with the corridor walls, which in turn can be reduced, by studying the problem in a coordinate frame attached to $\mathcal{P}$, to testing whether $\mathcal{P}$ intersects some resulting collection of algebraic arcs describing the relative motion of the corners and walls of the corridor. Our intersection detection algorithm can then be used to obtain an $O(\lambda_s(n)\log^2 n)$ motion planning algorithm (for some small $s > 0$), considerably improving the $O(n^2)$ algorithm given in [87]. More generally, we can use our procedure to obtain fast algorithms for testing a prescribed motion of a simple or "curved" polygon $\mathcal{P}$ for collision with a given collection of obstacles. Under reasonable assumptions on the simplicity of the motion of $\mathcal{P}$ (for example along algebraic paths of low degree) and on the shape of the boundaries of the obstacles and of $\mathcal{P}$, we can obtain close to linear collision detection procedures.

In the general case of the red-blue intersection detection problem, our algorithms are less efficient, but are still significantly better than quadratic. In a nutshell, our first algorithm is efficient because it can restrict the problem of detecting a red-blue intersection to a single face in the blue or the red arrangement. For arbitrary red and blue arrangements this is not possible in general, and one needs to calculate and search for an intersection in many such faces. We then face the problem of obtaining sharp bounds for the complexity of many such faces, and of their efficient calculation. For general arcs, when such procedures are not available, we present a deterministic algorithm that detects a red-blue intersection in time $O\left((m\sqrt{\lambda_{s+2}(n)} + n\sqrt{\lambda_{s+2}(m)})\log^{1.5}(m+n)\right)$. For certain special types of arcs, for which many faces in their arrangements can be computed efficiently, we obtain improved (albeit randomized) algorithms. For example, if all arcs are line segments or unit circles, then we can detect an intersection using a randomized algorithm whose expected running time is $O((m+n)^{4/3+\epsilon})$, for any $\epsilon > 0$, exploiting the results of [38], [59] and [55].

We believe our algorithms can be extended to obtain all red-blue intersections in an output-sensitive fashion; in Chapter 5 we show that if all arcs are line segments, then all K red-blue intersections can be reported (resp. counted) in time roughly $(m+n)^{4/3} + K$ (resp. $(m+n)^{4/3}$).

The chapter is organized as follows. Section 3.2 presents the efficient algorithm for the case of a simple (curved) polygon, and Section 3.3 gives applications of this algorithm to motion planning and collision detection. Section 3.4 describes algorithms for general red-blue intersection detection. Section 3.5 mentions a few additional applications of the general algorithm. Finally in Section 3.6 we conclude by mentioning some open problems.

3.2 Intersection between a Simple Polygon and Arcs

In this section, we give an efficient algorithm for the case when the red

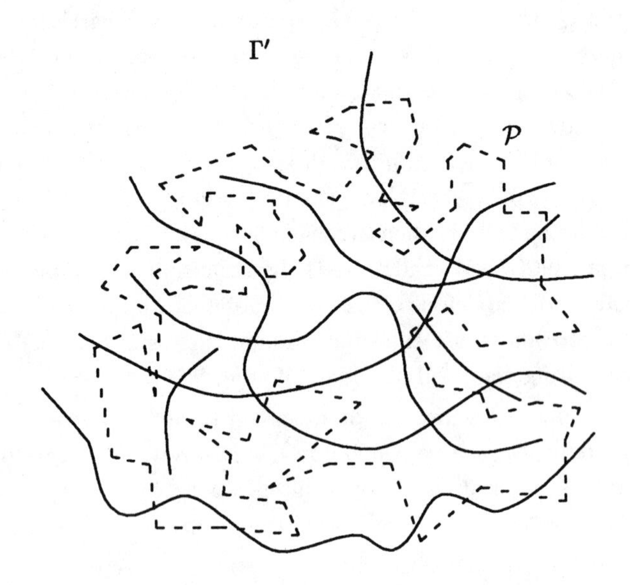

Figure 3.1: Red arcs and blue polygon

arcs in Γ are non-intersecting line segments forming the boundary of a simple polygon $\mathcal{P}$ (see figure 3.1) or, more generally, non-intersecting Jordan arcs whose union is the boundary of a simply-connected region. Let $\mathcal{A}(\Gamma')$ denote the arrangement of Γ', and let z denote a distinguished point on $\partial\mathcal{P}$. The following Lemma is a simple but key observation which allows us to obtain an efficient algorithm.

Lemma 3.1 *Let $\mathcal{F}$ be the face of $\mathcal{A}(\Gamma')$ containing the point z. Then $\partial\mathcal{P}$ and Γ' intersect if and only if $\partial\mathcal{F}$ and $\partial\mathcal{P}$ intersect.*

Proof: The "if" part is obvious. For the "only if" part, simply follow $\partial\mathcal{P}$ from z in, say, the clockwise direction until the first intersection with Γ' is encountered, which necessarily lies on $\partial\mathcal{F}$.

□

The above lemma suggests that we do not have to look at the entire arrangement of Γ', whose combinatorial complexity might be $\Omega(m^2)$ in the worst case, but instead we only need to compute a single face of $\mathcal{A}(\Gamma')$. Guibas, Sharir and Sifrony [75] (see also [105]) have studied the complexity of a single face in an arrangement of Jordan arcs. They have shown

Theorem 3.2 (Guibas et al. [75]) *Given a set $\Gamma' = \{\gamma_1, \gamma_2, \ldots, \gamma_m\}$ of Jordan arcs in the plane so that each pair of them intersect in at most s points, the complexity of a single face of $\mathcal{A}(\Gamma')$ is bounded by $O(\lambda_{s+2}(m))$ and it can be computed in $O(\lambda_{s+2}(m)\log^2 m)$ time, under an appropriate model of computation.*

Remark 3.3: If the arcs in Γ' are closed or bi-infinite Jordan curves, then the bound on the combinatorial complexity of (resp. the time to compute) a single face of their arrangement can be replaced by $O(\lambda_s(m))$ (resp. $O(\lambda_s(m)\log^2 m)$) (see [113] and [75]).

The combinatorial bound is obtained by showing that the circular sequence of arcs in the order in which they appear along the boundary of the given face can be written as a Davenport-Schinzel sequence of order $s+2$. The algorithm given by [75] uses a divide and conquer approach. It divides Γ' into two subsets $\mathbf{\Gamma'_1} = \{\gamma'_1, \ldots, \gamma'_{m/2}\}$ and $\mathbf{\Gamma'_2} = \{\gamma'_{m/2+1}, \ldots, \gamma'_m\}$ and recursively computes the face $\mathcal{F}_1$ of $\mathcal{A}(\Gamma'_1)$ and the face $\mathcal{F}_2$ of $\mathcal{A}(\Gamma'_2)$ containing the point z. The desired face $\mathcal{F}$ is the connected component of $\mathcal{F}_1 \cap \mathcal{F}_2$ that contains the point z. Using a relatively simple line-sweeping algorithm, a subsequent "red-blue merge" procedure then obtains $\mathcal{F}$ from $\mathcal{F}_1$ and $\mathcal{F}_2$ in $O(\lambda_{s+2}(m) \log m)$ time. Hence, the overall running time is $O(\lambda_{s+2}(m) \log^2 m)$.

Remark 3.4: Here and later we assume a model of computation in which various basic operations involving the given arcs are assumed to require $O(1)$ time. These include finding the intersection of a pair of arcs, or of an arc with a line, testing whether a point lies above or below an arc, and finding the points of vertical tangency along an arc. We are thus more interested in the combinatorial complexity than in the algebraic complexity of manipulating such arcs, which is an interesting subject in its own right.

Having computed $\mathcal{F}$, all we have to do is test whether $\partial\mathcal{F}$ and $\partial\mathcal{P}$ intersect. This is easily done by a variant of the Bentley-Ottmann algorithm [11] in time $O((\lambda_{s+2}(m) + n) \log(m + n))$. Thus, we can conclude

Theorem 3.5 *If the arcs in Γ form the boundary of a simple ("curved") polygon $\mathcal{P}$, then an intersection between Γ and Γ' can be detected in time*

$$O\left(\lambda_{s+2}(m) \log^2 m + (\lambda_{s+2}(m) + n) \log(m + n)\right)$$

where s is the maximum number of intersections between a pair of arcs in Γ'.

Proof: Immediate from Theorem 3.2 and the discussion given above.

□

Remark 3.6:

(i) If the arcs in Γ' are also line segments, then we can use the ray-shooting technique of [27] to detect an intersection of a blue arc with the red polygon in $O(\log n)$ time, after $O(n \log \log n)$ preprocessing.

(ii) If the arcs in Γ' are closed or bi-infinite Jordan curves, we can replace $s + 2$ by s in the preceding results.

The above technique can be extended further to detect an intersection between Γ and Γ' when each of $\mathcal{A}(\Gamma)$ and $\mathcal{A}(\Gamma')$ is a connected planar graph. Let p be an endpoint of an arc in Γ' and let $\mathcal{F}$ be the face of $\mathcal{A}(\Gamma)$ containing the point p. Then we have

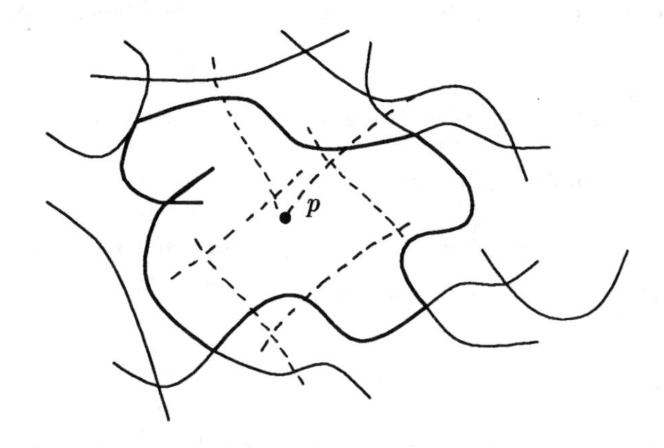

Figure 3.2: $\mathcal{A}(\Gamma)$ and $\mathcal{A}(\Gamma')$: solid (bold) arcs denote Γ (Γ'), and bold arcs denote $\mathcal{F}$

Lemma 3.7 *The arcs in Γ and Γ' intersect if and only if $\partial\mathcal{F}$ and Γ' intersect (see figure 3.2).*

Proof: The "if" part is obvious. For the "only if" part, let q be a red-blue intersection point. Since $\mathcal{A}(\Gamma')$ is a connected graph, there exists a connected path Π from p to q along the edges of $\mathcal{A}(\Gamma')$. Follow Π from p until its first intersection with $\mathcal{A}(\Gamma)$, which necessarily lies on $\partial\mathcal{F}$.

□

Thus, we can reduce the problem of detecting an intersection between Γ and Γ' to the problem of detecting an intersection between Γ' and a simply connected region $\mathcal{F}$, which we can solve using the preceding technique. Therefore, we obtain

Theorem 3.8 *If each of $\mathcal{A}(\Gamma)$ and $\mathcal{A}(\Gamma')$ is a connected planar graph, then an intersection between Γ and Γ' can be detected in $O(\lambda_{s+2}(m+n)\log^2(m+$*

n)) time, where s is the maximum number of intersections between a pair of arcs in Γ or in Γ'.

Proof: Using Theorem 3.2, we can compute the face $\mathcal{F}$ in $O(\lambda_{s+2}(n)\log^2 n)$ time. From Lemma 3.7 it follows that we only have to detect an intersection between $\partial\mathcal{F}$ and Γ', which, by Theorem 3.5 and the following remark, can be done in time

$$O\left(\lambda_{s+2}(m)\log^2 m + (\lambda_{s+2}(m)+\lambda_{s+2}(n))\log(m+n)\right)$$

and hence the overall running time is

$$\begin{aligned} &O\left(\lambda_{s+2}(m)\log^2(m+n) + \lambda_{s+2}(n)\log^2(m+n)\right) = \\ &O\left(\lambda_{s+2}(m+n)\log^2(m+n)\right). \end{aligned}$$

□

Remark 3.9:

(i) Note that we do not require any bound on the number of intersections between a given blue arc and a given red arc, because our algorithms stop as soon as one such intersection is detected. However, the algorithm still assumes that an intersection between a blue arc and a red arc can be detected in $O(1)$ time.

(ii) If the maximum number of intersection points between a pair of arcs in Γ (resp. Γ') is s (resp. s'), then the running time of the above algorithm is

$$O\left((\lambda_{s+2}(n)+\lambda_{s'+2}(m))\log^2(m+n)\right).$$

Moreover, if the arcs in Γ (resp. Γ') are closed or bi-infinite curves, then we can replace $s+2$ (resp. $s'+2$) by s (resp. s').

3.3 Applications to Collision Detection and Motion Planning

Using the intersection detection algorithm of the previous section, we present an efficient algorithm for the following problem:

Given a simple polygon $\mathcal{P}$ that is allowed to translate and rotate in the plane, a set of polygonal obstacles and a prescribed continuous motion of $\mathcal{P}$, check whether $\mathcal{P}$ collides with any obstacle during this motion.

We begin by reducing this problem to the one studied in the previous section. Let $\mathbf{O} = \{O_1, O_2, \ldots, O_l\}$ denote the given set of polygonal obstacles having pairwise disjoint interiors. We can represent their boundaries as a collection of "walls" $\mathbf{W} = \{W_1, \ldots, W_m\}$ where each W_i is an edge of some obstacle. We also define two coordinate systems: the *environment-frame* and the *object-frame*. The environment-frame is the usual coordinate system in which the obstacles are fixed and $\mathcal{P}$ moves. On the other hand, the object-frame is rigidly attached to $\mathcal{P}$, so in it $\mathcal{P}$ is fixed and the positions of the obstacles vary depending on the position of $\mathcal{P}$ in the environment-frame. The position of $\mathcal{P}$, in the environment-frame, can be described by a pair $Z = [X, \theta]$, where $X = (x, y)$ is the position of a fixed point p in $\mathcal{P}$, which is taken to be the origin of the object-frame, and θ is the orientation of the x-axis p_x of the object-frame, relative to the environment-frame (see figure 3.4). Formally, for any such placement Z of $\mathcal{P}$ we obtain an Euclidean transformation Ψ_Z which maps a point ξ in the object-frame to its position ω in the environment-frame as follows

$$x(\omega) = x + x(\xi) \cdot \cos\theta - y(\xi) \cdot \sin\theta \qquad (3.1)$$

$$y(\omega) = y + x(\xi) \cdot \sin\theta + y(\xi) \cdot \cos\theta. \qquad (3.2)$$

In the object-frame, the position of any obstacle O_i depends on the position Z of $\mathcal{P}$ in the environment-frame and is obtained by the corresponding inverse map Ψ_Z^{-1}.

Let Π denote a continuous motion of $\mathcal{P}$, given as a continuous path $\Pi : [0, 1] \to \mathcal{E}^2 \times \mathcal{S}^1$. We call Π a *basic* path if $\Pi([0, 1])$ is an algebraic curve in x, y and $\tan\frac{\theta}{2}$ having a fixed and small degree. Examples of basic paths are translation along a straight line or along a low-degree algebraic curve, rotation, sliding with a fixed pair of vertices of $\mathcal{P}$ touching a fixed pair of walls, etc. For the sake of exposition, we keep the notion of a basic path somewhat loose and informal, so for example we do not specify precisely how low do we want its degree to be, etc. We assume that Π is either a basic path or a concatenation of basic paths. The complexity of Π, denoted by $|\Pi| = k$, is the number of basic paths from which it is composed, and we write $\Pi = \pi_1 \| \pi_2 \| \cdots \| \pi_k$. We use the term Ψ_Π to denote the map that sends each point ξ in the object-frame to

$$\{\Psi_{\Pi(t)}(\xi) \mid t \in [0, 1]\}.$$

Ψ_Π^{-1} is defined symmetrically for the inverse transformation Ψ^{-1}.

Lemma 3.10 *Let q be a fixed point in the environment-frame and let $\mathcal{P}$ move along a basic path Π. Then $\Psi_\Pi^{-1}(q)$ is also a basic path (albeit its maximum degree might be higher). Similarly, if W is a wall, then $\Psi_\Pi^{-1}(W)$ is a region whose boundary consists of a constant number of algebraic curves of fixed degree.*

Proof: Let $q = (\rho, \sigma)$ be a fixed point in the environment-frame and let $\Pi(t) = [x(t), y(t), \theta(t)]$ denote the placement of $\mathcal{P}$ at time t as it moves along Π. Suppose $(\xi, \zeta) = \Psi_{\Pi(t)}^{-1}(q)$, then by (3.1) and (3.2)

$$\begin{aligned} \rho &= x(t) + \xi \cdot \cos\theta(t) - \zeta \cdot \sin\theta(t) \\ \sigma &= y(t) + \xi \cdot \sin\theta(t) + \zeta \cdot \cos\theta(t). \end{aligned}$$

Therefore,

$$\xi = (\rho - x(t)) \cdot \cos\theta(t) + (\sigma - y(t)) \cdot \sin\theta(t) \tag{3.3}$$

$$\zeta = -(\rho - x(t)) \cdot \sin\theta(t) + (\sigma - y(t)) \cdot \cos\theta(t). \tag{3.4}$$

Since Π is an algebraic curve of fixed degree, $x(t), y(t)$, $\sin\theta(t)$ and $\cos\theta(t)$ (or equivalently, $\tan\frac{\theta(t)}{2}$) are related by a pair of polynomial equations

$$P\left(x,\, y,\, \tan\frac{\theta}{2}\right) = 0, \tag{3.5}$$

$$Q\left(x,\, y,\, \tan\frac{\theta}{2}\right) = 0 \tag{3.6}$$

of some fixed degree. Eliminating these three variables from the four equations (3.3), (3.4), (3.5) and (3.6) (cf. [130]), we get an algebraic curve (in ξ and ζ) of fixed degree, showing that $\Psi_\Pi^{-1}(q)$ is a basic path.

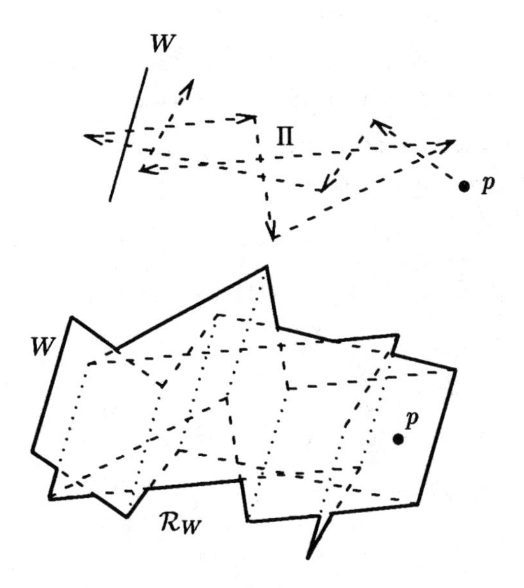

Figure 3.3: Area $\mathcal{R}_W$ swept by a wall W

Now let us consider the area swept by a wall W (see figure 3.3). Without loss of generality we can assume that W is a portion of the y-axis. It is easily checked that any point on the boundary of $\Psi_{\Pi}^{-1}(W)$ either lies on one of the paths $\Psi_{\Pi}^{-1}(q_1)$, $\Psi_{\Pi}^{-1}(q_2)$, where q_1, q_2 are the endpoints of W, or lies on the *envelope* of the parametric family of curves $\{\Psi_{\Pi(t)}^{-1}(\ell) : t \in [0,1]\}$, where ℓ is the y-axis (see [Co] for details). For each $t \in [0,1]$, let $\Pi(t) = [x(t), y(t), \theta(t)]$. Then, the points $(\xi, \zeta) \in \Psi_{\Pi(t)}^{-1}(\ell)$ in the object frame satisfy the following equation, as is implied by equations 3.1, 3.2

$$x(t) + \xi \cos\theta(t) - \zeta \sin\theta(t) = 0 \tag{3.7}$$

Let $f(\xi,\zeta,t)$ denote the left hand side of (3.7). The envelope Φ of $f(\xi,\zeta,t) = 0$ (with ξ, ζ as independent variables and t as a parameter) is found by eliminating t from the pair of equations $f(\xi,\zeta,t) = \frac{\partial}{\partial t} f(\xi,\zeta,t) = 0$ (see [46]). In our case these equations are

$$\begin{aligned} \xi\cos\theta(t) - \zeta\sin\theta(t) &= -x(t) \\ \xi\sin\theta(t) + \zeta\cos\theta(t) &= \frac{\dot{x}(t)}{\dot{\theta}(t)}. \end{aligned}$$

Rewriting these equations we obtain

$$\begin{aligned} \xi &= -x(t)\cos\theta(t) + \frac{\dot{x}(t)}{\dot{\theta}(t)}\sin\theta(t) \\ \zeta &= x(t)\sin\theta(t) + \frac{\dot{x}(t)}{\dot{\theta}(t)}\cos\theta(t), \end{aligned} \tag{3.8}$$

where both right-hand sides are algebraic in $x(t)$, $y(t)$, $\tan\frac{\theta}{2}$, as is easily checked. Eliminating these three variables as was done in the case of a point shows that the envelope Φ is algebraic (of degree generally larger than that of Π, but nevertheless having fixed maximum value, depending on the degree of Π). Thus, the boundary of $\Psi_{\Pi}^{-1}(W)$ consists of a constant number of connected subarcs of three basic paths $\Psi_{\Pi}^{-1}(q_1)$, $\Psi_{\Pi}^{-1}(q_2)$, Φ, where this constant depends on the pattern of intersections of these paths.

□

3.3.1 The algorithm

In this subsection we give an efficient algorithm for the collision detection problem, defined in the beginning of the section. A simple approach to solve this problem is to compute the region swept by every edge e_i of $\mathcal{P}$ as $\mathcal{P}$ moves along Π and test whether it intersects any wall W. Since for each basic path π_j, the region swept by e_i is of fixed complexity, we can check in $O(m)$ time if it intersects with any W. There are k basic paths and n edges, so the total time spent will be $O(k \cdot m \cdot n)$. Using the algorithm of Section 3.2, we show that we can do much better than this naive approach.

Without loss of generality, we can assume that the initial position of $\mathcal{P}$ in Π is a free position. In practical applications this can be expected to be the case; in general, we can run a standard line sweeping algorithm (such as that of [11]), to verify that $\mathcal{P}$ starts at a free placement, in $O((n+m)\log(n+m))$ time. Let $\mathcal{R}_j = \mathcal{R}_j(W)$ denote the region in the object-frame swept by the wall W when $\mathcal{P}$ moves along π_j, and let γ_j denote its boundary, that is $\gamma_j = \partial\Psi_{\pi_j}^{-1}(W)$, and let $\Gamma_W = \{\gamma_1, \gamma_2, \ldots, \gamma_k\}$; by Lemma 3.10, each γ_j is the union of $O(1)$ algebraic arcs.

Lemma 3.11 *$\mathcal{P}$ intersects a wall W while moving along Π if and only if $\partial\mathcal{P}$ intersects $\Psi_{\Pi}^{-1}(W)$ in the object frame. Therefore either $\partial\mathcal{P}$ intersects an arc in Γ_W, or $\partial\mathcal{P}$ lies "inside" one of the regions $\mathcal{R}_j$.*

Proof: We only prove the first statement of the Lemma because the second statement is an obvious consequence. Since the initial position of $\mathcal{P}$ is free, collision between W and $\mathcal{P}$ implies that there exists a placement $Z_0 \in \Pi$ such that $\partial\mathcal{P}$ and W intersect at some point $\zeta \in \Psi_{Z_0}(\partial\mathcal{P}) \cap W$. But then it follows by definition that

$$\Psi_{Z_0}^{-1}(\zeta) \in \partial\mathcal{P} \cap \Psi_{Z_0}^{-1}(W).$$

Hence, if $\mathcal{P}$ intersects a wall W while moving along Π, then $\partial\mathcal{P}$ intersects the area swept by W in the object frame.

□

We thus first test, in $O(mk)$ time, whether the origin p lies in any of the regions $\mathcal{R}_j$. If so, an intersection has been detected and we can stop right away. If not, the previous Lemma implies that it suffices to determine whether $\partial\mathcal{P}$ intersects an arc in $\Gamma = \bigcup_{W \in \mathbf{W}} \Gamma_W$. Hence we obtain

Theorem 3.12 *Given a polygon $\mathcal{P}$ with n vertices, a path $\Pi = \pi_1 \| \pi_2 \| \cdots \| \pi_k$, composed of k basic paths $\pi_1, \ldots, \pi_k$ and a set of polygonal obstacles* **O** *with a total of m walls, we can check in time*

$$O\left(\lambda_s(mk)\log^2(mk) + (n + \lambda_s(mk))\log(n + mk)\right)$$

if $\mathcal{P}$ collides with any of the obstacles in **O** *when it moves along Π. Here s is a constant depending on the maximum algebraic degree of the basic paths of Π.*

Proof: It requires only $O(mk)$ time to check if the origin p lies inside any of the regions $\mathcal{R}_j(W)$, the area swept the wall W when $\mathcal{P}$ moves along the basic path π_j. We can compute Γ, the collection of the boundary arcs of all the regions $\mathcal{R}_j$ in time $O(mk)$, and Theorem 3.5 implies that we can detect an intersection between Γ and $\partial\mathcal{P}$ in time

$$O\left(\lambda_s(mk)\log^2(mk) + (n + \lambda_s(mk))\log(n + mk)\right),$$

where $s - 2$ is the maximum number of intersections between a pair of arcs in Γ.

□

3.3.2 Applications to motion planning

In this section we describe some applications of the above collision-detection algorithm. This algorithm is useful when $mk \approx n$, because in this case it takes only $O(\lambda_s(n)\log^2 n)$ time to determine if the polygon $\mathcal{P}$ collides with any obstacle in **O** while moving along the prescribed path Π. Two such typical cases are: (i) There are $O(n)$ walls in the environment and the path of $\mathcal{P}$ consists of $O(1)$ basic paths, and (ii) The environment is of a small fixed size, i.e. has only $O(1)$ walls, but the path Π may consist of as many as $O(n)$ basic paths.

On the other hand if $mk \gg n$, $nk \approx m$, and the number of objects $l = O(1)$ (that is, there are not many obstacles though each obstacle can have many edges), then we can use a similar approach in the environment-frame instead of the object-frame. The basic idea is as follows. Let $\mathcal{R}_j$, for $1 \leq j \leq k$, denote the area swept by the edge e of $\mathcal{P}$, when $\mathcal{P}$ moves along the basic path π_j. Let $\gamma_j = \partial\mathcal{R}_j$ and $\Gamma_e = \{\gamma_1, \dots, \gamma_k\}$. Using Lemma 3.11 we can again show that $\mathcal{P}$ collides with an obstacle O_r if and only if either O_r lies inside any of $\mathcal{R}_j$ or ∂O_r intersects $\Gamma = \bigcup_{e\in\mathcal{P}} \Gamma_e$. Let m_r be the number of edges in O_r. Since it takes only $O(nk)$ time to check whether O_r lies inside any of the regions $\mathcal{R}_j$, it follows from Theorem 3.5 that an intersection between Γ and ∂O_r can be detected in time

$$O\left(\lambda_{s+2}(nk)\log^2 nk + (m_r + \lambda_{s+2}(nk))\log(nk + m_r)\right),$$

where s is the maximum number of intersections between any pair of arcs in Γ. Summing over all l obstacles, the total time required to detect a collision is

$$O\left(\lambda_{s+2}(nk)\log^2 nk + (m + \lambda_{s+2}(nk))\log(nk + m)\right) = O(\lambda_{s+2}(m)\log^2 m),$$

because $nk \approx m$ and $l = O(1)$. Notice that this argument assumes that each obstacle O_r is simply connected.

The main advantage of either of these two variants of our approach is therefore that the resulting complexity bound does not involve the term $O(mn)$, which seems to be unavoidable in any standard motion planning algorithm that explicitly constructs the configuration space.

Returning to our application, there are certain motion planning problems in which one can define a single canonical motion and show that if there exists a collision-free motion from the initial to the final position, then this canonical path is also collision-free. In such cases we can apply our collision detection technique to obtain an improved motion planning algorithm. Such a case is given by Maddila and Yap [87], who have studied the problem of moving a simple n-sided polygon $\mathcal{P}$ around a right-angle corner in a L-shaped corridor. They have presented an $O(n^2)$ algorithm for planning such a motion; using our techniques, we obtain an improved, almost linear algorithm.

A corridor with one right-angle corner is an infinite L-shaped region, where the two sides of the corridor extend in the $-x$ direction and the $+y$ direction, respectively (as shown in figure 3.4). The problem of moving a polygon $\mathcal{P}$ around the corner is to move $\mathcal{P}$ from a given initial position $Z_I = [x_I, y_I, \theta_I]$ in the left side of the corridor, to a final position $Z_F =$

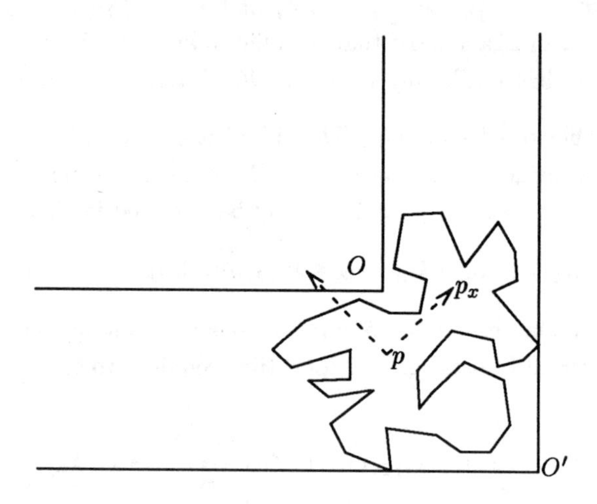

Figure 3.4: Right angle corridor and a simple polygon

$[x_F, y_F, \theta_F]$ in the other side of the corridor, without colliding with the corridor walls. In some applications Z_F may not be given and we only have to find out whether $\mathcal{P}$ can be moved around the corner, from some given initial placement Z_I. We will first analyze the former problem and then show how to handle the latter variant.

It is shown in [87] that a collision-free motion from Z_I to Z_F exists if and only if the following canonical motion of $\mathcal{P}$ is collision-free:

(i) Translate in the $-y$-direction, until $\mathcal{P}$ touches the horizontal wall of the convex corner of the corridor.

(ii) Translate in the x-direction until $\mathcal{P}$ touches the vertical wall of the convex corner.

(iii) Slide $\mathcal{P}$ while touching the two walls of the convex corner until the orientation becomes θ_F; this is the only non-trivial part of the motion, and we denote it by Π.

(iv) Translate in the y-direction until $y(p) = y_F$.

(v) Translate in the $-x$-direction until $x(p) = x_F$.

If Z_I and Z_F are free positions, then all of the translational motions are collision-free. Thus all we have to do is to calculate the path Π in (iii) and verify that it is also collision-free. This is done by applying our algorithm in the object frame to detect an intersection between $\mathcal{P}$ and the concave corner and adjacent walls of the corridor. The efficiency of the algorithm obviously depends on the complexity of Π, that is the number of basic paths composing it, which is analyzed as follows.

We can parameterize Π using the orientation of $\mathcal{P}$ as it changes monotonically and continuously along Π (see [87] for details). At any orientation, the rightmost vertex u of $\mathcal{P}$ touches the vertical wall of the convex corner and the bottommost vertex v of $\mathcal{P}$ touches the horizontal wall of the corner. Except at a finite number of *critical orientations*, where an edge of the convex hull of $\mathcal{P}$ adjacent to u (resp. v) becomes vertical (resp. horizontal), the *supporting pair* (u, v) is unique and does not change between any pair of consecutive critical orientations. The path formed by sliding $\mathcal{P}$ with a fixed pair of vertices touching a fixed pair of walls is known as a *glissette* [86]; each point on $\mathcal{P}$ traces an elliptic arc, and the path (as a mapping into $\mathcal{E}^2 \times \mathcal{S}^1$) is easily seen to be a basic path.

Lemma 3.13 (Maddila and Yap [87]) *The sequence (u_i, v_i) of supporting pairs encountered along Π changes in an orderly manner; (u_i, v_i) change in the order in which they appear along the convex hull of $\mathcal{P}$. Moreover, the number of supporting pairs is at most $2n$ and they can be found in $O(n)$ time.*

The above Lemma implies that Π consists of at most $2n$ basic paths, and by Lemma 3.10 the arcs describing the "inverse" motion of the walls in the object frame of $\mathcal{P}$ are algebraic curves of some fixed low degree. Thus, by Lemma 3.13 and Theorem 3.12 we obtain

Theorem 3.14 *It takes $O(\lambda_s(n) \log^2 n)$ time (for some small value of $s > 0$) to determine whether $\mathcal{P}$ can be moved around a right-angle corner in a corridor.*

If Z_F is not specified, then moving $\mathcal{P}$ around the corner amounts to sliding it as in the above canonical motion until it reaches an orientation

θ_F at which it can be translated vertically into the upper portion of the corridor. This is necessary and sufficient that the width of $\mathcal{P}$ in the direction perpendicular to θ_F should be less than the width of the upper portion of the corridor. This observation enables us to find the smallest such θ_F in $O(n)$ time using the standard "rotating calipers" method. Having found θ_F, the problem reduces to the one just studied, and the same algorithm can be applied.

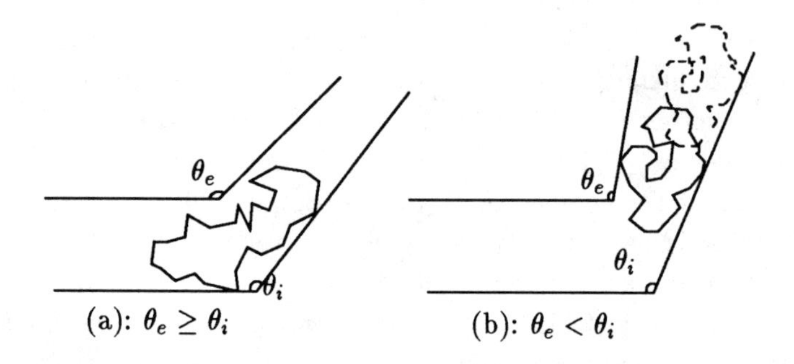

Figure 3.5: General corridors

We can extend our algorithm to a more general corridor in which the exterior angle of the concave corner is not less than the interior angle of the convex corner (see figure 3.5a). Indeed, our algorithm relies on the ability to reduce any collision-free motion of $\mathcal{P}$ around the corner to a canonical one in which $\mathcal{P}$ slides along the walls of the lower convex portion of the corridor. The conditions assumed are easily seen to imply that a "retraction" similar to that used above, that is first translate $\mathcal{P}$ to the right until it touches the right wall, and then slide it downwards along the wall until the lowest vertex of $\mathcal{P}$ touches the lower horizontal wall of the corridor, can be applied in this case. Once again, we can show that this canonical motion consists of at most $2n$ basic paths, leading to an algorithm similar to that presented above. Figure 3.5b gives an example where such a retraction is not continuous when the above conditions are not satisfied.

If $\mathcal{P}$ is a star-shaped polygon, we can somewhat improve the algorithm, because we can replace the calculation of a single face in $\mathcal{A}(\Gamma)$ (as in [75]) by the calculation of a certain lower envelope of Γ, where Γ is the collection of arcs describing the inverse motions of the walls as above. To this end, let ζ be a point in the kernel of $\mathcal{P}$, and consider each arc γ of Γ as a function $r = \gamma(\theta)$ in polar coordinates about ζ in the object frame; if necessary, we break each arc γ into a constant number of pieces, each monotone with respect to θ. The lower envelope $\mathcal{M}_\Gamma$ of Γ in these coordinates is defined as

$$\mathcal{M}_\Gamma(\theta) = \min_{\gamma \in \Gamma}\{\gamma(\theta)\}.$$

Lemma 3.15 *$\mathcal{P}$ collides with a wall forming the concave corner of the corridor while moving along the canonical path if and only if $\mathcal{P}$ intersects the lower envelope $\mathcal{M}_\Gamma$ defined above.*

Proof: Lemma 3.11 implies that $\mathcal{P}$ collides with a wall W while moving along the canonical path Π if and only if $\mathcal{P}$ and $\Psi_\Pi^{-1}(W)$ intersect. Therefore, intersection of $\mathcal{M}_\Gamma$ and $\mathcal{P}$ implies that $\mathcal{P}$ collides with some W while moving along Π.

Conversely, let ξ be an intersection point of $\mathcal{P}$ and $\partial\Psi_\Pi^{-1}(W)$. The point ξ lies on one of the curves in Γ, and if $\xi \notin \mathcal{M}_\Gamma$, then some other point σ on the segment $\overline{\zeta\xi}$ must lie on this lower envelope. Since $\mathcal{P}$ is star-shaped and ζ lies in its kernel, the entire segment $\overline{\zeta\xi}$, and thus also σ, lies inside $\mathcal{P}$. Therefore, $\mathcal{M}_\Gamma$ and $\mathcal{P}$ intersect.

□

An intersection between $\mathcal{M}_\Gamma$ and $\mathcal{P}$ can be easily detected as follows. Let a *breakpoint* of $\mathcal{M}_\Gamma$ denote either an endpoint of some arc $\gamma \in \Gamma$, a point of radial tangency on such an arc, or a point where $\mathcal{M}_\Gamma$ is simultaneously attained by two arcs in Γ. Sort the breakpoints of $\mathcal{M}_\Gamma$ in angular order about ζ and similarly sort the vertices of $\mathcal{P}$ in angular order about ζ. Note that the latter sorted list is the same as the order of the vertices of $\mathcal{P}$ along its boundary. Merge these two lists to obtain an overall list $\mathcal{L}$ of orientations about ζ, with the property that for each interval $\mathcal{I}$ between any pair of consecutive orientations in $\mathcal{L}$, $\mathcal{M}_\Gamma$ is attained over $\mathcal{I}$ by a single arc $\gamma_\mathcal{I}$ and the portion of $\partial\mathcal{P}$ seen from ζ in directions in $\mathcal{I}$ is a portion of a single edge $e_\mathcal{I}$. We can thus detect in constant time an intersection between $\mathcal{P}$ and $\mathcal{M}_\Gamma$ within the angular sector defined by $\mathcal{I}$. There are $O(\lambda_s(n))$ breakpoints in $\mathcal{M}_\Gamma$ (for some small fixed $s > 0$), so the length of $\mathcal{L}$ and the time required to detect an intersection between $\mathcal{M}_\Gamma$ and $\mathcal{P}$ are both $O(\lambda_s(n))$. $\mathcal{M}_\Gamma$ can be calculated (in the required sorted order) in time $O(\lambda_s(n)\log n)$, using a standard divide and conquer technique [6]. We thus have

Theorem 3.16 *It takes only $O(\lambda_s(n)\log n)$ time (for some small $s > 0$) to determine whether a star-shaped polygon $\mathcal{P}$ can be moved around a corner in a corridor.*

Remark 3.17: A similar technique applies if $\mathcal{P}$ is a monotone polygon, in which case the radial lower envelope is replaced by appropriate lower and upper envelopes in the direction of monotonicity of $\mathcal{P}$; we leave details of this easy extension to the reader.

3.4 Red-blue Intersection Detection in General

If the arcs in Γ or in Γ' do not form a simply connected region, or they are arbitrarily intersecting and their arrangements are disconnected, then Lemma 3.1 does not hold and we may have to search for an intersection in more than one face of $\mathcal{A}(\Gamma')$. Two difficulties can then arise. One is that the complexity of many faces in such an arrangement can be much higher than linear. Second, the algorithm of Guibas et al. [75] does not generalize to many faces, at least not for general arcs.

As discussed earlier, a sweep-line approach is doomed to be quadratic in the worst case, because it may have to sweep across many red-red or blue-blue intersections. A technique of Chazelle for point location in algebraic manifolds [18] (see also [39]) can be used in the case of algebraic arcs to obtain a slightly subquadratic algorithm, but otherwise we are not aware of any previous algorithm which achieves substantially subquadratic performance for this general red-blue intersection detection problem. In this section we present algorithms of this kind. We first describe a deterministic algorithm that works for general arcs, and then obtain in several restricted but useful special cases more efficient randomized algorithms.

Throughout this section we assume that the arcs in Γ and Γ' are x-monotonic. Note that if the arcs are algebraic curves, or intersect a vertical line in at most $O(1)$ points, this assumption does not pose any restriction because any non-monotonic arc can be split into $O(1)$ x-monotonic subarcs. Note that this assumption is also implicit in the algorithm of Guibas et al. [75] that we have used in Section 3.2.

3.4.1 Deterministic algorithm

Our deterministic algorithm consists of three phases. In the first phase, we partition Γ' into $\ell' < \frac{m}{r'}$ sets $\Gamma'_0, \Gamma'_1, \Gamma'_2, \ldots, \Gamma'_{\ell'}$, where r' is a parameter to be chosen later, so that (i) for $1 \le j \le \ell'$, every Γ'_j has $r' \le m_j < 2r'$ arcs and $\mathcal{A}(\Gamma'_j)$ forms a connected planar graph, (ii) Γ'_0 (possibly empty) contains the remaining arcs of Γ' and each connected component[1] of $\mathcal{A}(\Gamma'_0)$ has less than r arcs. For each $1 \le j \le \ell'$, we check whether an arc of Γ'_j intersects an arc of Γ, using a modified version of the technique of Section 3.2 (to be described below). In the second phase, we partition Γ into $\ell < \frac{n}{r}$ sets, Γ_0, $\Gamma_1, \Gamma_2, \ldots, \Gamma_\ell$, satisfying conditions analogous to those of the partition of Γ', and for each $1 \le j \le \ell$, we check whether Γ_j intersects Γ'. Finally, in the third phase we check whether Γ'_0 and Γ_0 intersect. Together, the three phases will detect an intersection between Γ and Γ', if one exists.

The first phase of the algorithm proceeds as follows. We partition Γ' into $\Gamma'_0, \Gamma'_1, \ldots, \Gamma'_{\ell'}$ using a line sweep approach [11]. We sweep Γ' from left to right with a vertical line; every time we sweep through an intersection point of two arcs, we merge the connected components of $\mathcal{A}(\Gamma')$ containing them, and whenever a component is found to contain more than r' arcs, we remove all these arcs from Γ'. In more detail, as in any standard line sweep algorithm we store the arcs currently intersecting the sweep line in a list Q, sorted in increasing order of y coordinate of their intersections with the sweep line. In addition we maintain the connected components of $\mathcal{A}(\overline{\Gamma'} \cap h^-)$, where $\overline{\Gamma'}$ is the set of arcs that have been encountered by the sweep line so far but have not yet been deleted by the algorithm, and h^- is the half plane lying to the left of the sweep line. For each arc $\gamma' \in \Gamma'$, we store the connected component $C(\gamma')$ that contains it. If an arc γ' has not been encountered by the sweep line, $C(\gamma')$ is not defined. At every event point of the sweep, i.e.,

[1] In what follows we use the term "connected component" of an arrangement to refer to a component of the union of its arcs, whereas the term "face" will continue to refer to a connected component of the complement of the union of these arcs.

an endpoint of one arc or an intersection point between two arcs in $\overline{\Gamma'}$, we perform the standard list and priority queue updating operations as well as the following additional tasks.

(i) If we reach the left endpoint of some arc γ', then we create a new connected component $C(\gamma')$ containing only γ'. Sweeping through the right endpoint of an arc requires no special action.

(ii) If we reach an intersection point of two arcs γ' and γ'', and $C(\gamma')$ is different from $C(\gamma'')$, then we merge the two connected components containing γ' and γ'' respectively.

(iii) If any connected component C is found to have more than r' arcs, then we output it as a set Γ'_j, and delete C and all of its arcs from $\overline{\Gamma'}$ as well as Q. Note that C need not be a full component of $\mathcal{A}(\Gamma')$, but may be only a subset of such a component. We also remove all endpoints and intersection points along these arcs from the priority queue. The priority queue is also updated by inserting into it all the intersection points between newly adjacent pairs of arcs in Q.

It is easy to see that the number of arcs in each connected component processed during the sweep is between r' and $2r'$. The arcs not deleted by the sweep are put into the set Γ'_0. See below for analysis the time complexity of this sweep.

Next for each $1 \leq j \leq \ell'$, we check whether Γ'_j intersects Γ using the following procedure.

We first construct the arrangement $\mathcal{A}(\Gamma'_j)$ of the arcs in Γ'_j using a line sweep technique. If the two endpoints of an arc $\gamma \in \Gamma$ lie in two different faces of $\mathcal{A}(\Gamma'_j)$, then obviously γ intersects some arc of Γ'_j. If they lie in the same face f_i of $\mathcal{A}(\Gamma'_j)$, then γ need not intersect Γ'_j; however if it does, it has to intersect a subarc lying on the boundary of f_i (as in Lemma 3.1). Therefore the next step is to determine for each arc $\gamma \in \Gamma$, the faces that contain the endpoints of γ. This can be easily done during the line sweeping procedure that produces $\mathcal{A}(\Gamma'_j)$ (see also [106]), in overall time $O((r'^2 + n)\log(r' + n))$. To reiterate, our sweep performs the following steps for each $\gamma \in \Gamma$.

(i) If the two endpoints of γ lie in two different faces of $\mathcal{A}(\Gamma'_j)$, then we have found a red-blue intersection and we stop.

(ii) If the two endpoints of γ lie in the same face f_i of $\mathcal{A}(\Gamma'_j)$, we assign γ to f_i.

Thus if the sweep does not detect an intersection, it produces a partition $\Gamma_{f_1}, \ldots, \Gamma_{f_p}$ of Γ, for $p = O(r'^2)$, such that all arcs in Γ_{f_i} have both of their endpoints in the face f_i of the arrangement. Let $n_i = |\Gamma_{f_i}|$. As argued above, it suffices to check the arcs of each Γ_{f_i} for intersection with the edges of f_i. A simple way of doing this is to take an arc of Γ_{f_i} and check its intersection with all edges of f_i. But in the worst case f_i may be bounded by $\Omega(r')$ arcs and all arcs of Γ may lie in this face, in which case this naive procedure will be too expensive. The following Lemma suggests a way to improve this procedure by exploiting the property that the boundary of each face in $\mathcal{A}(\Gamma'_j)$ is connected.

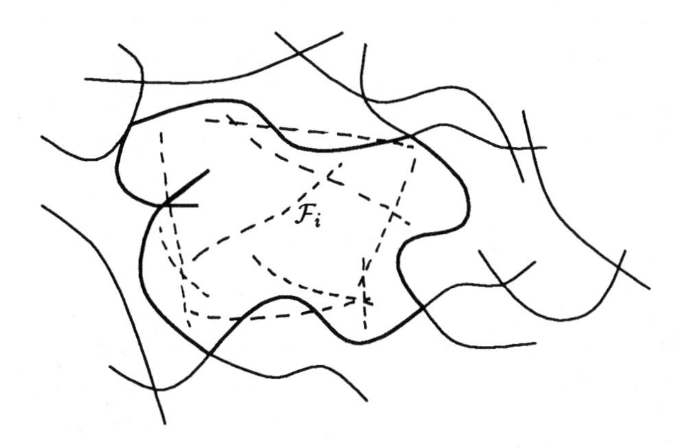

Figure 3.6: Illustration for Lemma 3.18

Lemma 3.18 *Let f_i be a simply connected face of $\mathcal{A}(\Gamma'_j)$, and let $\mathcal{F}_i$ be the unbounded face of $\mathcal{A}(\Gamma_{f_i})$. An arc $\gamma \in \Gamma_{f_i}$ intersects Γ'_j if and only if $\partial\mathcal{F}_i$ and ∂f_i intersect (see figure 3.6). For unbounded faces f_i of $\mathcal{A}(\Gamma'_j)$, take $\mathcal{F}_i$ to be the face of $\mathcal{A}(\Gamma_{f_i})$ that contains some point of ∂f_i.*

Proof: Using the argument of Lemma 3.1 we can prove that Γ_{f_i} and Γ'_j intersect if and only if ∂f_i and Γ_{f_i} intersect. By assumption, ∂f_i is a simple closed Jordan curve, or a simple unbounded Jordan arc. Take any point $z \in \partial f_i$, and let $\mathcal{F}_i$ be the face of $\mathcal{A}(\Gamma_{f_i})$ that contain z; it is easily verified, using the x-monotonicity of the arcs in Γ_{f_i}, that if f_i is bounded then $\mathcal{F}_i$ is the unbounded face of $\mathcal{A}(\Gamma_{f_i})$. The claim now follows from another application of Lemma 3.1, in which the roles of Γ and Γ' are interchanged.

□

We thus compute one face $\mathcal{F}_i$ for each Γ_{f_i}, as prescribed by Lemma 3.18, and then detect an intersection between any ∂f_i and $\partial \mathcal{F}_i$ by performing another line sweep.

This completes the description of the first phase of our algorithm. The second phase proceeds in a completely symmetric manner, where we partition Γ into connected components of size roughly r (to be specified later) instead of r'. For the final phase, we first construct the full arrangements $\mathcal{A}(\Gamma_0')$ and $\mathcal{A}(\Gamma_0)$. Then by sweeping the two arrangements with a vertical line we check whether an edge of $\mathcal{A}(\Gamma_0')$ intersects an edge of $\mathcal{A}(\Gamma_0)$.

The correctness of the algorithm follows easily from the analysis given above. As to analysis of its time complexity, we first need the following simple lemma.

Lemma 3.19 *If each connected component of $\mathcal{A}(\Gamma)$ has at most r arcs, then $\mathcal{A}(\Gamma)$ has only $O(nr)$ vertices.*

Proof: Let $\Gamma_1, \ldots, \Gamma_k$ denote the sets of arcs constituting each connected component of $\mathcal{A}(\Gamma)$. If two sets Γ_i and Γ_j have at most $\frac{r}{2}$ arcs each, we merge them together. We repeat this process until there are no two such subsets. Then each Γ_i (except possibly one) has at least $\frac{r}{2}$ but fewer than r arcs, so the number of these sets is at most $\frac{2n}{r} + 1$ and each $\mathcal{A}(\Gamma_i)$ has $O(r^2)$ vertices. Since $\mathcal{A}(\Gamma_i)$ and $\mathcal{A}(\Gamma_j)$, for $j \neq i$, do not intersect, there are only $\left(\frac{2n}{r} + 1\right) \times O(r^2) = O(nr)$ vertices in $\mathcal{A}(\Gamma)$.

□

Lemma 3.20 *Partitioning Γ' into $\Gamma_1', \ldots, \Gamma_{\ell'}'$ in the first phase of the algorithm can be accomplished in $O(mr' \log m)$ time.*

Proof: Inspecting the line sweeping algorithm used to produce the partitioning, we see that most of its steps can be performed quite efficiently.

For example, maintaining the connected components of $\mathcal{A}(\overline{\Gamma'})$ (creating new connected components and merging pairs of components) can be easily done in time $O(m\alpha(m))$ using the union-find data structure [129]. Also, the time needed to delete arcs from Q every time a component is produced is $O(m \log m)$ because each arc is deleted only once. The most expensive part of the algorithm is the maintenance of the priority queue, which takes time $O(K \log m)$, where K is the number of endpoints and intersection points encountered by the algorithm. Therefore, to prove the Lemma, we only have to show that $K = O(mr')$.

Let $\Gamma'_0, \Gamma'_1, \ldots, \Gamma'_{\ell'}$ be the connected components produced by the algorithm. Since the algorithm merges connected components whenever it sweeps through an intersection point of two arcs lying in two different components, it never encounters an intersection point between arcs of two different Γ'_i and Γ'_j; more precisely, it may have added such points into the priority queue, but it never reaches these points during the sweep. Moreover, at every intersection swept by the line, only a constant number of events are added to the priority queue. Hence, the total number K of intersections encountered by the algorithm is proportional to the number of intersections swept through, which in turn is at most the total number of vertices in all the subarrangements for $\mathcal{A}(\Gamma'_j)$ for $0 \leq j \leq \ell'$. For $j \geq 1$, Γ'_j has less than $2r'$ arcs, therefore $\mathcal{A}(\Gamma'_j)$ has $O(r'^2)$ vertices. Furthermore, each connected component of $\mathcal{A}(\Gamma'_0)$ has less than r' arcs, so by Lemma 3.19 $\mathcal{A}(\Gamma'_0)$ has only $O(mr')$ vertices. Since $\ell' < \left\lceil \frac{m}{r'} \right\rceil$, the total number of intersection points K is $O(mr')$. Hence, it takes only $O(mr' \log m)$ time to partition Γ' into $\Gamma'_0, \ldots, \Gamma'_{\ell'}$.

□

Lemma 3.21 *The time spent in the first phase of the algorithm is bounded by*

$$O\left(mr' \log(m+n) + \frac{m}{r'}\lambda_{s+2}(n) \log^2 n\right).$$

Proof: It follows from the previous Lemma that $\Gamma'_0, \ldots, \Gamma'_{\ell'}$ can be obtained in time $O(mr' \log m)$. Therefore, we only have to bound the time spent in detecting an intersection between Γ'_j, for $1 \leq j \leq \ell'$, and Γ.

If $|\Gamma_{f_i}| = n_i$, then Lemma 3.2 implies that $\partial\mathcal{F}_i$ can be computed in time $O(\lambda_{s+2}(n_i)\log^2 n_i)$. By using a line sweep technique we can determine in time $O((m_i + \lambda_{s+2}(n_i))\log(n_i + m_i))$ whether $\partial\mathcal{F}_i$ and f_i intersect, where m_i is the number of edges in ∂f_i. Summing these costs over all faces f_i of $\mathcal{A}(\Gamma'_j)$, we get

$$\begin{aligned}
&\sum_{i=1}^{p} O\left(m_i \log(n_i + m_i) + \lambda_{s+2}(n_i)\log^2 n_i\right) \\
= \;& O\left(\sum_{i=1}^{p} m_i \log(n_i + m_i) + \sum_{i=1}^{p} \lambda_{s+2}(n_i)\log^2 n_i\right).
\end{aligned}$$

The total number of edges in $\mathcal{A}(\Gamma'_j)$ is bounded by $O(r'^2)$, so $\sum_{i=1}^{p} m_i = O(r'^2)$. Since every arc of Γ is in exactly one Γ_{f_i}, $\sum_{i=1}^{p} n_i = n$, and $\mathcal{A}(\Gamma'_j)$ and $\Gamma_{f_1}, \ldots, \Gamma_{f_p}$ can be computed in time $O(r'^2 \log r')$ and $O((r'^2+n)\log(r'+n))$ respectively. Thus, the total time spent in detecting an intersection between Γ and Γ'_j is

$$O\left(r'^2 \log(r' + n) + \lambda_{s+2}(n)\log^2 n\right).$$

Since, $\ell' < \left\lceil \frac{m}{r'} \right\rceil$, the overall time spent in the first phase, including the overhead of partitioning Γ, is at most

$$\begin{aligned}
&O(mr' \log m) + O\left(\frac{m}{r'} r'^2 \cdot \log(n + r') + \frac{m}{r'} \lambda_{s+2}(n) \cdot \log^2 n\right) \\
= \;& O\left(m \cdot r' \log(n + m) + \frac{m}{r'} \cdot \lambda_{s+2}(n)\log^2 n\right).
\end{aligned}$$

□

Using a symmetric analysis, it follows that the second phase requires

$$O\left(n \cdot r \log(m + n) + \frac{n}{r} \cdot \lambda_{s+2}(m)\log^2 m\right)$$

time. As to the third phase, Lemma 3.19 implies that $\mathcal{A}(\Gamma'_0)$ has only $O(mr')$ vertices, therefore we can easily compute it in time $O(mr' \log m)$. Similarly, $\mathcal{A}(\Gamma_0)$ has $O(nr)$ vertices and can be computed in time $O(nr \log n)$. Once we have $\mathcal{A}(\Gamma'_0)$ and $\mathcal{A}(\Gamma_0)$, we can easily check in time $O((mr' + nr)\log(m + n))$ whether they intersect, which bounds the running time of the third phase. We thus obtain the main result of this section

Theorem 3.22 *Given a set of n "red" Jordan arcs Γ, and another set of m "blue" Jordan arcs Γ', we can detect in time*

$$O\left((m\sqrt{\lambda_{s+2}(n)} + n\sqrt{\lambda_{s+2}(m)})\log^{1.5}(m+n)\right)$$

whether Γ and Γ' intersect, where s is the maximum number of intersections between a pair of arcs in Γ or in Γ'.

Proof: Summing the cost of all the three phases of the algorithm and choosing $r' = \sqrt{\lambda_{s+2}(n)\log(m+n)}$, $r = \sqrt{\lambda_{s+2}(m)\log(m+n)}$, the overall running time becomes

$$O\left((m\sqrt{\lambda_{s+2}(n)} + n\sqrt{\lambda_{s+2}(m)})\log^{1.5}(m+n)\right).$$

□

Remark 3.23: In some special cases we can do somewhat better as follows.

(i) If the maximum number of intersections between a pair of arcs in Γ (resp. Γ') is s (resp. s'), then the running time of the above algorithm becomes

$$O\left((m\sqrt{\lambda_{s+2}(n)} + n\sqrt{\lambda_{s'+2}(m)})\log^{1.5}(m+n)\right).$$

(ii) If Γ' can be partitioned into subsets of the appropriate size r so that the boundary of each face in the arrangement of each subset of Γ is simply connected, then we do not need the last two phases of the above algorithm. Therefore, it easily follows that the running time of the algorithm becomes $O(m\sqrt{\lambda_{s+2}(n)}\log^{1.5} n)$, which is certainly better than the running time of the above algorithm for $n \gg m$. A case in which we can ensure that each f_i is simply connected is when the arcs in Γ' are concatenated to one another at their endpoints to form a single connected (possibly self-intersecting) chain. This situation typically arises in collision-detection problems of the sort studied in Section 3.2 and 3.

(iii) Another special case in which we can get a slightly improved bound is when the arcs in Γ are non-intersecting. Again, we do not need the last two phases. We partition Γ' arbitrarily into $\left\lceil \frac{m}{r'} \right\rceil$ sets Γ'_j, each of size at most r', and then, for each Γ'_j, as in the first phase of the general algorithm, we compute $\Gamma_{f_1}, \ldots, \Gamma_{f_p}$. Since arcs in Γ are non-intersecting we do not have to compute any specific face of Γ_{f_i}, but instead we can detect an intersection between ∂f_i and Γ_{f_i} by performing a line sweep. By choosing $r' = \sqrt{n}$, one can easily check that the overall running time is bounded by $O(m\sqrt{n}\log n)$.

(iv) Again, if all the arcs in Γ (resp. Γ') are closed or bi-infinite Jordan curves, then we can replace $s + 2$ (resp. $s' + 2$) by s (resp. s') in the preceding results.

3.4.2 Relation between many faces and red-blue intersection detection

In this subsection we explore the close relationship between the problem of computing many faces in the arrangement of a given set of Jordan arcs and the red-blue intersection detection problem for those arcs. That is, we show that an efficient algorithm for the former problem can be used to obtain an efficient algorithm for the latter one. The basic idea is that a red-blue intersection can be detected by knowing only a few faces of $\mathcal{A}(\Gamma)$ and $\mathcal{A}(\Gamma')$. We formalize this idea in the following lemma.

Lemma 3.24 *Let $\mathcal{F} = \{\partial f_1, \partial f_2, \ldots, \partial f_k\}$ (resp. $\mathcal{F}' = \{\partial f'_1, \partial f'_2, \ldots, \partial f'_{k'}\}$) be the collection of the (connected components of the) boundaries of the faces in $\mathcal{A}(\Gamma)$ (resp. $\mathcal{A}(\Gamma')$) that are either unbounded or contain at least one endpoint of a bounded arc either in Γ or Γ'. Then Γ and Γ' intersect if and only if $\mathcal{F}$ and $\mathcal{F}'$ intersect.*

Proof: The "if" part is obvious. For the "only if" part, suppose first that all arcs in Γ and Γ' are bounded, and again recall our assumption that all these arcs are x-monotonic. Let σ be a red-blue intersection point. Let $\mathcal{A}(\Gamma_\sigma)$ (resp. $\mathcal{A}(\Gamma'_\sigma)$) denote the connected component of $\mathcal{A}(\Gamma)$ (resp. $\mathcal{A}(\Gamma')$) containing the point σ (see figure 3.7). It is obvious that the rightmost point r' of $\mathcal{A}(\Gamma'_\sigma)$ lies in its unbounded face f', therefore its boundary $\partial f'$ is in the set $\mathcal{F}'$. Let r be the rightmost endpoint of $\mathcal{A}(\Gamma_\sigma)$; again, r must lie in the

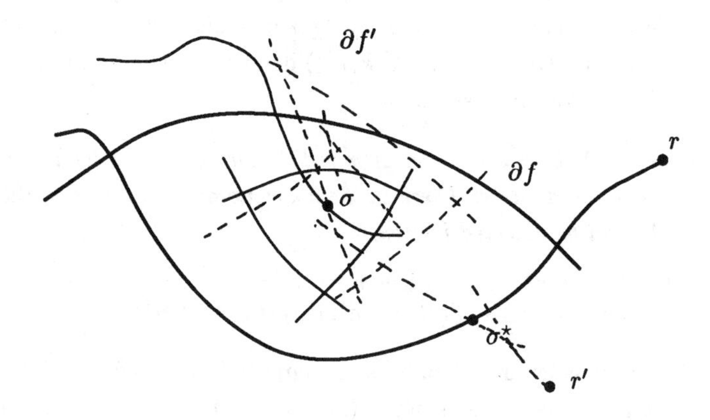

Figure 3.7: Illustration for Lemma 3.24

unbounded face f of $\mathcal{A}(\Gamma_\sigma)$, so that ∂f is in $\mathcal{F}$. Without loss of generality assume that r is to the right of r', which implies that r lies in f'.

There exists a connected path Π in $\mathcal{A}(\Gamma_\sigma)$ from r to σ. Follow $\partial f'$ from r' until we encounter a red-blue intersection point $\sigma^\star$. Such an intersection must be encountered because either σ lies on $\partial f'$ or $\partial f'$ intersects Π. Since $\sigma^\star \in \partial f'$ and also on the boundary of the face of $\mathcal{A}(\Gamma)$ containing r', $\mathcal{F}$ and $\mathcal{F}'$ intersect.

If Γ or Γ' contains unbounded arcs, we reduce the analysis to the bounded arcs by drawing a sufficiently large circle that contains all bounded arcs and intersection points of $\Gamma \cup \Gamma'$, cuts each unbounded arc in two points, and traverses only unbounded faces of $\mathcal{A}(\Gamma)$, $\mathcal{A}(\Gamma')$. The "clipped" arrangement now consists only of bounded arcs, and the unbounded face of each of them is the union of all unbounded faces in the original arrangement. Applying the preceding analysis, the claim follows easily in this case too.

□

If $\Gamma \cup \Gamma'$ has t unbounded arcs, then there are at most $2(m + n - t)$ endpoints and the total number of unbounded faces is at most $2t$. Therefore, the above lemma shows that it is enough to compute only $2(m + n - t) + 2t = 2(m + n)$ faces of each of $\mathcal{A}(\Gamma)$ and $\mathcal{A}(\Gamma')$ instead of computing the whole arrangement which can have $\Omega(n^2)$ (or $\Omega(m^2)$) faces. Let $\beta(k,n)$ (resp. $\beta'(k,m)$) denote the maximum number of edges in k distinct faces of $\mathcal{A}(\Gamma)$ (resp. $\mathcal{A}(\Gamma')$), and let $\Lambda(k,n)$ (resp. $\Lambda'(k,m)$) denote the time required to compute these faces. We can easily show

Theorem 3.25 *Let Γ be a collection of n red Jordan arcs and let Γ' be a collection of m blue Jordan arcs. Then, in the above notation, an intersection between Γ and Γ' can be detected in time*

$$O(\Lambda(2(m+n),n)+\Lambda'(2(m+n),m)+ \\ (\beta(2(m+n),n)+\beta'(2(m+n),m))\cdot\log(m+n)).$$

Proof: By the previous lemma, it suffices to compute $2(m+n)$ faces of $\mathcal{A}(\Gamma)$ and of $\mathcal{A}(\Gamma')$, which can be done in time $\Lambda(2(m+n),n)$ and $\Lambda(2(m+n),m)$, respectively. Once having computed these faces, an intersection between them can be detected in time $O((\beta(2(m+n),n)+\beta'(2(m+n),m))\log(m+n))$ by using a line sweep. Thus, the overall running time is bounded by

$$O(\Lambda(2(m+n),n)+\Lambda'(2(m+n),m)+ \\ (\beta(2(m+n),n)+\beta'(2(m+n),m))\cdot\log(m+n)).$$

□

[38], [59] and [55] have studied the complexity of many faces in arrangements of certain classes of Jordan curves and for some special cases they have obtained bounds that are either optimal or optimal within logarithmic factors, together with comparably efficient randomized algorithms for calculating many such faces, which are based on the random sampling technique of [36] and [78] (see also Chapter 5). Using these results and Theorem 3.25, we obtain

Corollary 3.26 *If the arcs in Γ and Γ' are line segments, unit circles, or pseudo lines (that is, unbounded arcs, each pair of which intersect at most once), then an intersection between Γ and Γ' can be detected by a randomized algorithm whose expected running time is $O((m+n)^{4/3+\epsilon})$, for any $\epsilon>0$ (where the constant of proportionality depends on ϵ). If Γ or Γ' is a set of arbitrary circles, and the other set is either a set of curves in one of the above classes or is also a set of arbitrary circles, then an intersection between Γ and Γ' can be detected in randomized expected time $O((m+n)^{7/5+\epsilon})$, for any $\epsilon>0$.*

Proof: The proof follows immediately from the results of [59], [38] and [75]. For unit circles, it is shown in [38] that m faces in an arrangement of n unit circles have $O\left(m^{2/3}n^{2/3}\alpha(n)^{1/3} + n\alpha(n)\right)$ edges, and the results of [75], [59] imply that they can be computed in randomized expected time

$$O\left(m^{2/3-\epsilon}n^{2/3+2\epsilon} + n\log^2 n \log m\right),$$

for any $\epsilon > 0$. In this case we have

$$\begin{aligned} \Lambda(m+n,n) &= \Lambda'(m+n,m) \\ &= (\beta(2(m+n),n) + \beta'(2(m+n),m))\log(m+n) \\ &= O((m+n)^{4/3+\epsilon}) \end{aligned}$$

for any $\epsilon > 0$, so Theorem 3.25 implies the claim. Similar bounds for many faces in arrangements of the other kind of curves, as established in [38] and [75], imply the claim in all other cases.

□

Remark 3.27: In Chapter 5 we show that in the more special case, where all arcs are line segments, we can count all red blue intersections in time $O(n^{4/3}\log^{\beta} n)$ time, where β is a constant < 1.77.

If the arcs in Γ are non-intersecting, then $\mathcal{A}(\Gamma)$ has only one face and there are only n edges in it. Moreover, it is enough to detect an intersection between Γ and the faces of $\mathcal{A}(\Gamma')$ that are unbounded or contain the endpoints of bounded arcs in Γ. There are at most $2n$ faces of $\mathcal{A}(\Gamma')$ containing endpoints of arcs in Γ; the unbounded faces of $\mathcal{A}(\Gamma')$ can be regarded as a single connected face if we clip each unbounded arc of Γ' at two sufficiently distant points, as in the proof of Lemma 3.24. We thus obtain

Corollary 3.28 *If the arcs in Γ are non-intersecting, then an intersection between Γ and Γ' can be detected in time $O(\Lambda'(2n,m)+\beta'(2n,m)\log(m+n))$, where $\Lambda'(n,m)$ is the time needed to compute n faces of $\mathcal{A}(\Gamma')$, and $\beta'(n,m)$ is the maximum number of edges in n such faces.*

3.5 Further Applications

In this section, we mention some applications of the general red-blue intersection detection algorithms. We describe only two applications and leave it to the imagination of the reader to roam free in search of further applications.

Consider the following problem:

> *Given a billiard table in the shape of an arbitrary k-sided convex polygon, and n billiard balls of equal size lying on the table. Suppose we shoot one of these balls B in a given direction, and allow it to bounce m times off the sides of the table. Will B hit any of the other balls?*

This problem can be solved using our red-blue intersection detection algorithm as follows. Since all balls have the same radius, it is easy to see that if B collides with any other ball B_i, it does so on the plane that passes through the centers of all the balls. Without loss of generality let us assume that this is the xy-plane. Therefore we can consider this problem as a two dimensional collision-detection problem, in which the moving object is a circle, all obstacles are also circles, and we want to determine whether the moving circle collides any other circle while following the path of B.

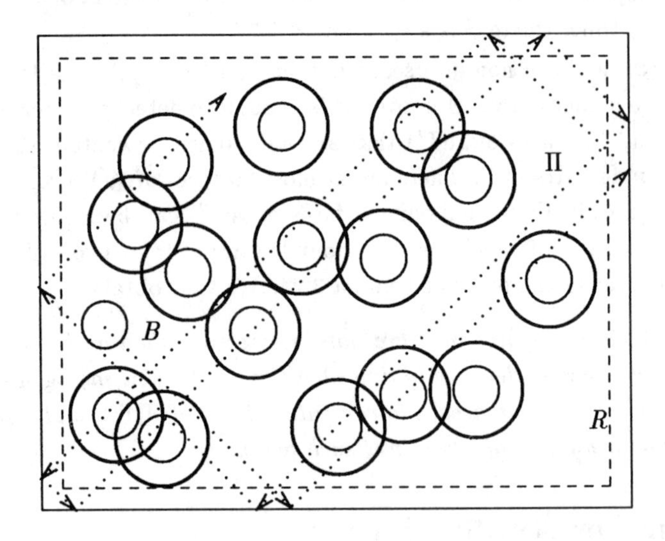

Figure 3.8: Detecting intersection between billiards balls

Let γ_i^* denote the circle obtained by projecting B_i on the xy-plane and let $\gamma^\star$ be the projection of B. We expand each circle $\gamma_i^\star$ by the radius of $\gamma^\star$ to obtain a set of n possibly intersecting circles, Γ (see figure 3.8). Next we shrink the table by the radius of B. Let R denote this convex polygon, which has at most k edges. It is easy to see that this procedure reduces the moving ball to a single point z in the sense that B collides with any other ball B_i if and only if z intersects any circle of Γ while moving on the path followed by the center of B (strictly speaking, the projection of the path on the xy-plane). The path Π followed by z can be determined in time $O(m \log k)$ after $O(k)$ preprocessing, using a trivial ray-shooting algorithm in a convex polygon. Let Γ' denote the set of segments forming Π. It follows from the above discussion that B collides with any other ball B_i if and only if Γ and Γ' intersect.

Hence, we can solve the above problem by applying our general algorithm for red-blue intersection detection. Since the arcs in Γ are closed curves such that no two of them intersect in more than two points, and the arcs in Γ' intersect in at most one point, it follows from Theorem 3.22 that we can detect an intersection between Γ and Γ' in time

$$O\left(\left(n\sqrt{m\alpha(m)} + m\sqrt{n}\right)\log^{3/2}(m+n)\right).$$

However in this problem the arcs in Γ and Γ' have some additional properties that improve the running time of the algorithm. For example the segments in Γ' form a connected path, so Γ' can be partitioned into subsets of appropriate size with the property that, the boundary of every face in the arrangement of each subset is connected. Hence by the remark following Theorem 3.22, the running time of the above algorithm is $O(m\sqrt{n}\log^{3/2} n)$.

On the other hand observe that B collides with any other obstacle if and only if the unbounded face $\mathcal{F}$ of $\mathcal{A}(\Gamma)$ and Π intersect. Kedem et al. [80] have proved that the complexity of $\mathcal{F}$ is $O(n)$, and that it can be computed in time $O(n\log^2 n)$; as a matter of fact, this can be improved to $O(n \log n)$, using the Voronoi diagram of the centers of the circles (see [100], [84]). Since the edges in $\mathcal{F}$ are non-intersecting, by the remark following Theorem 3.22 we can detect an intersection between $\mathcal{F}$ and Γ' in time $O(m\sqrt{n}\log n + n\log n)$.

Finally, the circles in Γ have the same radius, Corollary 3.26 implies that we can detect an intersection between Γ and Γ' in randomized expected time $O((m+n)^{4/3+\epsilon})$, for any $\epsilon > 0$. Therefore, we can conclude

Theorem 3.29 *The "billiard ball" problem defined above can be solved*

(i) deterministically in time $O(m\sqrt{n}\log n + n\log n)$,

(ii) in randomized expected time $O((m+n)^{4/3+\epsilon})$, *for any* $\epsilon > 0$.

□

Remark 3.30: We can extend our deterministic algorithm in several ways. For example we can generalize our algorithm for balls with different radii. In fact our algorithm works for balls of other shapes as well.

Another application, more related to computer graphics, is the following multiple ray-shooting, or the "death squad" problem:

> *Given m rays in the plane and a collection of objects bounded by n algebraic arcs (of small degree), does any ray hit any object?*

Again this problem can be reduced to an instance of the red-blue intersection detection problem by considering the boundaries of the objects as a collection of red arcs and the rays as a collection of blue arcs. Hence, assuming the objects are non-intersecting, we can solve this problem in (deterministic) time $O(m\sqrt{n}\log n)$ by Remark (ii) following Theorem 3.22, or in randomized expected time $O(n^{2/3-\epsilon}m^{2/3+2\epsilon} + m\alpha(m)\log^2 m \log n)$, for any $\epsilon > 0$ using Corollary 3.28.

If instead of rays we shoot identical bullets of some convex shape having small and fixed complexity, we expand each object by the bullet shape to reduce the problem to the original ray shooting problem. In this case we can detect an intersection, using our general algorithm, in time $O((m\sqrt{\lambda_{s+2}(n)} + n\sqrt{m\alpha(m)})\log^{1.5}(m+n))$, where s is the maximum number of intersections between any two expanded objects. Again, the running time can be improved by exploiting several special properties of the problem structure. Notice that the arcs in Γ' are rays rather than segments. It has been proved in [3] that a single face in an arrangement of m rays has only $O(m)$ complexity and can be calculated in $O(m\log m)$ time, therefore the running time becomes $O((m\sqrt{\lambda_{s+2}(n)}\log^{1.5}(m+n) + n\sqrt{m}\log(m+n))$. Finally, if all the objects being shot at are convex, and all bullets are identical and convex, any pair of arcs in Γ intersect in at most two points (cf. [80]), and the union of all expanded objects can be computed in time $O(n \log n)$. As in the billiard-ball problem it suffices to detect an intersection between the unbounded face of $\mathcal{A}(\Gamma)$ and Γ', therefore we can detect an intersection in additional $O(m\sqrt{n}\log n)$ time. Hence, we can conclude

Theorem 3.31 *(i) The multiple ray shooting can be solved deterministically in $O(m \cdot \sqrt{n} \cdot \log n + n\log(m+n))$ time, or in randomized expected time $O(n^{2/3-\epsilon}m^{2/3+2\epsilon} + m\alpha(m)\log^2 m \log n)$.*
(ii) The multiple bullet shooting problem can be solved deterministically in time

$$O(m\sqrt{\lambda_{s+2}(n)}\log^{1.5}(m+n) + n\sqrt{m}\log(m+n)),$$

which reduces to $O(m\sqrt{n}\log n + n\log n)$ when all objects are convex.

Remark 3.32: Our techniques also allow us to solve efficiently various other extensions of the above problem. For example, if the bullets being shot have a finite range, then we need to detect an intersection between a collection of segments and the given objects. Similarly, in the case of a "stone-throwing squad", where all the stones move along parabolic trajectories in say, the $x - z$ plane, or along other (algebraic) trajectories, we need to detect an intersection between those trajectories and the objects.

3.6 Discussion and Open Problems

We obtained several efficient algorithms for the red-blue intersection detection problem. Our algorithm for the case when each of $\mathcal{A}(\Gamma)$ and $\mathcal{A}(\Gamma')$ is a connected planar graph is close to optimal, and our general (deterministic and randomized) algorithms are faster than previously known algorithms. We applied these algorithms to obtain fast algorithms for several problems in motion planning and collision detection. However, there are still several open questions, as listed below:

(i) The most important is whether the running time of our deterministic algorithm for the general case can be improved. Theorem 3.25 shows that an efficient algorithm for computing many faces in the arrangement of a given set of Jordan arcs also yields a fast algorithm for the red-blue intersection detection problem, therefore one way of improving the running time is to find an efficient deterministic algorithm to compute many faces in arrangement of Jordan arcs.

(ii) We also obtained randomized algorithms for several special cases, that are faster than our deterministic algorithm but at present we do not have any such algorithm for the general case.

(iii) Another open question is whether a red-blue intersection can be detected in randomized expected time $O((n+m)\log^{O(1)}(m+n))$ if all the arcs are line segments. This appears to be a very difficult problem, because it is closely related to Hopcroft's problem that calls for detecting an incidence between a set of points and a set of lines, for which no solution better than roughly $(m+n)^{4/3}$ (see Chapter 5, [59]) is known.

(iv) Finally, can one report all red-blue intersections in an output-sensitive fashion for general arcs as well?

Chapter 4

Partitioning Arrangements of Lines

4.1 Introduction

In the last few years several randomized divide and conquer algorithms for a variety of geometric problems have been developed ([34], [36], [37], [39], [40], [59], [73], [74], etc.), using ϵ-nets [78] or the random sampling technique of [36] (see also [109]). The ϵ-net theory shows that for a given set X of n objects and a set $\mathcal{R} \subseteq 2^X$ of *ranges* with finite *Vapnik-Chervonenkis dimension* (see [78] for definition), there exists a subset $N \subset X$ of size r, for any $r > 0$, such that if $\tau \in \mathcal{R}$ and $\tau \cap N = \emptyset$, then $|\tau| = O(\frac{n}{r} \log r)$. A similar result has been independently obtained by Clarkson [36]. This property allows us to split the original problem into subproblems of small size, which can then be solved recursively. For example, if the objects are "lines in the plane" and an element of $\mathcal{R}$ is the "set of lines intersecting a given triangle", then these results imply that for a given set $\mathcal{L}$ of n lines in the plane, there exists a subset $N \subset \mathcal{L}$ of size r such that any triangle which misses all lines in N intersects at most $O(\frac{n}{r} \log r)$ lines of $\mathcal{L}$. A problem involving the set $\mathcal{L}$ of lines can then be split into subproblems by computing the arrangement of the lines in N and by triangulating each face of the arrangement; each resulting triangle Δ induces a subproblem involving only $O(\frac{n}{r} \log r)$ lines of $\mathcal{L}$ intersecting Δ.

Unfortunately the random sampling technique or the ϵ-net theory only proves the existence of such a subset N, but does not show how to construct it deterministically (and efficiently). This has forced all algorithms based on these techniques to use randomization to obtain N (the only randomized step in most of these algorithms is that of choosing a good sample N). The randomization is justified because the above theories show that most subsets N of size r satisfy the desired properties, so a random choice of such a subset will succeed with high probability. Still, a major open problem in this area is to find an efficient deterministic algorithm to obtain a "good" sample.

Recently Chazelle and Friedman [26] gave a general framework which unifies the results of Haussler-Welzl [78] and Clarkson [36], [37], and yields a deterministic algorithm to construct a good sample. Although their algorithm runs in polynomial time, it is not very efficient (in particular its time complexity is too high an overhead for most of the applications). This has motivated researchers to look for special cases, where faster deterministic constructions are possible. Woeginger [134] gave an $O(n)$ algorithm to construct an ϵ-net for a very special case, where the objects are points in the plane and the ranges are half-planes. A much more significant result in this direction has been obtained by Matoušek [89], who showed that, given a set of n lines in the plane, the plane can be partitioned into $O(r^2)$ triangles in time $O(nr^2 \log^2 r)$ so that no triangle intersects more than $O(\frac{n}{r})$ lines (however, these triangles in general are not formed by an arrangement of a sample subset of the lines). An important property of the algorithm is that each triangle it produces is cut by only $O(\frac{n}{r})$ lines, rather than the $O(\frac{n}{r} \log r)$ promised by the probabilistic techniques. If r is constant, then this algorithm is optimal, and can be used to remove randomization from most of the "random-sampling" based algorithms involving lines or segments ([59], [73], [74], [78] etc.) However, it is not efficient if r is large, in particular if $r = n^\alpha$, for some $0 < \alpha < 1$.

Let us be more specific concerning this efficiency issue. Even though Matoušek's algorithm is much faster than that of [26], it is still inefficient when compared to the following "lower bound". Since the algorithm produces $O(r^2)$ triangles and each is met by $O(\frac{n}{r})$ lines, the total input size of all the corresponding subproblems (that is, the total number of line-triangle crossings) is $O(nr)$ (one can show that the bound is tight in the worst case).

Thus it is natural to seek an improved algorithm whose complexity is close to $O(nr)$. As we will see later, this improvement does make a difference, because in many applications we do want to choose r to be n^α, for some $0 < \alpha < 1$.

In this chapter we achieve such an improvement, and obtain an $O(nr \log n \cdot \log^\omega r)$ deterministic algorithm which, given a set $\mathcal{L}$ of n lines, partitions the plane into $O(r^2)$ triangles so that no triangle intersects more than $O(\frac{n}{r})$ lines (here ω is some constant < 3.33). Then we apply this algorithm to remove randomization and to improve the time complexity of solutions to several problems. A common characteristic of the time complexity of the previously known algorithms for these problems is that it is worse by a factor of $O(n^\delta)$, for any $\delta > 0$ (with a constant of proportionality depending on δ), than the worst case output size, or than some other natural measure of complexity that one can associate with the problem. For example, the running time of the best known (randomized) algorithm to compute the incidences between m points and n lines is $O(m^{2/3-\delta}n^{2/3+2\delta} + (m+n)\log n)$, for any $\delta > 0$ [59], which is worse by a factor of $O((\frac{n^2}{m})^\delta)$ than the maximum possible number of such incidences. (A slightly faster but still randomized algorithm for this problem, which runs in $O(m^{2/3}n^{2/3}\log^4 n + (m + n^{3/2})\log^2 n)$ expected time, has been given in [55]. The difference between these two algorithms is that the first algorithm can be made deterministic without increasing its running time, using Matoušek's algorithm, while the second algorithm is not yet known to admit a similarly efficient determinization.) Factors like $O(n^\delta)$ appear in the bound of these algorithms because they use small values of r (in most cases constant values). Although Matoušek's algorithm makes most of these algorithms deterministic, it also cannot choose a large value of r, because its running time is quadratic in terms of r, in which case the overhead of constructing the partition dominates substantially the time complexity, resulting in an inefficient algorithm. On the other hand, by applying our algorithm, we can use a sufficiently large value of r to remove the $O(n^\delta)$ factor from the bounds. Another disadvantage of using a small value of r is that the algorithm then becomes recursive and more complex. In contrast, by choosing an appropriate large value of r we partition the plane just once, because then the subproblems are sufficiently small and can be solved directly by other means; this makes the algorithms much simpler.

In the next chapter we describe several applications of our partitioning algorithm, including an $O(m^{2/3}n^{2/3}\log^{2/3} n \log^{\omega/3} \frac{m}{\sqrt{n}} + (m+n)\log n)$ algorithm to compute incidences between m points and n lines, an $O(m^{2/3}n^{2/3} \cdot \log^{5/3} n \log^{\omega/3} \frac{m}{\sqrt{n}} + (m+n)\log n)$ algorithm to compute m distinct faces in an arrangement of n lines, an $O(n^{4/3}\log^{(\omega+2)/3} n)$ algorithm to compute intersection in a set of n segments and an $O(n^{3/2}\log^{\omega+1} n)$ algorithm to compute a family of $O(\log n)$ spanning trees of a set of n points, with low stabbing number.

Our algorithm for partitioning the plane works in two phases. In the first phase it partitions the plane into $O(r^2 \log^{\omega} r)$ triangles so that no triangle intersects more than $O(\frac{n}{r})$ lines. In the second phase we reduce the number of triangles to $O(r^2)$ using a technique similar to that of Matoušek [89]. We believe that the second phase is redundant, i.e. that the first phase or some appropriate variant of it produces only $O(r^2)$ triangles, but we have not been able to prove it so far. Even if this is not the case, in most of the applications we can stop after the first phase and solve the subproblems directly within each of the resulting $O(r^2 \log^{\omega} r)$ triangles, without increasing the asymptotic complexity of the algorithm.

The chapter is organized as follows. In Section 4.2 we describe the geometric concepts. Section 4.3 gives an algorithm for a subproblem that is interesting in its own right, namely that of computing the κth leftmost intersection point induced by a set of n lines inside a given convex quadrilateral. Section 4.4 is the heart of our algorithm; it presents an algorithm to partition a given convex quadrilateral $\mathcal{Q}$ intersecting m lines and containing K of their intersection points into $O(\frac{m}{\zeta} + \frac{K}{\zeta^2})$ convex quadrilaterals (for an arbitrary parameter ζ) so that both the number of lines crossing any subquadrilateral $\mathcal{Q}'$ and the number of their intersections inside $\mathcal{Q}'$ are small (the precise conditions are given in Theorem 4.22). In Section 4.5 we describe the first phase of the partitioning algorithm, which basically consists of applying the algorithm of Section 4.4 recursively, and Section 4.6 describes the second phase of the algorithm. In earlier sections we assume that the lines are in general position; we show in Section 4.7 how to modify the analysis to handle degenerate cases as well. In Section 4.8 we conclude with some final remarks and open problems.

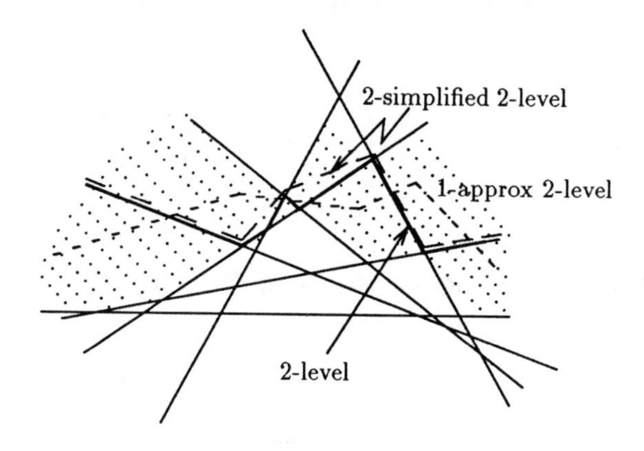

Figure 4.1: 2-level, 1-approximate 2-level, 2-simplified 2-level

4.2 Geometric Preliminaries

This section defines the geometric concepts, and formalizes the notation that we will be using. Let $\mathcal{H} = \{\ell_1, \ell_2, \ldots, \ell_n\}$ denote a set of n lines in the plane. These lines induce a planar map called the *arrangement* $\mathcal{A}(\mathcal{H})$ of $\mathcal{H}$, whose *vertices* are the intersection points of lines in $\mathcal{H}$, *edges* are maximal connected portions of lines in $\mathcal{H}$ not containing a vertex, and *faces* are maximal connected portions of the plane not meeting any edge or vertex of $\mathcal{A}(\mathcal{H})$. See [52] for more details. For simplicity we assume that the lines in $\mathcal{H}$ are in general position, that is, no three lines are concurrent, and that no line in $\mathcal{H}$ is vertical. For any point $p \in \mathbb{R}^2$, we define its *level* to be the number of lines in $\mathcal{H}$ lying above it (not counting the lines passing through p). For any $0 \leq k < n$, the *k-level* of $\mathcal{A}(\mathcal{H})$ is the set of (closure of) edges of $\mathcal{A}(\mathcal{H})$ whose level is k (see figure 4.1). A k-level of $\mathcal{A}(\mathcal{H})$ is a monotone polygonal chain with two unbounded rays. More details on k-levels can be found in [52], [65], [131].

Lemma 4.1 *Let p_1 and p_2 be two points in the plane whose levels are k_1 and k_2 respectively. Then the line segment $\overline{p_1p_2}$ intersects at least $|k_1 - k_2|$ lines of $\mathcal{H}$.*

Proof: Obvious from the definition of k-levels.

□

Lemma 4.2 (Matoušek [89]) *Let $\mathcal{U}_1, \mathcal{U}_2, \ldots, \mathcal{U}_s$ denote pairwise disjoint sets of levels of $\mathcal{A}(\mathcal{H})$ with $|\mathcal{U}_i| \geq \varphi$ for all $i \leq s$, then there exists a level $U_i \in \mathcal{U}_i$ such that $\sum_{i=1}^{s} |U_i| \leq \dfrac{n^2}{\varphi}$.*

Proof: Let U_i be the shortest level in $\mathcal{U}_i$. Since no two levels share an edge,

$$\sum_{U \in \mathcal{U}_i} |U| \geq \varphi |U_i|.$$

But there are n^2 edges in $\mathcal{A}(\mathcal{H})$, therefore

$$n^2 \geq \sum_{i=1}^{s} \sum_{U \in \mathcal{U}_i} |U| \geq \varphi \sum_{i=1}^{s} |U_i|.$$

□

We call an x-monotone polygonal path Π (not necessarily formed by the edges of $\mathcal{A}(\mathcal{H})$) an *ϵ-approximate k-level*, for $\epsilon \leq k$, if it lies in the strip bounded by the $k - \epsilon$ and $k + \epsilon$ levels of $\mathcal{A}(\mathcal{H})$ (see figure 4.1). A set of ϵ-approximate $2\epsilon i$-levels, for $i \leq \lfloor \frac{n}{2\epsilon} \rfloor$, is called an *$\epsilon$-approximate leveling* of $\mathcal{A}(\mathcal{H})$. Let $p_1, p_2, \ldots, p_m$ be the vertices of a k-level of $\mathcal{A}(\mathcal{H})$. For any $\delta < m$, we define the *δ-simplified k-level* to be the polygonal path formed by connecting p_1 to $p_{\delta+1}$, $p_{\delta+1}$ to $p_{2\delta+1}$, ..., $p_{\lfloor \frac{m}{\delta} \rfloor \delta + 1}$ to p_m, and concatenated with the left and right rays of the k-level incident to p_0 and p_m, respectively. Edelsbrunner and Welzl [65] proved that

Lemma 4.3 (Edelsbrunner and Welzl [65]) *Given set $\mathcal{H}$ of n lines and $0 \leq k < n$, a δ-simplified k-level is a $\lceil \frac{\delta}{2} \rceil$-approximate k-level of $\mathcal{A}(\mathcal{H})$.*

Proof: Let $e = ab$ be an edge of the δ-simplified k-level. Let ℓ denote a line of $\mathcal{H}$ properly intersecting e, that is, it intersects the relative interior of e. It is easily seen that at least one vertex of the k-level on ℓ is in the sequence of the vertices replaced by e. Thus at most $\delta - 1$ lines intersect e.

If lemma were not true, then there would be a point $p \in e$ whose level is either less than $k - \lceil \delta/2 \rceil$ or more than $k + \lceil \delta/2 \rceil$, which implies that both

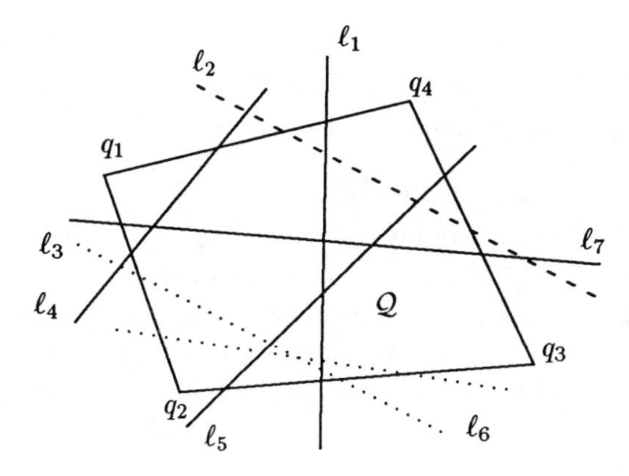

Figure 4.2: $\mathcal{Q}$ and $\mathcal{L}$: solid lines denote red lines, dashed lines denote green lines and dotted lines denote blue lines.

ap and bp would intersect at least $\lceil \delta/2 \rceil + 1$ lines of $\mathcal{H}$, contradicting the fact that e intersects at most $\delta - 1$ lines. □

Let $\mathcal{Q}$ be a convex quadrilateral with vertices q_1, q_2, q_3, q_4 ordered counterclockwise and let $\partial\mathcal{Q}$ denote the boundary of $\mathcal{Q}$. Let $\mathcal{L} = \{\ell_1, \ell_2, \ldots, \ell_m\}$ denote a set of lines passing through $\mathcal{Q}$, and let K be the number of intersection points of $\mathcal{L}$ contained inside $\mathcal{Q}$. The intersection points of $\partial\mathcal{Q}$ and $\mathcal{L}$ are called the *endpoints* of $\mathcal{L}$. For simplicity let us assume that every line of $\mathcal{L}$ intersects $\partial\mathcal{Q}$ at two points, i.e. does not touch $\mathcal{Q}$ at one of its vertices (or at an edge).

The set $\mathcal{L}$ of lines can be partitioned into three different classes according to the location of their endpoints (see figure 4.2).

(i) "Red" lines $\mathcal{L}_r$: a line ℓ belongs to $\mathcal{L}_r$ if one of its endpoints lies on $\overline{q_1q_2}$ or $\overline{q_2q_3}$ (referred to as the left endpoint) and the other lies on $\overline{q_1q_4}$ or $\overline{q_4q_3}$ (referred to as the right endpoint), i.e. ℓ crosses the diagonal $\overline{q_1q_3}$; let $|\mathcal{L}_r| = m_r$.

(ii) "Blue" lines $\mathcal{L}_b$: a line ℓ belongs to $\mathcal{L}_b$ if one of its endpoints lies on $\overline{q_1q_2}$ (referred to as the left endpoint) and the other lies on $\overline{q_2q_3}$ (referred to as the right endpoint); let $|\mathcal{L}_b| = m_b$.

(iii) "Green" lines $\mathcal{L}_g$: a line ℓ belongs to $\mathcal{L}_g$ if one of its endpoints lies on $\overline{q_1 q_4}$ (referred to as the left endpoint) and the other lies on $\overline{q_4 q_3}$ (referred to as the right endpoint); let $|\mathcal{L}_g| = m_g$.

These three sets are pairwise disjoint and $\mathcal{L}_r \cup \mathcal{L}_g \cup \mathcal{L}_b = \mathcal{L}$. Let K_{xy} denote the number of intersection points between $\mathcal{L}_x$ and $\mathcal{L}_y$ lying inside $\mathcal{Q}$, where $x, y \in \{r, b, g\}$. It is easily seen that $\mathcal{L}_b$ and $\mathcal{L}_g$ do not intersect inside $\mathcal{Q}$, that is $K_{bg} = 0$. Without loss of generality we can assume that $m_r \geq \frac{m}{2}$ because otherwise we can cyclically renumber the vertices of $\mathcal{Q}$ so that the above inequality is satisfied. For the sake of exposition we assume that q_1 is the upper left corner of $\mathcal{Q}$.

Let $A = \{a_1, \ldots, a_{m_r}\}$ (resp. $B = \{b_1, \ldots, b_{m_r}\}$) be the left (resp. right) endpoints of $\mathcal{L}_r$ appearing in counter-clockwise (resp. clockwise) direction around $\partial\mathcal{Q}$. For each line $\ell_i \in \mathcal{L}_r$ let $a_{\pi(i)}$ (resp. $b_{\sigma(i)}$) denote its left (resp. right) endpoint.

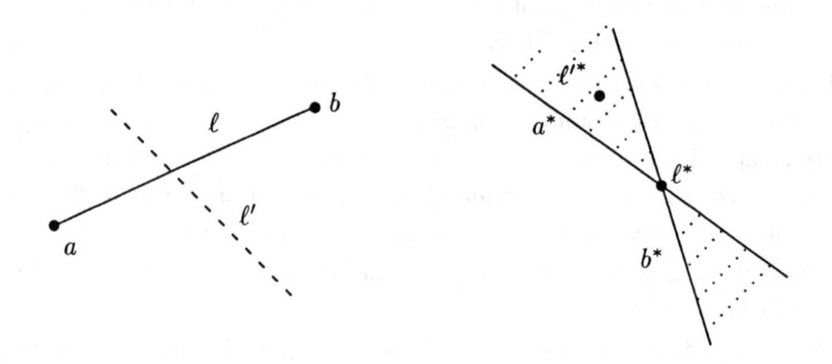

Figure 4.3: A segment $e = \overline{ab}$ and its dual e^*

Another geometric concept we use is *duality* (see [14], [52]). In $\mathbb{R}^2$, the dual of a line is a point, and the dual of a point is a line. The duality transformation can be chosen in such a way that it preserves the "above-below" relationship between points and lines. The dual of a line segment $\overline{pq}$ is a double wedge formed between the dual lines p^*, q^* of p, q respectively, and not containing the vertical line through their intersection point. We denote the dual of a feature (point, line or segment) γ by $\gamma^\star$.

4.3 Selecting the κth Leftmost Intersection Point

Let $p_1, \ldots, p_K$ denote the intersection points of the lines in $\mathcal{L}$ lying inside $\mathcal{Q}$ and ordered by increasing x-coordinate. In this section we consider the problem of calculating the κth leftmost point p_κ of these intersections, for any given $\kappa \leq K$. The algorithm extends the recent algorithm of Cole et al. [43] for calculating the kth leftmost intersection point of n lines in the entire plane. One of the key steps of the algorithm is to count the number of intersection points of $\mathcal{L}$ lying inside a given convex quadrilateral. We first give an algorithm that counts the number of intersections inside a convex quadrilateral in time $O(m \log m)$, where m is the number of lines passing through $\mathcal{Q}$.

4.3.1 Counting intersection points inside a convex quadrilateral

We partition the set of lines $\mathcal{L}$ into $\mathcal{L}_r$, $\mathcal{L}_b$ and $\mathcal{L}_g$ as defined earlier, and count each of K_{xy}, for $x, y \in \{r, b, g\}$, separately. First, we show how to count the number of red-red intersection points lying inside $\mathcal{Q}$, i.e. K_{rr}. It is obvious that two red lines ℓ_i and ℓ_j in $\mathcal{L}$ intersect if and only if $\pi(i) < \pi(j)$ and $\sigma(i) > \sigma(j)$ or vice versa. We thus obtain K_{rr} by counting the number of inversions in σ with respect to π, which is easily done in time $O(m_r \log m_r)$ by sorting the endpoints of $\mathcal{L}_r$ along the boundary of $\mathcal{Q}$. Similarly, we can compute K_{bb} (resp. K_{gg}) in time $O(m_b \log m_b)$ (resp. $O(m_g \log m_g)$).

As for determining K_{rg}, let e_1 (resp. e_2) be the left (resp. right) endpoint of a green line ℓ. The line ℓ intersects a red line ℓ_j if the right endpoint of ℓ_j lies on either $\overline{e_1 q_4}$ or $\overline{q_4 e_2}$ (see figure 4.4). Therefore using the same sorted list B we can compute the number of red lines that ℓ intersects by locating e_1 and e_2 in B, which implies that K_{rg} can be computed in time $O(m_g \log m_r)$. Similarly, K_{rb} can be computed in time $O(m_b \log m_r)$. Thus, we have shown

Lemma 4.4 *For a given convex quadrilateral $\mathcal{Q}$ and a set $\mathcal{L}$ of m lines*

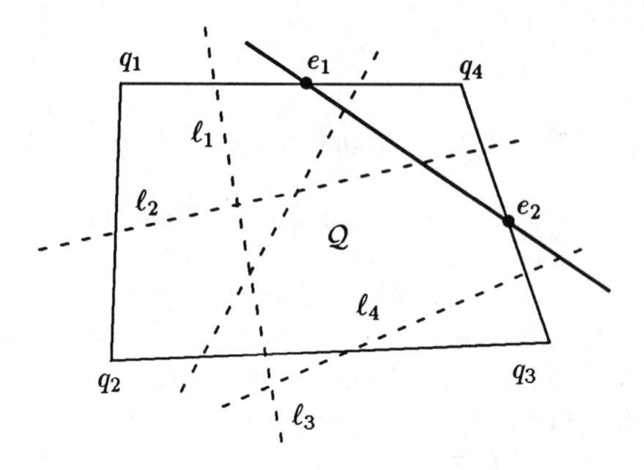

Figure 4.4: Green line ℓ and its endpoints e_1, e_2; ℓ_1, ℓ_4 do not intersect ℓ

passing through it, we can count the number of intersections of $\mathcal{L}$ lying inside $\mathcal{Q}$ in time $O(m \log m)$. □

Remark 4.5: The above algorithm can be extended to count the number of intersections lying inside any k-gon $\mathcal{B}$ in $O(km \log m)$ time, by triangulating $\mathcal{B}$, and by counting the number of intersections in each triangle separately.

Next we give an algorithm for finding the κth leftmost intersection point inside $\mathcal{Q}$ using the counting procedure just described.

4.3.2 Computing the κth leftmost intersection

Cole et al. [43] have given an $O(m \log m)$ algorithm that returns the κth leftmost intersection point in a set $\mathcal{L}$ of m lines in the plane. Their algorithm cannot be applied directly to our case because it counts all intersection points of $\mathcal{L}$ lying to the left of the κth one, while we count only those intersection points that lie inside $\mathcal{Q}$. However, we will show that the algorithm of [43] can be easily adapted to our case. We assume that the reader is familiar with this algorithm, so we describe it here only briefly, and show how to modify it for our case; for more details, see [43].

Let $\pi(a)$ denote the permutation of lines in $\mathcal{L}$ sorted in decreasing order of their intercepts with the line $x = a$, assuming that no two lines in $\mathcal{L}$ intersect at $x = a$. The *rank* of a, denoted by $\psi(a)$, is defined to be the number of intersection points of $\mathcal{L}$ lying inside $\mathcal{Q}$ and to the left of $x = a$. The κth leftmost intersection point p_κ is such that $\psi(a^*) = \kappa$, where $a^* = x(p_\kappa) + \epsilon$, for some sufficiently small $\epsilon > 0$ (in this description we assume that no two points p_i, p_j have the same x-coordinate).

Cole et al.'s algorithm [43] is based on Megiddo's technique [96], further improved by Cole [41], for implicit parametric searching. Using a sorting network, e.g. that of Ajtai et al. [2], $\pi(a^*)$ can be obtained by performing $O(m \log m)$ comparisons, performed in $O(\log m)$ batches of $O(m)$ independent comparisons each, where each comparison answers a question of the form: "Does ℓ_i lie above ℓ_j at $x = a^*$?". Although we do not know a^*, we can answer a question of this form by computing the rank of the abscissa $x(q_{ij})$, where q_{ij} is the intersection point of ℓ_i and ℓ_j. If $\psi(x(q_{ij})) > \kappa$, then q_{ij} lies to the left of $x = a^*$, so the order of ℓ_i and ℓ_j there coincides with their order at $x = -\infty$; otherwise the order is reversed. In such a sorting network there are $O(\log m)$ levels and at each level $s = O(m)$ questions are asked; let $z_1, z_2, \ldots, z_s$ denote the x-coordinates of the intersection points corresponding to the questions to be answered at a level j. If we resolve the question for the median value z_{med} of $z_1, \ldots, z_s$, then we can resolve the question for at least half of the above points, e.g. if $z_{\text{med}} < a^*$, then $z_i < a^*$ for all $z_i < z_{\text{med}}$. For the remaining half of the points, we can resolve the question by applying the same method recursively, thus allowing us to resolve all m questions by actually computing the rank of only $O(\log m)$ points. Once all comparisons are resolved at level j, we can continue to level $j+1$ of the network. At the end, all comparisons have been resolved and we have obtained the sorted order of the lines in $\mathcal{L}$ along $x = a^*$, still without knowing the exact value of a^*. However, this order is easily seen to determine uniquely how many intersection points of $\mathcal{L}$ lie to the left of $x = a^*$, thus the set of these intersections, as well as the subset of these points lying inside $\mathcal{Q}$, are fully determined at the end of the algorithm. By keeping track of all inequalities of the form $z_{\text{med}} < a^*$ or $z_{\text{med}} > a^*$ we finally obtain an interval $\alpha < a^* < \beta$, and the logic of the procedure is easily seen to imply that α is the x-coordinate of p_κ. The point p_κ itself is also easy to obtain by recording, for each z_{med}, the two lines that induce it.

To resolve a comparison at $x = z_{\text{med}}$, we proceed as follows. Let Q^- denote the portion of Q lying to the left of the vertical line $x = z_{\text{med}}$. By Lemma 4.4 and the subsequent Remark, we can count the number of intersection points of $\mathcal{L}$ lying inside Q^- (which is at most a pentagon) in $O(m \log m)$ time. Thus the rank of z_{med} can be computed in $O(m \log m)$ time, so the overall running time of the algorithm is $O(m \log^3 m)$. By using Cole's improvement [41], the time can be reduced to $O(m \log^2 m)$. Cole et al. [43] observed that it is not necessary to compute the exact rank of a point because we are only interested in knowing whether z_{med} lies to the left or to the right of p_κ. They proved that it suffices to compute an approximate rank of a point, which can be done in $O(m)$ amortized time, yielding an $O(m \log m)$ algorithm. Using the same idea we can also compute our version of approximate rank in amortized time $O(m)$, and therefore we obtain

Theorem 4.6 *Given a convex quadrilateral Q, a set of m lines $\mathcal{L}$ and an integer κ, we can determine, in time $O(m \log m)$, a vertical line, so that Q contains exactly κ intersection points to the left of that line.*

□

Remark 4.7:

(i) The remark following Lemma 4.4 implies that we can extend the above algorithm to any k-gon with $O(km \log m)$ running time.

(ii) The above algorithm uses the sorting network of Ajtai et al. [2], which involves a very large constant, therefore making our algorithm impractical. A possible solution to circumvent this problem is to use the much simpler Batcher's sorting network [10]; this network has depth $O(\log^2 m)$ and therefore the overall running time of the resulting algorithm becomes $O(m \log^2 m)$. However, to obtain a better asymptotic complexity, we will be using the former algorithm.

4.4 Partitioning a Convex Quadrilateral

This section describes an algorithm to partition the convex quadrilateral Q into convex quadrilateral subcells so that on the average only few lines pass through each cell, and the maximum number of lines passing through each cell is also low. To be more precise, for any given positive integer $\zeta < m$, we want to partition $\mathcal{Q}$ into $M = O(\frac{m}{\zeta} + \frac{K}{\zeta^2})$ cells so that the following conditions are satisfied (where m_i is the number of lines passing through the ith cell, and K_i is the number of intersections within the ith cell):

$$\text{(i)} \quad \sum_{i=1}^{M} m_i = O\left(m + \frac{K}{\zeta}\right), \qquad \max_i m_i = O(\sqrt{K})$$

$$\text{(ii)} \quad \sum_{i=1}^{M} K_i = K, \qquad \max_i K_i \le \zeta\sqrt{K}.$$

4.4.1 A special case

Before solving this general problem we consider a special case in which $\mathcal{Q}$ contains only "red" lines and we do not care how many intersection points lie in a cell. Moreover, although most cells in the construction to follow will be quadrilaterals, some could have up to six edges.

First, we compute the left and right endpoints of all lines in $\mathcal{L}$ and sort them to obtain the lists A and B. Next, for all $1 \le i < t = \lceil \frac{m}{\zeta} \rceil$ we connect $a_{i\zeta}$ to $b_{i\zeta}$. The segments $\overline{a_{i\zeta} b_{i\zeta}}$ are called *pseudo edges* (see figure 4.5). The pseudo edges partition $\mathcal{Q}$ into cells $\mathcal{Q}_1, \mathcal{Q}_2, \ldots, \mathcal{Q}_t$. The following sequence of lemmas show that this partitioning has the desired properties.

Lemma 4.8 *A line $\ell_i \in \mathcal{L}$ intersects at least $|\pi(i) - \sigma(i)|$ lines of $\mathcal{L}$ inside $\mathcal{Q}$.*

Proof: Without loss of generality assume that $\pi(i) \ge \sigma(i)$. Since only $\sigma(i) - 1$ lines have their right endpoints preceding the right endpoint of ℓ_i, at least $\pi(i) - \sigma(i)$ of the lines whose left endpoints precede that of ℓ_i, have their right endpoints after $b_{\sigma(i)}$ in B. But every such line crosses the line ℓ_i inside $\mathcal{Q}$, showing that at least $|\pi(i) - \sigma(i)|$ lines of $\mathcal{L}$ intersect ℓ_i inside $\mathcal{Q}$. □

Let δ_i denote the number of pseudo edges intersected by the line ℓ_i.

Lemma 4.9 $$\sum_{i=1}^{m} \delta_i \le \frac{2K}{\zeta} + m.$$

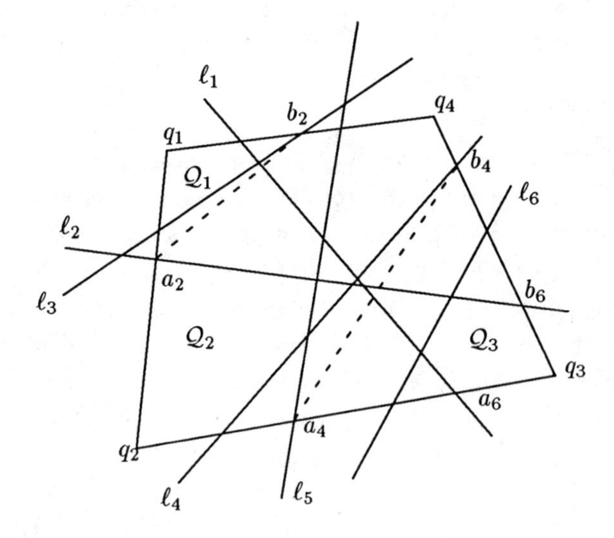

Figure 4.5: Partitioning $\mathcal{Q}$ into $\mathcal{Q}_1, \ldots, \mathcal{Q}_t$, with $m = 6$ and $\zeta = 2$. The dashed edges are pseudo edges.

Proof: If a line ℓ_i intersects δ_i pseudo edges, then it is easily seen that

$$|\pi(i) - \sigma(i)| \geq (\delta_i - 1)\zeta.$$

Summing over all lines, we obtain

$$\sum_{i=1}^{m} |\pi(i) - \sigma(i)| \geq \sum_{i=1}^{m} (\delta_i - 1)\zeta.$$

Let ν_i denote the number of lines of $\mathcal{L}$ that ℓ_i intersects inside $\mathcal{Q}$. We have $\sum_{i=1}^{m} \nu_i = 2K$, because there are only K intersection points inside $\mathcal{Q}$ and each intersection is counted twice. By Lemma 4.8, $|\pi(i) - \sigma(i)| \leq \nu_i$, therefore

$$\sum_{i=1}^{m} (\delta_i - 1)\zeta \leq 2K \quad \text{or} \quad \sum_{i=1}^{m} \delta_i \leq \frac{2K}{\zeta} + m.$$

□

Let m_i denote the number of lines passing through the cell $\mathcal{Q}_i$. Since the line ℓ_j lies in $\delta_j + 1$ cells, we have

Corollary 4.10

$$\sum_{i=1}^{t} m_i = \sum_{j=1}^{m} (\delta_j + 1) \leq 2\left(\frac{K}{\zeta} + m\right).$$

□

Lemma 4.11 *Let μ_i denote the number of lines passing through the cell $\mathcal{Q}_i$ but not having any of their endpoints on $\partial\mathcal{Q}_i$. Then $\mu_i \leq 2\sqrt{K}$, for all $i \leq t$.*

Proof: Let $\mathcal{L}_i$ denote the set of lines passing through the cell $\mathcal{Q}_i$ but not having any endpoint in $\mathcal{Q}_i$. Partition the set $\mathcal{L}_i$ into disjoint subsets $\mathcal{L}_{xy}$, for $x = 1, \ldots, t-i$, $y = 1, \ldots, i-1$, such that a line ℓ_j belongs to $\mathcal{L}_{xy}$ if one of its endpoint lies in $\mathcal{Q}_{i+x}$ and the other in $\mathcal{Q}_{i-y}$. Let $\mu_{xy} = |\mathcal{L}_{xy}|$. Note that

$$\forall \ell_j \in \mathcal{L}_{xy}, \; |\pi(j) - \sigma(j)| \geq \zeta(x + y - 1).$$

By Lemma 4.8 a line $\ell_j \in \mathcal{L}_{xy}$ intersects at least $(x + y - 1)\zeta$ lines (see figure 4.6). Summing over all x and y we get

$$\begin{aligned} 2K &\geq \sum_x \sum_y \mu_{xy}(x + y - 1) \cdot \zeta \\ \text{or} \quad \frac{2K}{\zeta} &\geq \sum_x \sum_y x \cdot \mu_{xy} + \sum_x \sum_y y \cdot \mu_{xy} - \sum_x \sum_y \mu_{xy}. \end{aligned} \tag{4.1}$$

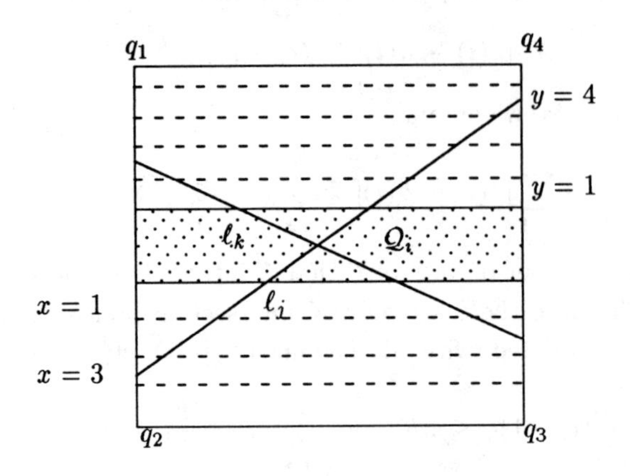

Figure 4.6: ℓ_j intersects at least 6ζ lines of $\mathcal{L}$.

Let $\mathcal{L}'_x \subseteq \mathcal{L}_i$ denote the set of all lines having one endpoint in Q_{i+x}. Then

$$\sum_x \sum_y x \cdot \mu_{xy} = \sum_x x\Big(\sum_y \mu_{xy}\Big) = \sum_x x|\mathcal{L}'_x|. \tag{4.2}$$

Similarly, if $\mathcal{L}''_y \subseteq \mathcal{L}_i$ denotes the set of lines having one endpoint in Q_{i-y}, then

$$\sum_x \sum_y y \cdot \mu_{xy} = \sum_y y|\mathcal{L}''_y|.$$

Moreover $\sum_x \sum_y \mu_{xy} = \mu_i$. Therefore (4.1) can be written as

$$\sum_x x \cdot |\mathcal{L}'_x| + \sum_y y \cdot |\mathcal{L}''_y| - \mu_i \leq \frac{2K}{\zeta}. \tag{4.3}$$

Since $\sum_x |\mathcal{L}'_x| = \mu_i$ and for all x, $|\mathcal{L}'_x| \leq 2\zeta$, the term $\sum_x x|\mathcal{L}'_x|$ is minimized when

$$|\mathcal{L}'_x| = \begin{cases} 2\zeta & \text{for } 1 \leq x \leq \left\lfloor \dfrac{\mu_i}{2\zeta} \right\rfloor \\ \mu_i - 2\zeta \left\lfloor \dfrac{\mu_i}{2\zeta} \right\rfloor & \text{for } x = \left\lfloor \dfrac{\mu_i}{2\zeta} \right\rfloor + 1 \\ 0 & \text{otherwise.} \end{cases}$$

Hence,

$$\begin{aligned}\sum_x x \cdot |\mathcal{L}'_x| &\geq \sum_{x=1}^{\lfloor \frac{\mu_i}{2\zeta} \rfloor} x \cdot 2\zeta + \left(\mu_i - 2\zeta\left\lfloor\frac{\mu_i}{2\zeta}\right\rfloor\right)\left(\left\lfloor\frac{\mu_i}{2\zeta}\right\rfloor + 1\right) \\ &\geq \frac{2\zeta}{2}\left\lfloor\frac{\mu_i}{2\zeta}\right\rfloor\left(\left\lfloor\frac{\mu_i}{2\zeta}\right\rfloor + 1\right) + \left(\mu_i - 2\zeta\left\lfloor\frac{\mu_i}{2\zeta}\right\rfloor\right)\left(\left\lfloor\frac{\mu_i}{2\zeta}\right\rfloor + 1\right) \\ &= \left(\left\lfloor\frac{\mu_i}{2\zeta}\right\rfloor + 1\right)\left(\mu_i - \zeta\left\lfloor\frac{\mu_i}{2\zeta}\right\rfloor\right). \qquad (4.4)\end{aligned}$$

Similarly,

$$\sum_y y \cdot |\mathcal{L}''_y| \geq \left(\left\lfloor\frac{\mu_i}{2\zeta}\right\rfloor + 1\right)\left(\mu_i - \zeta\left\lfloor\frac{\mu_i}{2\zeta}\right\rfloor\right). \qquad (4.5)$$

By substituting (4.4) and (4.5) in (4.3), we obtain

$$\frac{2K}{\zeta} \geq 2\left(\left\lfloor\frac{\mu_i}{2\zeta}\right\rfloor + 1\right)\left(\mu_i - \zeta\left\lfloor\frac{\mu_i}{2\zeta}\right\rfloor\right) - \mu_i. \qquad (4.6)$$

Let $\left\lfloor\frac{\mu_i}{2\zeta}\right\rfloor = a$ and $\frac{\mu_i}{2\zeta} = a + \nu$, for some $0 \leq \nu < 1$, then (4.6) becomes

$$\begin{aligned}\frac{2K}{\zeta} &\geq 2(a+1)(2\zeta(a+\nu) - \zeta a) - 2\zeta(a+\nu) \\ \text{or} \quad \frac{K}{\zeta^2} &\geq (a+1)(a+2\nu) - (a+\nu) \\ &= (a+\nu)^2 + \nu(1-\nu) \\ &\geq (a+\nu)^2 \qquad \text{(because } 0 \leq \nu < 1\text{)} \\ &= \frac{\mu_i^2}{4\zeta^2}.\end{aligned}$$

Therefore, $\mu_i \leq 2\sqrt{K}$.

□

Concerning running time, notice that the sets A and B, and therefore $\mathcal{Q}_1, \ldots, \mathcal{Q}_t$ can be computed in time $O(m \log m)$. Moreover, for each line ℓ_j we can easily compute the cells it crosses— they are simply all the cells lying between (and including) the cell containing $a_{\pi(j)}$ and the cell containing $b_{\sigma(j)}$. Thus the total time spent in computing the lines passing through each cell is bounded by $O(m \log m + K/\zeta)$. Now it follows from Lemma 4.4 that number K_i of intersections within $\mathcal{Q}_i$ can be computed in time $O(m_i \log m_i)$. Therefore the total time spent in computing intersections is at most

$$O\Big(\sum_i m_i \log m_i\Big) = O\Big((\sum_i m_i) \log m\Big) = O\Big((m + \frac{K}{\zeta}) \log m\Big).$$

Hence, we can conclude that

Lemma 4.12 *The above algorithm partitions the cell $\mathcal{Q}$ into $t = \lceil \frac{m}{\zeta} \rceil$ subcells $\mathcal{Q}_1, \ldots, \mathcal{Q}_t$ in time $O(m \log m)$ such that (i) $\sum_{i=1}^{t} m_i \le 2\left(m + \frac{K}{\zeta}\right)$, and (ii) $\max_i m_i \le 2\sqrt{K} + 2\zeta$. Moreover, we can compute $\mathcal{L}_i$ and K_i (as defined above) for all $\mathcal{Q}_i$, in time $O((m + \frac{K}{\zeta}) \log m)$.*

□

Remark 4.13: Note that in the special case at hand we have ignored the issue of making the numbers K_i small. Also, the resulting partitioning can have one cell with 6 edges or two cells with 5 edges each. These issues will be addressed in the following general algorithm.

4.4.2 General algorithm

Next we give an algorithm for the general case using the above procedure as a subroutine. Our algorithm consists of five steps.

The first step applies the preceding algorithm to the collection of red lines, to partition the quadrilateral $\mathcal{Q}$ into a set of cells $\mathbf{Q}^r = \{\mathcal{Q}_1^r, \ldots, \mathcal{Q}_t^r\}$, for $t = \lceil \frac{m_r}{\zeta} \rceil$, so that every cell contains the left as well as right endpoints of at most ζ red lines. Let $R = \{r_1, \ldots, r_{t-1}\}$ denote the set of pseudo edges that bound the quadrilaterals of $\mathbf{Q}^r$; they will be referred to as "red" pseudo edges.

The second step applies the preceding algorithm to the collection of green lines (where this time $\partial\mathcal{Q}$ is divided into a left and a right portions at q_4 and q_2), to partition $\mathcal{Q}$ into a collection of cells $\mathbf{Q}^g = \{\mathcal{Q}_1^g, \ldots, \mathcal{Q}_u^g\}$, for $u = \lceil \frac{m_g}{\zeta} \rceil$, so that every cell contains the left as well as right endpoints of at most ζ green lines. Let $G = \{g_1, \ldots, g_{u-1}\}$ denote the set of pseudo edges bounding the quadrilaterals of $\mathbf{Q}^g$; these pseudo edges will be referred to as "green" pseudo edges. Since the endpoints of all green lines lie on $\overline{q_1 q_4}$

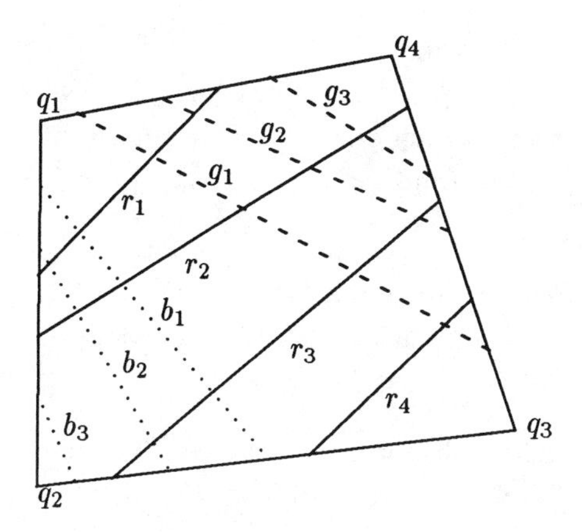

Figure 4.7: Red, green and blue pseudo edges

and $\overline{q_3q_4}$, all green pseudo edges extend from $\overline{q_1q_4}$ to $\overline{q_3q_4}$. Assume that the quadrilaterals $\mathcal{Q}_i^g$ are sorted in their order along $\overrightarrow{q_1q_4}$, and that $\mathcal{Q}_i^g$ is bounded by the two pseudo edges g_{i-1}, g_i, for $1 < i < u$.

In the third step we apply the same algorithm to the collection of blue lines (again, partitioning $\partial\mathcal{Q}$ at q_2 and q_4), to partition $\mathcal{Q}$ into a collection of cells $\mathbf{Q}^b = \{\mathcal{Q}_1^b, \ldots, \mathcal{Q}_v^b\}$, for $v = \lceil \frac{m_b}{\zeta} \rceil$, so that every cell contains the left as well as right endpoints of at most ζ blue lines. Let $B = \{b_1, \ldots, b_{v-1}\}$ denote the set of pseudo edges bounding the quadrilaterals of $\mathbf{Q}^b$; we refer to them as "blue" pseudo edges. This time all pseudo edges extend from $\overline{q_1q_2}$ to $\overline{q_2q_3}$. Assume that the quadrilaterals $\mathcal{Q}_i^b$ are sorted in their order along $\overrightarrow{q_1q_2}$, and that $\mathcal{Q}_i^b$ is bounded by the two pseudo edges b_{i-1}, b_i, for $1 < i < v$. See figure 4.7 for an illustration of the resulting pseudo edges.

Let $\mathcal{S} = R \cup B \cup G$, and let $\mathcal{E}$ denote the set of endpoints of segments in $\mathcal{S}$. Let $\mathbf{F}$ denote the set of bounded faces in the arrangement of $\partial\mathcal{Q} \cup \mathcal{S}$. If a face $\mathcal{F}_\alpha \in \mathbf{F}$ touches the boundary of $\mathcal{Q}$, then it is called a *boundary cell*, otherwise it is called an *internal cell* (see figure 4.8). The boundary of an internal cell is formed by four pseudo edges, two of which are always red. On the basis of the color of their edges we partition the cells into three categories:

(i) red-green cells: cells formed by the intersection of $\mathcal{Q}_i^g$ for $1 < i \leq u$ and $\mathcal{Q}_j^r$ for $1 \leq j \leq t$. A red-green cell formed by $\mathcal{Q}_i^g \cap \mathcal{Q}_j^r$ is denoted by $\mathcal{F}_{ij}$. If $\mathcal{F}_{ij}$ is an internal cell, then its edges are portions of g_{i-1}, $g_i \in G$ and r_{j-1}, $r_j \in R$ (and $i < u$, $1 < j < t$).

(ii) red-blue cells: cells formed by the intersection of $\mathcal{Q}_i^b$ for $1 < i \leq v$ and $\mathcal{Q}_j^r$ for $1 \leq j \leq t$. A red-blue cell formed by $\mathcal{Q}_i^b \cap \mathcal{Q}_j^r$ is denoted by $\mathcal{F}_{ij}$. If $\mathcal{F}_{ij}$ is an internal cell, then its edges are portions of b_{i-1}, $b_i \in B$ and r_{j-1}, $r_j \in R$ (and $i < v$, $1 < j < t$).

(iii) red-blue-green cells: cells formed by the intersection of $\mathcal{Q}_1^g$, $\mathcal{Q}_1^b$ and $\mathcal{Q}_i^r$ for $1 \leq i \leq t$. We denote such a cell by $\mathcal{F}_{1i}$. If $\mathcal{F}_{1i}$ is an internal cell, then its edges are portions of b_1, g_1, r_{i-1} and r_i (and $1 < i < t$).

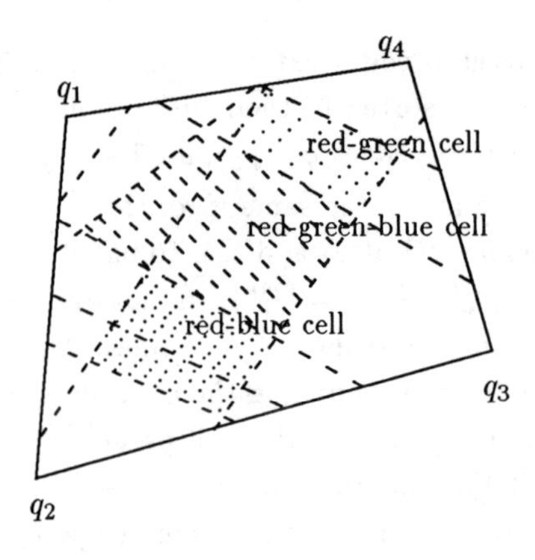

Figure 4.8: Internal and boundary cells: shaded regions denote internal cells

The problem with the boundary cells produced by the arrangement of $\partial\mathcal{Q} \cup \mathcal{S}$ is that some of them may have more than four edges. However, since every cell has at most four pseudo edges, it cannot have more than eight edges in total. At the fourth step of our algorithm, we partition all boundary cells with more than four edges into two or three convex quadrilaterals and triangles by appropriate diagonals, which we also call pseudo edges (see figure 4.9).

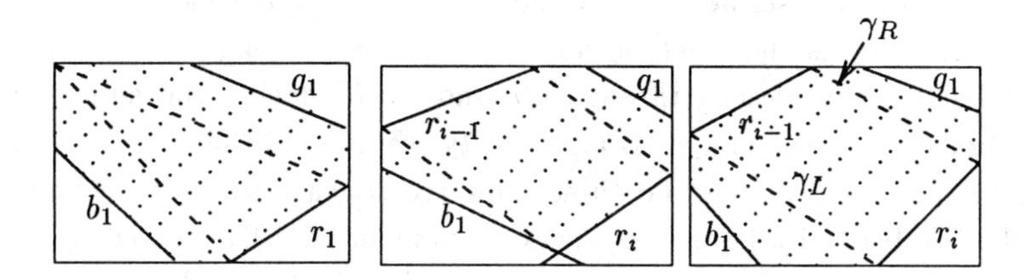

Figure 4.9: Cells with seven or eight edges: dashed lines are the pseudo edges added to partition the cell into convex quadrilaterals and triangles

A cell with five or six edges is partitioned arbitrarily into two convex quadrilaterals, but a cell with seven or eight edges is partitioned in such a way so that one of the new pseudo edges intersects only red and green lines, and the other intersects only red and blue lines. It is easily seen that such a partition is always possible. For example, an 8-cell lying between the red pseudo edges r_{i-1}, r_i is partitioned by connecting the left endpoint of r_{i-1} to the left endpoint of r_i, and the right endpoint of r_{i-1} to the right endpoint of r_i (see figure 4.9). 7-cells are partitioned in a similar manner.

Note that there can be at most one cell having 7 or 8 edges. The existence of an 8-cell implies that no pair of pseudo edges intersect, and the existence of a 7-cell allows only very limited patterns of pseudo-edge intersections. Moreover the existence of a 7 or 8-cell implies that the first step creates at least one red pseudo edge.

Next, for each of the cells produced so far, we compute the lines passing through it, and count the number of intersection points lying in it. If $K \geq 6\zeta^2$ and a cell $\mathcal{F}_\alpha$ has $K_\alpha > \zeta\sqrt{K}$ intersection points, then we partition it into $\lceil \frac{K_\alpha}{\zeta\sqrt{K}} \rceil$ cells, each containing at most $\zeta\sqrt{K}$ intersections, using the algorithm of Section 4.3. With some care we can obtain such a partition with at most one 5-cell, which we then split by a diagonal into two subcells.

This yields a final collection of quadrilaterals $\{\mathcal{Q}_1, \mathcal{Q}_2, \ldots, \mathcal{Q}_M\}$, which is the output of the algorithm.

4.4.3 Analysis of the algorithm

The algorithm just described for partitioning a convex quadrilateral is fairly simple. Its analysis, however, is not. For $i = 1, \ldots, M$, let m_i denote the number of lines passing through $\mathcal{Q}_i$, and let K_i denote the number of intersections contained in $\mathcal{Q}_i$. For simplicity we assume that the lines are in general position, that is, no three lines are concurrent, and no line is vertical. We begin by bounding the total number of cells created by the algorithm and the number of lines meeting each cell. As we shall see later, the constants appearing in the bounds for $\max_i m_i$ and $\sum_i m_i$ control the exponent of the logarithmic factor in the bound for the time complexity of the overall partitioning algorithm, therefore we try to obtain as small constants as possible. If we do not worry about constants, the proofs can be simplified a lot. To bound the total number of cells produced by our algorithm, we first estimate the number of cells formed by overlapping the original red, green and blue pseudo edges.

Lemma 4.14 *Let N be the number of cells produced by overlapping the red, green and blue pseudo edges, then*

$$N \le \frac{2}{\zeta}m + \frac{K}{\zeta^2} - 2. \tag{4.7}$$

Proof: We bound the number of boundary and internal cells separately. The number of boundary cells is obviously equal to the number of endpoints of pseudo edges, namely $2|S| = 2(u + v + t - 3)$.

Next, consider the number of internal cells. To bound the number of red-green and red-blue-green cells, we use the following charging scheme. Let w be the intersection point of a red line ℓ and a green line ℓ'. Suppose that the right endpoint of ℓ lies on $\overline{q_1q_4}$ in a cell $\mathcal{Q}_j^r$, and that the left endpoint of ℓ' (on $\overline{q_1q_4}$) lies in $\mathcal{Q}_i^g$. If $i = 1$ and $1 < j < t$, then we charge w to the red-blue-green cell $\mathcal{F}_{1j}$ (e.g. w_4 in figure 4.10) , and if $1 < i < u$ and $1 < j < t$, then we charge w to the red-green cell $\mathcal{F}_{ij}$ (e.g. w_2 in figure 4.10). If $\mathcal{Q}_i^g \cap \mathcal{Q}_j^r$ is a boundary cell, then we do not charge w to any cell (e.g. w_3

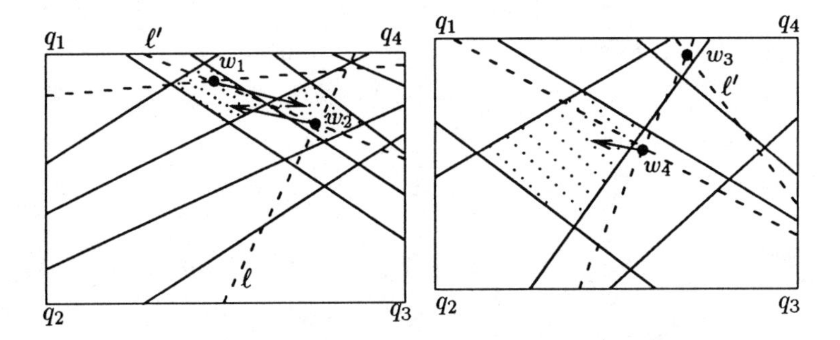

Figure 4.10: Charging of a red-green intersection point

in figure 4.10). Similarly, if the right endpoint of ℓ lies on $\overline{q_3q_4}$, then we charge w to the cell $\mathcal{F}_{ij}$ (e.g. w_1 in figure 4.10), where $\mathcal{Q}_i^g$ contains the right endpoint of ℓ' (on $\overline{q_3q_4}$) and $\mathcal{Q}_j^r$ contains, as before, the right endpoint of ℓ. Again, no charge is made if $\mathcal{Q}_i^g \cap \mathcal{Q}_j^r$ is a boundary cell.

Clearly, each internal red-green intersection is charged to at most one cell. Moreover, it is easily checked that each internal red-green or red-blue-green cell is charged exactly ζ^2 red-green intersections, namely those between the red lines whose right endpoint lies in $\mathcal{Q}_j^r$ and the green lines whose right or left endpoint (as the case might be) lies in $\mathcal{Q}_i^g$ (by construction, all those intersections points do lie inside $\mathcal{Q}$). Hence there are at most $\frac{K_{rg}}{\zeta^2}$ red-green and red-blue-green cells. Similarly we can prove that there are at most $\frac{K_{rb}}{\zeta^2}$ red-blue cells. Thus, the overall number N of cells after the third step is bounded by

$$\begin{aligned}
N &\leq 2(t+u+v) + \frac{K_{rg}}{\zeta^2} + \frac{K_{rb}}{\zeta^2} - 6 \\
&\leq 2\left(\left\lceil\frac{m_r}{\zeta}\right\rceil + \left\lceil\frac{m_g}{\zeta}\right\rceil + \left\lceil\frac{m_b}{\zeta}\right\rceil\right) + \frac{1}{\zeta^2}(K_{rg} + K_{rb}) - 6 \\
&\leq \frac{2}{\zeta}(m+2) + \frac{1}{\zeta^2}(K_{rg} + K_{rb}) - 6
\end{aligned}$$

$$\leq \frac{2}{\zeta}m + \frac{K}{\zeta^2} - 2.$$

□

Lemma 4.15 *Let M be the total number of cells produced by the above algorithm. Then*

$$M \leq \frac{3}{\zeta}m + \frac{2K}{\zeta^2}.$$

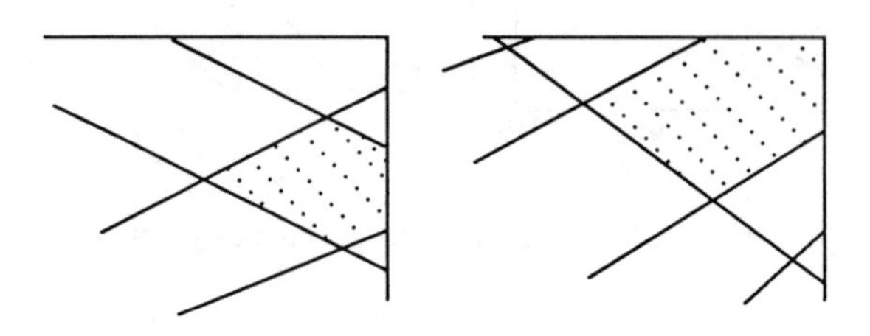

Figure 4.11: No two adjacent cells have more than four edges

Proof: It is easy to see that for any two adjacent boundary cells, one of them has at most four edges, therefore there are at most $|S|$ cells with five or more edges (see figure 4.11). Moreover there is at most one cell with seven or eight edges. Since each 5 or 6-cell is partitioned into two convex quadrilaterals and the 7 or 8-cell into three convex quadrilaterals, the fourth step creates at most $|S|+1$ convex quadrilaterals, which in conjunction with (4.7) implies that after the fourth step the number of cells is at most

$$\frac{2m}{\zeta} + \frac{K}{\zeta^2} - 2 + (t+u+v-3) \leq \frac{3m}{\zeta} + \frac{K}{\zeta^2} - 5.$$

Finally, the fifth step adds at most $\lceil \frac{K_\alpha}{\zeta\sqrt{K}} \rceil$ new cells at each quadrilateral containing $K_\alpha > \zeta\sqrt{K}$ intersection points. Since there can be at most $\frac{K}{\zeta\sqrt{K}}$ such quadrilaterals, the total number of new cells is at most

$$\sum_\alpha \left(\frac{K_\alpha}{\zeta\sqrt{K}} + 1 \right) \leq \frac{2\sqrt{K}}{\zeta} \leq \frac{K}{\zeta^2},$$

because we apply this step only when $K > 6\zeta^2$. Thus

$$M \leq \frac{3m}{\zeta} + \frac{2K}{\zeta^2}.$$

□

Next, we bound the maximum number of lines passing through any cell and the total number of line-cell crossings.

Lemma 4.16 *For each* $1 \leq i \leq M$, *the number* m_i *of lines passing through the cell* Q_i *satisfies*

$$m_i \leq \max\{2\sqrt{2}\sqrt{K} + 4\zeta, 2\sqrt{K} + 6\zeta\}. \tag{4.8}$$

Proof: Note that a cell Q_i produced by our algorithm is a subset of some cell $\mathcal{F}_\alpha \in \mathbf{F}$ (the cells obtained by overlapping red, blue and green pseudo edges). Therefore it suffices to bound the number of lines passing through a cell in $\mathbf{F}$. First, let us consider a red-green cell $\mathcal{F}_{ij} = Q_i^g \cap Q_j^r$. By Lemma 4.12, Q_i^g meets at most $2\sqrt{K_{gg}} + 2\zeta$ green lines and Q_j^r meets at most $2\sqrt{K_{rr}} + 2\zeta$ red lines, and since no blue line passes through a red-green cell, there are at most $2(\sqrt{K_{rr}} + \sqrt{K_{gg}} + 2\zeta)$ lines passing through $\mathcal{F}_{ij}$. Similarly, we can show that a red-blue cell meets at most $2(\sqrt{K_{rr}} + \sqrt{K_{bb}} + 2\zeta)$ lines.

Finally, if $\mathcal{F}_{1i}$ is a red-blue-green cell, then $\mathcal{F}_{1i}$ is the intersection of Q_1^g, Q_1^b and Q_i^r. Since Q_1^g (resp. Q_1^b) meets at most 2ζ green (resp. blue) lines, $\mathcal{F}_{1i}$ has at most $2\sqrt{K_{rr}} + 6\zeta$ lines. Now the lemma follows from the fact that $\max\{\sqrt{K_{rr}} + \sqrt{K_{bb}}, \sqrt{K_{rr}} + \sqrt{K_{gg}}\} \leq \sqrt{2K}$.

□

For $x, y \in \{r, b, g\}$, let δ_i^{xy} denote the number of pseudo edges of color x intersected by the line ℓ_i of color y. It follows from Lemma 4.9 that

$$\sum_{i=1}^{m_x} \delta_i^{xx} \leq \frac{2K_{xx}}{\zeta} + m_x.$$

In the next two lemmas we bound the "mixed" sums $\sum_{i=1}^{m_g} \delta_i^{rg}$, $\sum_{i=1}^{m_b} \delta_i^{rb}$, $\sum_{i=1}^{m_r} \delta_i^{gr}$ and $\sum_{i=1}^{m_r} \delta_i^{br}$.

Lemma 4.17 $$\sum_{i=1}^{m_g} \delta_i^{rg} \leq \frac{K_{rg}}{\zeta} + m_g \quad \text{and} \quad \sum_{i=1}^{m_b} \delta_i^{rb} \leq \frac{K_{rb}}{\zeta} + m_b.$$

Proof: We will only prove the first part of the lemma; the second part can be proved in a symmetric way. Let ℓ_i be a green line intersecting δ_i^{rg} red pseudo edges, $r_{k+1}, \ldots, r_{k+\delta_i^{rg}}$, and let ℓ' be a red line having its right endpoint in Q_l^r, for $k+1 < l < k+\delta_i^{rg}$ (see figure 4.12a). It is easily checked that ℓ' intersects ℓ_i inside $\mathcal{Q}$. Since each $\mathcal{Q}_l^r$ contains the right endpoints of ζ lines, ℓ_i intersects at least $(\delta_i^{rg} - 1)\zeta$ lines inside $\mathcal{Q}$. But there are at most K_{rg} red-green intersection points, therefore summing over all green lines, we obtain

$$\sum_{i=1}^{m_g}(\delta_i^{rg} - 1)\zeta \leq K_{rg} \quad \text{or} \quad \sum_{i=1}^{m_g}\delta_i^{rg} \leq \frac{K_{rg}}{\zeta} + m_g.$$

□

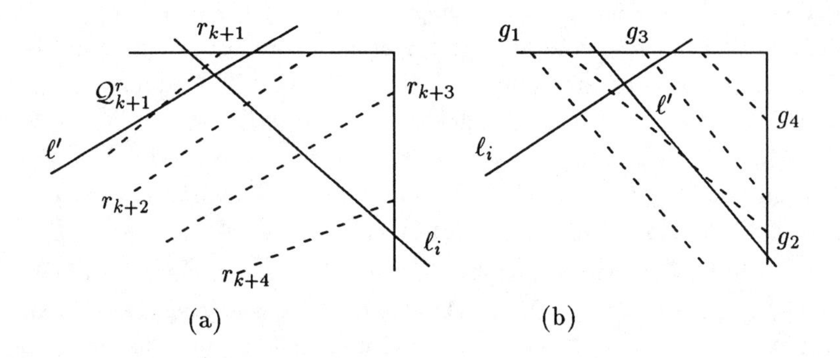

Figure 4.12: Illustration for Lemma 4.17 and Lemma 4.18

Lemma 4.18 $$\sum_{i=1}^{m_r}\delta_i^{gr} \leq \frac{K_{rg}}{\zeta} \quad \text{and} \quad \sum_{i=1}^{m_r}\delta_i^{br} \leq \frac{K_{rb}}{\zeta}.$$

Proof: Let ℓ_i be a red line intersecting δ_i^{gr} green pseudo edges, $g_1, \ldots, g_{\delta_i^{gr}}$. Suppose that the right endpoint of ℓ_i lies on $\overline{q_1q_4}$. Let ℓ' be a green line whose left endpoint (on $\overline{q_1q_4}$) lies in $\mathcal{Q}_l^g$, for $l \leq \ell_i^{gr}$ (see figure 4.12b). It is easily seen that ℓ' intersects ℓ_i inside $\mathcal{Q}$. Similarly, if the right endpoint of ℓ_i is on $\overline{q_3q_4}$, then ℓ_i intersects any green line whose right endpoint (on $\overline{q_3q_4}$) lies in $\mathcal{Q}_l^g$, for $l \leq \ell_i^{gr}$. Since each $\mathcal{Q}_l^g$ contains ζ left endpoints, and also ζ right endpoints, of green lines, ℓ_i intersects at least $\delta_i^{gr} \cdot \zeta$ green lines inside $\mathcal{Q}$. Summing over all red lines, we obtain

$$\sum_{i=1}^{m_r}\delta_i^{gr} \cdot \zeta \leq K_{rg} \quad \text{or} \quad \sum_{i=1}^{m_r}\delta_i^{gr} \leq \frac{K_{rg}}{\zeta}.$$

Similarly, we can prove that $\sum_{i=1}^{m_r} \delta_i^{br} \leq \frac{K_{rb}}{\zeta}$.

□

Lemma 4.19 *Let m'_i denote the number of lines passing through the ith cell of* **F**, *then*

$$\sum_{i=1}^{N} m'_i \leq \frac{2}{\zeta}K + \frac{5}{2}m. \tag{4.9}$$

Proof: For $x \in \{r, b, g\}$, let η_j^x denote the number of cells crossed by a line $\ell_j \in \mathcal{L}_x$. For each cell $\mathcal{F}$ meeting ℓ_j, we charge $\mathcal{F}$ to the leftmost point $w_{\mathcal{F}}$ of $\partial\mathcal{F} \cap \ell_j$. For all cells crossed by ℓ_j, except the leftmost one, $w_{\mathcal{F}}$ is an intersection point of a pseudo edge and ℓ_j (see figure 4.13). For red lines it follows from Lemma 4.9 and Lemma 4.18 that

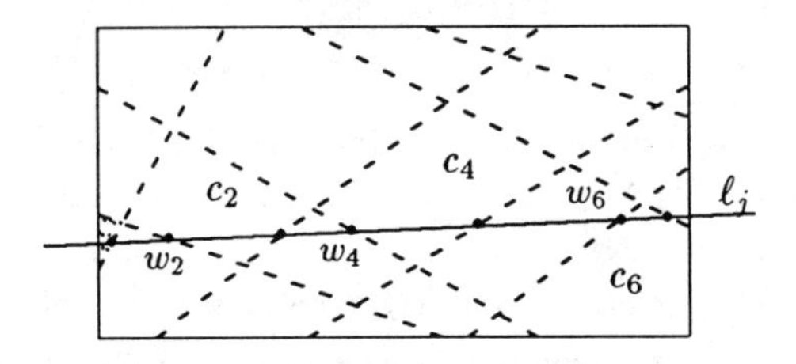

Figure 4.13: Charging line cell crossings; cell c_i is charged to w_i

$$\begin{aligned}\sum_{i=1}^{m_r} \eta_i^r &= \sum_{i=1}^{m_r} (\delta_i^{rr} + \delta_i^{gr} + \delta_i^{br} + 1) \\ &\leq 2\frac{K_{rr}}{\zeta} + \frac{K_{rg}}{\zeta} + \frac{K_{rb}}{\zeta} + 2m_r. \end{aligned} \tag{4.10}$$

Similarly, using Lemma 4.9 and Lemma 4.17 we can show that

$$\begin{aligned}\sum_{i=1}^{m_b} \eta_i^b &= \sum_{i=1}^{m_b} (\delta_i^{bb} + \delta_i^{rb} + \delta_i^{gb} + 1) \\ &\leq 2\frac{K_{bb}}{\zeta} + \frac{K_{rb}}{\zeta} + 3m_b, \end{aligned} \tag{4.11}$$

$$\text{and} \quad \sum_{i=1}^{m_g} \eta_i^g \leq 2\frac{K_{gg}}{\zeta} + \frac{K_{rg}}{\zeta} + 3m_g. \tag{4.12}$$

Since $\sum_{i=1}^{N} m'_i = \sum_{\substack{x\in\\ \{r,b,g\}}} \sum_{j=1}^{m_x} \eta_j^x$, summing (4.10), (4.11) and (4.12), we obtain

$$\begin{aligned}\sum_{i=1}^{N} m'_i &\leq \frac{2}{\zeta}(K_{rr} + K_{bb} + K_{gg} + K_{rg} + K_{rb}) + 3(m_g + m_b) + 2m_r \\ &\leq \frac{2}{\zeta}K + \frac{5}{2}m \qquad \text{(because } m_r \geq \tfrac{m}{2}\text{; see Section 4.2).}\end{aligned}$$

□

Next, we bound $\sum_{i=1}^{M} m_i$, that is, the total number of line-cell crossings, for the final cells produced by the algorithm.

Lemma 4.20 *Let m_i denote the number of lines in cell Q_i; then*

$$\sum_{i=1}^{M} m_i \leq (4 + 4\sqrt{2})\frac{K}{\zeta} + (5 + 4\sqrt{2})m.$$

Proof: Let m_α denote the number of lines meeting the αth cell after the fourth step. First consider the case when $\mathbf{F}$ has a cell $\mathcal{F}_\alpha$ with seven or eight edges, say eight edges (see figure 4.9). Let r_{i-1} and r_i denote the red pseudo lines bounding $\mathcal{F}_\alpha$. In this case the fourth step adds only two pseudo edges, γ_L and γ_R. Since γ_L does not intersect any green lines and intersects only those red lines whose left endpoints lie in Q_i^r, it adds at most $\zeta + m_b$ line-cell crossings. Similarly γ_R adds at most $\zeta + m_g$ line-cell crossings. Moreover, if $\mathbf{F}$ has a 7-cell, then it can also have one 5-cell. Thus by Lemma 4.19,

$$\sum_{\alpha} m_\alpha \leq \frac{2}{\zeta}K + \frac{5}{2}m + (2\zeta + m_g + m_b) + m.$$

Note that in this case the first three steps create at least three pseudo edges, which implies that $m > 3\zeta$. Since $m_g + m_b \leq \frac{m}{2}$, we have

$$\sum_{\alpha} m_\alpha \leq \frac{2}{\zeta}K + 5m.$$

On the other hand if all cells after the third step have at most six edges, then every cell is partitioned into at most two cells. Thus by Lemma 4.19, the total number of line-cell crossing after the fourth step is

$$\sum_{\alpha} m_\alpha \leq \frac{4}{\zeta}K + 5m.$$

The final step creates at most $\frac{2\sqrt{K}}{\zeta}$ new cells. Moreover, in that step a cell $\mathcal{F}_\alpha$ is partitioned only when $K > 6\zeta^2$, in which case $m_\alpha < 2\sqrt{2}\sqrt{K} + 4\zeta$ (cf. Lemma 4.16). Hence, we obtain

$$\begin{aligned}\sum_{i=1}^{M} m_i &\leq \frac{4}{\zeta}K + 5m + \frac{2\sqrt{K}}{\zeta}(2\sqrt{2}\sqrt{K} + 4\zeta) \\ &= (4 + 4\sqrt{2})\frac{K}{\zeta} + 8\sqrt{K} + 5m \\ &\leq (4 + 4\sqrt{2})\frac{K}{\zeta} + (5 + 4\sqrt{2})m\end{aligned}$$

because $K \leq \frac{m^2}{2}$.

□

Finally, we analyze the running time of the algorithm.

Lemma 4.21 *It requires $O\left((m + \frac{K}{\zeta})\log m\right)$ time to compute all subquadrilaterals $\mathcal{Q}_1, \ldots, \mathcal{Q}_M$, to determine the lines passing through each $\mathcal{Q}_i$, and to count the number of intersection points in each $\mathcal{Q}_i$.*

Proof: It follows from Lemma 4.12 that $\mathbf{Q}^r$, $\mathbf{Q}^g$ and $\mathbf{Q}^b$ can be computed in time $O((m_r + m_b + m_g)\log m) = O(m \log m)$. Once we have computed all pseudo edges, the set $\mathbf{F}$ of cells, obtained by overlaying these pseudo edges, can be computed, for example, by sweeping a vertical line through the quadrilateral $\mathcal{Q}$. The time spent in the sweep is $O((|\mathcal{S}| + |\mathbf{F}|)\log \mathcal{S})$ $= O\left((\frac{m}{\zeta} + \frac{K}{\zeta^2})\log m\right)$, because it encounters at most $O(|\mathbf{F}|)$ intersection points. Since step 4 simply scans all boundary cells and splits those having more than four edges, it can be easily done in $O(|\mathcal{S}|)$ time.

As for the final step, let m_α (resp. K_α) be the number of lines meeting (resp. intersection points contained in) the αth cell produced by the first four steps. The lines passing through each cell $\mathcal{F}_\alpha$ can be computed in time $O(m \log m + \sum_\alpha m_\alpha \log m) = O((m + \frac{K}{\zeta})\log m)$ (cf. Lemma 4.20) by tracing every line through the cells it intersects and spending $O(\log m)$ time at each cell. It follows from Lemma 4.4 that the quantities K_α, for all cells $\mathcal{F}_\alpha$, can be calculated in time $O((m + \frac{K}{\zeta})\log m)$. Moreover, step 5 of the algorithm partitions the αth cell into at most $\lceil \frac{K_\alpha}{\zeta\sqrt{K}} \rceil$ convex quadrilaterals, which is done applying the algorithm of Section 4.3 at most $\lceil \frac{K_\alpha}{\zeta\sqrt{K}} \rceil - 1$ times, each application taking $O(m_\alpha \log m_\alpha)$ time (cf. Theorem 4.6). Therefore, the total time spent in the final step is bounded by

$$
\begin{aligned}
& O\Big(\sum_\alpha \Big(\Big\lceil \frac{K_\alpha}{\zeta\sqrt{K}} \Big\rceil - 1\Big) m_\alpha \log m_\alpha\Big) \\
= \; & O\Big(\frac{1}{\zeta\sqrt{K}} \sum_\alpha K_\alpha \cdot m_\alpha \log m_\alpha\Big) \\
= \; & O\Big(\frac{\sqrt{K}+\zeta}{\zeta\sqrt{K}} \log m \Big(\sum_\alpha K_\alpha\Big)\Big) \\
& (\text{because } m_\alpha \leq 2\sqrt{2}\sqrt{K} + 4\zeta) \\
= \; & O\Big(\Big(\frac{K}{\zeta} + \sqrt{K}\Big) \log m\Big) \\
= \; & O\Big(\Big(\frac{K}{\zeta} + m\Big) \log m\Big) \\
& (\text{because } K \leq \frac{m^2}{2}).
\end{aligned}
$$

Thus, the overall running time of our algorithm is $O\left((m + \frac{K}{\zeta}) \log m\right)$.

□

Hence, we can conclude that

Theorem 4.22 *We can partition the convex quadrilateral Q into $M \leq \frac{3m}{\zeta} + \frac{2K}{\zeta^2}$ convex subquadrilaterals $Q_1, \ldots, Q_M$, with the property that each Q_i is crossed by m_i lines and contains K_i intersection points, so that the following conditions are satisfied:*

$$
\begin{array}{llll}
(i) & \sum_{i=1}^M m_i \leq A_1 m + A_2 \frac{K}{\zeta}; & \max m_i \leq A_3\sqrt{K} + A_4\zeta \\
(ii) & \sum_{i=1}^M K_i = K; & \max K_i \leq \zeta\sqrt{K}
\end{array}
$$

where $A_1 = (5 + 4\sqrt{2})$, $A_2 = (4 + 4\sqrt{2})$, $A_3 = 2\sqrt{2}$ *and* $A_4 = 6$.

□

Remark 4.23: If some other simpler sorting network of $O(\log^2 m)$ depth is used in the final step, the running time will be $O((m + K/\zeta) \log^2 m)$ instead of $O((m + K/\zeta) \log m)$.

4.5 Partitioning the Plane into Quadrilaterals

We now obtain the main result of this chapter. Let $\mathcal{L} = \{\ell_1, \ell_2, \ldots, \ell_n\}$ be a set of n lines in the plane, and $\mathbf{R}$ an "enclosing rectangle" of $\mathcal{L}$, i.e. one that contains all $\binom{n}{2}$ intersection points of $\mathcal{L}$. For a given integer $r > 0$, we want to partition $\mathbf{R}$ into $O(r^2 \log^\beta r)$ convex quadrilaterals, for some constant $\beta > 0$ to be determined later, so that each of them is crossed by at most $\frac{n}{r}$ lines.

The idea is to use the algorithm of the previous section recursively. Fix $\zeta = \frac{n}{\gamma r}$, where γ is a constant to be chosen later (for simplicity let us assume that n is a multiple of γr). At each step of the algorithm we have a convex quadrilateral $\mathcal{Q}$ whose interior meets m of the given lines and contains K of their intersection points. Initially $\mathcal{Q} = \mathbf{R}$, $m = n$ and $K = \binom{n}{2}$ (assuming that the lines are in general position). At each recursive step, the algorithm proceeds as follows: If $m \le \frac{n}{r}$, there is nothing to do, so we stop; otherwise we partition $\mathcal{Q}$ into M convex quadrilaterals using the algorithm of the previous section (for the initial fixed value of ζ).

Let m_i denote the number of lines meeting the interior of $\mathcal{Q}_i$ and let K_i denote the number of intersection points of $\mathcal{L}$ contained in the interior of $\mathcal{Q}_i$. Let $\mathcal{C}(m, K)$ denote the maximum number of cells into which such a $\mathcal{Q}$ will be partitioned by all subsequent recursive applications of the algorithm, and let $\mathcal{T}(m, K)$ denote the maximum time required for such a partitioning. It follows from Theorem 4.22 that

$$\mathcal{C}(m,K) \le \begin{cases} \sum_{i=1}^{M} \mathcal{C}(m_i, K_i) & \text{if } m > \frac{n}{r} \\ 1 & \text{otherwise} \end{cases} \tag{4.13}$$

$$\mathcal{T}(m,k) \le \begin{cases} \sum_{i=1}^{M} \mathcal{T}(m_i, K_i) + D\left(m + \frac{K}{\zeta}\right)\log m & \text{if } m > \frac{n}{r} \\ D & \text{otherwise} \end{cases} \tag{4.14}$$

where D is some constant > 0 and M, m_i and K_i satisfy the bounds given in Theorem 4.22.

Next we bound the values of $\mathcal{C}(m, K)$ and $\mathcal{T}(m, K)$ using (4.13) and (4.14). In what follows by Log$x$ we mean $\max\{\log x, 1\}$, and similarly, later on, by LogLog$x$ we mean $\max\{\log\log x, \log\log\sqrt{6}\}$.

Lemma 4.24 *There exists a constant $E > 0$ such that*

$$\mathcal{C}(m,K) \le E\left(\frac{m}{\zeta} + \frac{K}{\zeta^2}\right)\operatorname{Log}^{\beta}\frac{K}{\zeta^2}, \tag{4.15}$$

where $\beta = \max\{\log A_1, \log(1+A_2)\} < 3.33$.

Proof: We prove the above inequality by induction on K. If we choose $\gamma \ge 4(\sqrt{3}+1)$, then for $K \le 6\zeta^2$, after applying the algorithm once, we have

$$\begin{aligned}\max m_i &\le \max\{2\sqrt{2}\sqrt{K} + 4\zeta, 2\sqrt{K} + 6\zeta\}\\ &= \max\{(4\sqrt{3}+4)\zeta, (2\sqrt{6}+6)\zeta\}\\ &= 4(\sqrt{3}+1)\zeta \le \frac{n}{r}\end{aligned}$$

and therefore, the algorithm stops. By Theorem 4.22 this step partitions $\mathcal{Q}$ into at most $3\left(\frac{m}{\zeta} + \frac{K}{\zeta^2}\right)$ cells. Thus, if we choose $E > 3$, then for $K \le 6\zeta^2$, (4.15) holds trivially.

For $K > 6\zeta^2$, suppose inductively that (4.15) holds for all $K' < K$. Since $K_i \le \zeta\sqrt{K}$ and $K > 6\zeta^2$, $K_i \le \sqrt{\frac{K}{6}}\cdot\sqrt{K} = \frac{K}{\sqrt{6}} < K$. Therefore by induction hypothesis, (4.15) holds for all $\mathcal{Q}_i$, so (4.13) implies

$$\mathcal{C}(m,K) \le \sum_{i=1}^{M} E\left(\frac{m_i}{\zeta} + \frac{K_i}{\zeta^2}\right)\operatorname{Log}^{\beta}\frac{K_i}{\zeta^2}. \tag{4.16}$$

Since $\forall i$, $K_i \le \zeta\sqrt{K}$, we have

$$\operatorname{Log}\frac{K_i}{\zeta^2} \le \operatorname{Log}\frac{\zeta\sqrt{K}}{\zeta^2} = \max\{\log\sqrt{\frac{K}{\zeta^2}}, 1\}.$$

$K \ge 6\zeta^2$ implies that $\log\sqrt{\frac{K}{\zeta^2}} > 1$, therefore (4.16) can be written as

$$\begin{aligned}\mathcal{C}(m,K) &\le \sum_{i=1}^{M} E\left(\frac{m_i}{\zeta} + \frac{K_i}{\zeta^2}\right)\log^{\beta}\sqrt{\frac{K}{\zeta^2}}\\ &= E\left(\frac{1}{\zeta}\sum_{i=1}^{M} m_i + \frac{1}{\zeta^2}\sum_{i=1}^{M} K_i\right)\cdot\frac{1}{2^{\beta}}\log^{\beta}\frac{K}{\zeta^2}\\ &\le \frac{E}{2^{\beta}}\left(A_1\frac{m}{\zeta} + (A_2+1)\frac{K}{\zeta^2}\right)\log^{\beta}\frac{K}{\zeta^2}\\ &\qquad \text{(By Theorem 4.22).}\end{aligned}$$

Let $\beta = \max\{\log A_1, \log(1 + A_2)\}$, then

$$\frac{1}{2^\beta}\left(A_1\frac{m}{\zeta} + (A_2+1)\frac{K}{\zeta^2}\right) \le \left(\frac{m}{\zeta} + \frac{K}{\zeta^2}\right).$$

Therefore

$$\begin{aligned} \mathcal{C}(m, K) &\le E\left(\frac{m}{\zeta} + \frac{K}{\zeta^2}\right)\log^\beta\frac{K}{\zeta^2} \\ &\le E\left(\frac{m}{\zeta} + \frac{K}{\zeta^2}\right)\mathrm{Log}^\beta\frac{K}{\zeta^2}. \end{aligned}$$

□

Remark 4.25: The actual value of β depends on the constants appearing in Theorem 4.22(i). We believe that these constants are not optimal, so β is likely to be smaller than $\log_2(5 + 4\sqrt{2}) = 3.33$.

Next, we bound the running time $\mathcal{T}(m, K)$ of the algorithm.

Lemma 4.26 *There exists a constant $F > 0$ such that*

$$\mathcal{T}(m, K) \le F\left(m + \frac{K}{\zeta}\right)\log m \cdot \mathrm{Log}^\beta\frac{K}{\zeta^2} \cdot \mathrm{LogLog}\frac{K}{\zeta^2}. \tag{4.17}$$

Proof: Again we prove the inequality by induction on K. In Lemma 4.24 we showed that, for $K \le 6\zeta^2$, the algorithm stops after one step of recursion. By Theorem 4.22, this step requires $O((m + \frac{K}{\zeta})\log m)$ time, therefore if we choose F sufficiently large, then (4.17) holds trivially.

For $K > 6\zeta^2$, suppose inductively that (4.17) holds for all $K' < K$. Since $K_i < K$ (cf. Lemma 4.24), by induction hypothesis, (4.17) holds for all $\mathcal{Q}_i$, so (4.14) implies

$$\begin{aligned} \mathcal{T}(m, K) \le\ & \sum_{i=1}^{M} F\left(m_i + \frac{K_i}{\zeta}\right)\log m_i \cdot \mathrm{Log}^\beta\frac{K_i}{\zeta^2} \cdot \mathrm{LogLog}\frac{K_i}{\zeta^2} + \\ & D\left(m + \frac{K}{\zeta}\right)\log m. \end{aligned} \tag{4.18}$$

Note that, for all i, $K_i \le \zeta\sqrt{K}$, therefore

$$\mathrm{Log}^\beta\frac{K_i}{\zeta^2}\mathrm{LogLog}\frac{K_i}{\zeta^2} \le \max\{\log^\beta\sqrt{\frac{K}{\zeta^2}}, 1\} \cdot \max\{\log\log\sqrt{\frac{K}{\zeta^2}}, \log\log\sqrt{6}\}.$$

But $K \geq 6\zeta^2$, therefore it is easily seen that

$$\mathrm{Log}^{\beta}\sqrt{\frac{K}{\zeta^2}}\mathrm{LogLog}\sqrt{\frac{K}{\zeta^2}} = \log^{\beta}\sqrt{\frac{K}{\zeta^2}}\log\log\sqrt{\frac{K}{\zeta^2}}. \tag{4.19}$$

Substituting it in (4.18), we obtain

$$\begin{aligned}
\mathcal{T}(m,K) &\leq \sum_{i=1}^{M} F\left(m_i + \frac{K_i}{\zeta}\right)\log m \cdot \log^{\beta}\sqrt{\frac{K}{\zeta^2}} \cdot \log\log\sqrt{\frac{K}{\zeta^2}} + \\
&\quad D\left(m + \frac{K}{\zeta}\right)\log m \\
&\leq \left(\sum_{i=1}^{M} m_i + \frac{1}{\zeta}\sum_{i=1}^{M} K_i\right)\frac{F}{2^{\beta}} \cdot \log^{\beta}\frac{K}{\zeta^2}\left(\log\log\frac{K}{\zeta^2} - 1\right)\log m + \\
&\quad D\left(m + \frac{K}{\zeta}\right)\log m \\
&\leq \left(A_1 m + (A_2+1)\frac{K}{\zeta}\right)\frac{F}{2^{\beta}} \cdot \log^{\beta}\frac{K}{\zeta^2}\left(\log\log\frac{K}{\zeta^2} - 1\right)\log m + \\
&\quad D\left(m + \frac{K}{\zeta}\right)\log m \qquad \text{(by Theorem 4.22)} \\
&\leq \left(m + \frac{K}{\zeta}\right)\left[F\log^{\beta}\frac{K}{\zeta^2} \cdot \log\log\frac{K}{\zeta^2} + \left(D - F\log^{\beta}\frac{K}{\zeta^2}\right)\right]\log m
\end{aligned}$$

because $\beta = \max\{\log A_1, \log(A_2+1)\}$. Since $K > 6\zeta^2$, we have $\log^{\beta}\frac{K}{\zeta^2} > 2^{\beta}$, and therefore, if we choose $F = \frac{D}{2^{\beta}}$, we obtain

$$\begin{aligned}
\mathcal{T}(m,K) &\leq F\left(m + \frac{K}{\zeta}\right)\log m \cdot \log^{\beta}\frac{K}{\zeta^2} \cdot \log\log\frac{K}{\zeta^2} \\
&\leq F\left(m + \frac{K}{\zeta}\right)\log m \cdot \mathrm{Log}^{\beta}\frac{K}{\zeta^2} \cdot \mathrm{LogLog}\frac{K}{\zeta^2}.
\end{aligned}$$

□

These lemmas imply that

Theorem 4.27 *Given a set $\mathcal{L}$ of n lines in the plane and a parameter $1 < r < n$, we can partition an enclosing rectangle $\mathbf{R}$ into $O(r^2\log^{\beta} r)$ convex quadrilaterals in time $O(nr\log n \cdot \log^{\beta} r \cdot \log\log r)$ so that each cell meets at most $\frac{n}{r}$ lines of $\mathcal{L}$.*

Proof: Choose $\zeta = \frac{n}{\gamma r}$, where $\gamma = 4(\sqrt{3}+1)$. At the top level of the recursion $m = n$ and $K = \binom{n}{2} < \frac{n^2}{2}$. Therefore

$$\frac{K}{\zeta^2} < \frac{n^2/2}{n^2/\gamma^2 r^2} = \frac{\gamma^2 r^2}{2}$$

and $\dfrac{m}{\zeta} = \dfrac{n}{n/\gamma r} = \gamma r$. Substituting these values in (4.15) we obtain

$$\begin{aligned} \mathcal{C}(n, n^2/2) &\leq E\left(\gamma r + \frac{\gamma^2 r^2}{2}\right) \text{Log}^\beta \frac{\gamma^2 r^2}{2} \\ &= O\left(r^2 \log^\beta r\right) \end{aligned}$$

and by substituting $\dfrac{K}{\zeta} = \dfrac{n^2/2}{2n/\gamma r} = \dfrac{\gamma r}{4} n$ in (4.17) we get

$$\begin{aligned} \mathcal{T}(n, n^2) &\leq F\left(n + \frac{\gamma r}{4} n\right) \log n \cdot \text{Log}^\beta \frac{\gamma^2 r^2}{2} \text{LogLog} \frac{\gamma^2 r^2}{2} \\ &= O\left(nr \log n \log^\beta r \log\log r\right). \end{aligned}$$

□

Remark 4.28:

(i) As a side product our algorithm outputs, for each cell, the lines intersecting its interior.

(ii) To make the notation easier to follow, we shall henceforth replace the term $\log^\beta r \cdot \log\log r$ by $\log^\omega r$, for some fixed constant ω slightly larger than β. Since we do not know what the best value for β is, this convention involves no real loss of information.

4.6 Constructing Approximate Levels

In this section we describe the second phase of our algorithm, which reduces the number of triangles from $O(r^2 \log^\beta r)$ to $O(r^2)$, while maintaining the property that each triangle meets at most $O(\frac{n}{r})$ lines of $\mathcal{L}$. As mentioned in the introduction, this second phase is not required in most of the applications. As an intermediate step, the algorithm constructs an $\frac{n}{2r}$-approximate leveling of $\mathcal{A}(\mathcal{L})$ (as defined in Section 4.2), with $O(r^2 \log^\beta r)$ edges in total, from the partition obtained in the first phase of the algorithm. Once an $\frac{n}{2r}$-approximate leveling has been constructed, we proceed in the same way as Matoušek [89]. We first describe how to obtain an $\frac{n}{2r}$-approximate leveling.

Let $\overline{\mathcal{Q}}$ denote the planar map induced by the preceding partition of the enclosing rectangle $\mathbf{R}$. We assume that all lines in $\mathcal{L}$ intersect $\partial\mathbf{R}$ at its vertical edges. Let $A = \{a_1, a_2, \ldots, a_n\}$ (resp. $B = \{b_1, b_2, \ldots, b_n\}$) denote the intersection points of the lines in $\mathcal{L}$ and the left (resp. right) vertical edge of $\partial\mathbf{R}$, sorted in decreasing order of their y-coordinates. For $1 \le i \le r$, add $a_{in/r}$ and $b_{in/r}$ to the set of vertices of $\overline{\mathcal{Q}}$. We triangulate all faces of $\overline{\mathcal{Q}}$, in time $O(r^2 \log^\beta r)$. The triangulated map $\mathcal{G}$ also has $O(r^2 \log^\beta r)$ edges. The following observations enable us to compute approximate levels efficiently.

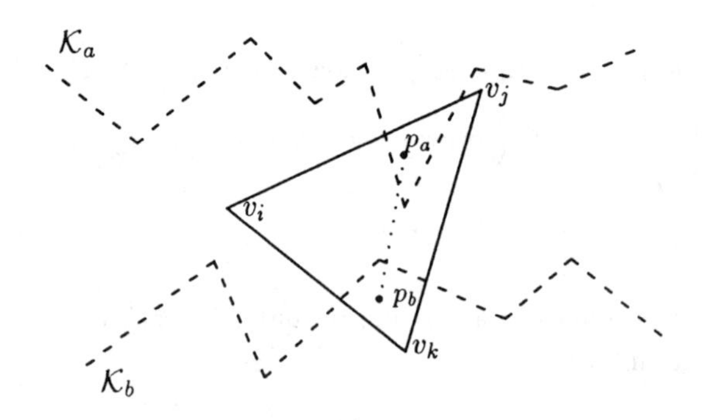

Figure 4.14: A triangle intersecting both $\mathcal{K}_a$ and $\mathcal{K}_b$.

Lemma 4.29 *Let $\mathcal{K}_a$ (resp. $\mathcal{K}_b$) denote the $k - \frac{n}{2r}$ (resp. $k + \frac{n}{2r}$) level of $\mathcal{A}(\mathcal{L})$. Then the interior of a triangle in $\overline{\mathcal{Q}}$ does not intersect both $\mathcal{K}_a$ and $\mathcal{K}_b$.*

Proof: Suppose to the contrary that there is such a triangle $\Delta = v_i v_j v_k$ that crosses $\mathcal{K}_a$ and $\mathcal{K}_b$, then Δ contains a point p_a of the level $k - \frac{n}{2r} - 1$ and another point p_b of level $k + \frac{n}{2r}$ (see figure 4.14). Obviously, the segment $\overline{p_a p_b}$ lies entirely in Δ, but by Lemma 3.10 it intersects at least $\frac{n}{r} + 1$ lines of $\mathcal{L}$, which contradicts the property that no triangle in $\mathcal{G}$ intersects more than $\frac{n}{r}$ lines of $\mathcal{L}$.

□

Lemma 4.30 *Let $\mathcal{K}_a$ (resp. $\mathcal{K}_b$) denote the $k - \frac{n}{2r}$ (resp. $k + \frac{n}{2r}$) level of $\mathcal{A}(\mathcal{L})$. Let v_0 be a vertex of $\mathcal{G}$ on the left vertical edge of* **R**, *whose level is between $k - \frac{n}{2r}$ and $k + \frac{n}{2r}$. There exists a path Π in $\mathcal{G}$ from v_0 to the right vertical edge of* **R** *that lies between $\mathcal{K}_a$ and $\mathcal{K}_b$. Moreover Π can be converted into a monotone path without increasing its number of edges, such that the new path also lies between $\mathcal{K}_a$ and $\mathcal{K}_b$.*

Proof: Let $\mathbf{\Delta} = \Delta_1, \Delta_2, \ldots, \Delta_t$ denote the sequence of triangles visited if we follow $\mathcal{K}_b$ from left to right (if a portion of $\mathcal{K}_b$ lies on an edge of a triangle, then we pick up the triangle lying below that edge). If a triangle appears more than once in $\mathbf{\Delta}$, then retain only its first occurrence. It is easily seen that $\mathbf{\Delta}$ forms a connected region from the left vertical edge to the right vertical edge of $\partial\mathbf{R}$, and its boundary is formed by the edges of $\mathcal{G}$ (see figure 4.15); we call it a *corridor*. Let Π denote the top portion of $\partial\mathbf{\Delta}$. Π is a connected polygonal path from the left vertical edge of $\partial\mathbf{R}$ to its right vertical edge, formed by the edges of $\mathcal{G}$. Obviously Π does not intersect $\mathcal{K}_b$. Since each triangle of $\mathbf{\Delta}$ intersects $\mathcal{K}_b$, Π cannot intersect $\mathcal{K}_a$ (cf. Lemma 4.29), thus Π lies between $\mathcal{K}_a$ and $\mathcal{K}_b$.

To prove the second part, let Γ denote any polygonal path lying between $\mathcal{K}_a$ and $\mathcal{K}_b$. Let v_1 be the first vertex on Γ at which Γ turns backwards. Let ξ denote the first point on Γ after v_1 (along Γ), which has the same x-coordinate as v_1. Since v_1 lies above $\mathcal{K}_b$, and $\mathcal{K}_b$ is a x-monotone path, the vertical segment $\overline{v_1\xi}$ does not cross $\mathcal{K}_b$. The same argument implies that $\overline{v_1\xi}$ does not intersect $\mathcal{K}_a$, and therefore the new path also lies between $\mathcal{K}_a$ and $\mathcal{K}_b$. Whenever we add a new edge, we remove at least one edge of Γ, which proves that shortcutting Γ does not increase its number of edges.

□

Since the path Π lies between the $k - \frac{n}{2r}$ and $k + \frac{n}{2r}$ levels of $\mathcal{A}(\mathcal{L})$, it is an $\frac{n}{2r}$-approximate k-level of $\mathcal{A}(\mathcal{L})$. Therefore $\mathcal{G}$ contains an $\frac{n}{2r}$-approximate $\frac{in}{r}$-level, for each $i \leq r$. Let $\mathcal{G}_k$ denote the subgraph of $\mathcal{G}$ consisting of all edges that fully lie between the levels $k - \frac{n}{2r}$ and $k + \frac{n}{2r}$. To obtain $\mathcal{G}_k$, we

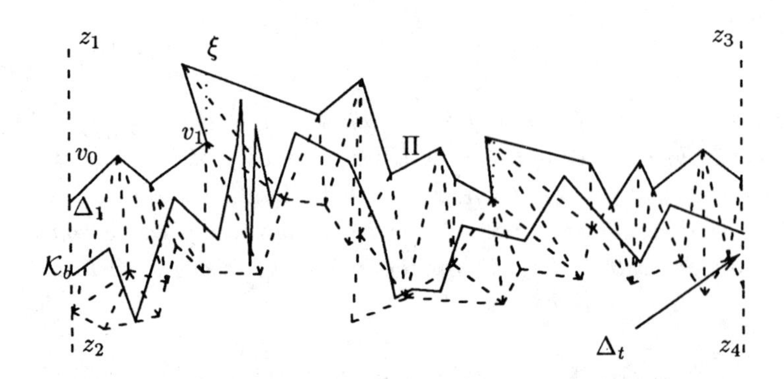

Figure 4.15: Level $\mathcal{K}_b$ contained in the corridor $\mathbf{\Delta}$; Π is the upper boundary of $\mathbf{\Delta}$.

need to compute the level of each vertex of $\mathcal{G}$ and the highest and the lowest levels crossed by each edge of $\mathcal{G}$. Let $\psi(v)$ denote the level of a vertex $v \in \mathcal{G}$. Let $\overline{v_i v_j}$ denote an edge of $\mathcal{G}$ with $x(v_i) < x(v_j)$. If there are h lines of $\mathcal{L}$ intersecting $\overline{v_i v_j}$ and k of these lines lie above v_i, then

$$\psi(v_j) \;=\; \psi(v_i) - k + (h - k) \;=\; \psi(v_i) - 2k + h. \tag{4.20}$$

We can compute the level of each vertex of $\mathcal{G}$ lying on the left vertical edge of $\partial\mathbf{R}$ by counting the number of lines of $\mathcal{L}$ lying above it. The partitioning algorithm also produces the subset of lines of $\mathcal{L}$ that crosses each edge of $\mathcal{G}$, so it is trivial to count how many of them lie above the left endpoint of e. The levels of all vertices of $\mathcal{G}$ are now easy to determine by propagating levels from left to right along the edges of $\mathcal{G}$, using (4.20). As for determining the levels crossed by an edge e of $\mathcal{G}$, we sort the lines intersecting e along the edge. Once we know the level of the endpoints of e, we can easily compute the levels crossed by e. Next we partition $\mathcal{G}$ into $\mathcal{G}_1, \ldots, \mathcal{G}_r$, and find a path from $a_{in/r}$ to $b_{in/r}$ in $\mathcal{G}_i$ using a depth first search from $a_{in/r}$, for $i \le r$. Finally, if any of the resulting paths is not x-monotone, we make it monotone by shortcutting all edges that turn backwards. Since the edges in $\mathcal{G}_i$, for $i \le r$, are pairwise disjoint, there are at most $O(r^2 \log^\beta r)$ edges in the resulting $\frac{n}{2r}$-approximate leveling of $\mathcal{A}(\mathcal{L})$.

The correctness of the algorithm follows immediately from the above discussion, so we only have to analyze the running time of the algorithm.

Lemma 4.31 *Given a set $\mathcal{L}$ of n lines, we can construct, in $O(nr \log n \log^{\omega} r)$ time, an $\frac{n}{2r}$-approximate leveling of $\mathcal{A}(\mathcal{L})$ having only $O(r^2 \log^{\beta} r)$ edges in total.*

Proof: By Theorem 4.27, the planar graph $\mathcal{G}$ can be constructed in time $O(nr \log n \cdot \log^{\omega} r)$ and it has only $O(r^2 \log^{\beta} r)$ triangles. It follows from the above discussion that it takes $O(nr \log n \log^{\omega} r)$ time to compute the level of each vertex and the levels crossed by each edge of $\mathcal{G}$. Therefore, we can obtain $\mathcal{G}_i$, for $i \leq r$, in $O(nr \log n \log^{\omega} r)$ time. The depth first search takes only $O(r^2 \log^{\beta} r)$ time, and to converting the computed paths into monotone paths takes the same amount of time. Hence, the lemma follows.

□

Remark 4.32: Matoušek [89] also constructs an approximate leveling as an intermediate step, but in a direct and much simpler (albeit inefficient) manner. His algorithm works roughly as follows.

Partition $\mathbf{R}$ into r^2 vertical slabs each containing at most $O(\frac{n}{r^2})$ intersection points of $\mathcal{L}$. Let $V_0, \ldots, V_{r^2}$ denote the vertical edges of these slabs, and let $a_{i,1}, \ldots, a_{in}$ denote the intersection points of V_i and $\mathcal{L}$ sorted in decreasing order of their y-coordinates. For each $j \leq \frac{r}{2}$ and $0 \leq i < r^2$ connect $a_{i,2jn/r}$ to $a_{i+1,2jn/r}$. These polygonal paths are shown to form an $\frac{n}{r}$-approximate leveling of $\mathcal{A}(\mathcal{L})$.

The problem with this approach is that the approximate leveling has $O(r^3)$ edges in total, and the time needed to obtain it is $O(nr^2 \log^2 r)$, which is substantially dominated by the partitioning of $\mathbf{R}$ into vertical slabs. Partitioning $\mathbf{R}$ into r^2 slabs is done to ensure that no segment $\overline{a_{i,2jn/r}a_{i+1,2jn/r}}$ crosses too many levels. Using our improved partitioning technique we are able to obtain an $\frac{n}{2r}$-approximate partitioning that has almost an order of magnitude fewer edges (in terms of r), and the running time of our algorithm is also about an order of magnitude faster. If r is small, e.g. $O(1)$, then Matoušek algorithm is better (it runs in optimal linear time), but for large values of r it becomes very inefficient. As we will see in Chapter 5, in most of the applications it is desirable to use a large value of r.

After computing a $\frac{n}{2r}$-approximate leveling of $\mathcal{A}(\mathcal{L})$, we apply the same technique of Matoušek to partition R into $O(r^2)$ triangles. Let $K_1, K_2, \ldots, K_r$ denote the set of $\frac{n}{2r}$-approximate levels of $\mathcal{A}(\mathcal{L})$. Using Lemma 4.2 and Lemma 4.3, Matoušek proved that

Lemma 4.33 (Matoušek [89]) *For $i \le r/3$, there exists a polygonal path Π_i between K_{3i-1} and K_{3i+1} such that $\sum_{i=1}^{r/3} |\Pi_i| = O(r^2)$.*

Proof: By the definition of approximate levels, the region between K_{3i-1} and K_{3i+1} must contain all the levels between $(6i-1)\frac{n}{2r}$ and $(6i+1)\frac{n}{2r}$. Applying Lemma 4.2 with $\omega = n/2r$, we obtain levels $U_i \in [(6i-1)\frac{n}{2r}, (6i+1)\frac{n}{2r}]$ such that $\sum_{i \le r/3} |U_i| = O(nr)$. If we take Π_i to be the $\frac{n}{2r}$-simplified U_i, then by Lemma 4.3 Π_i also lies in the region bounded by $(6i-1)\frac{n}{2r}$ and $(6i+1)\frac{n}{2r}$. Since Π_i is obtained by shortcutting every $\frac{n}{2r}^{th}$ vertex, $|\Pi_i| \le \frac{|U_i|}{n/2r} + 2$, therefore $\sum |\Pi_i| = O(r^2)$.

□

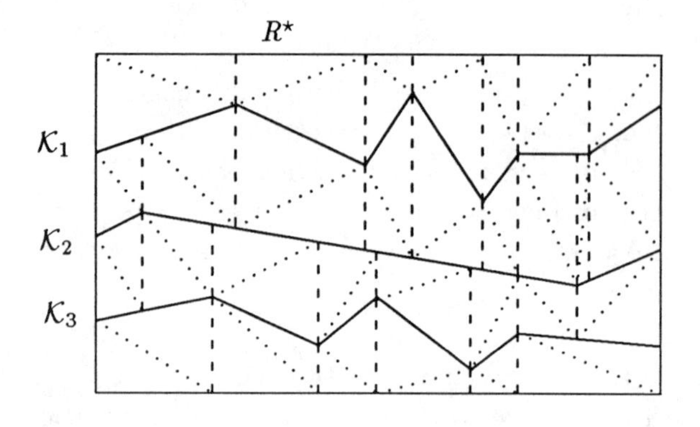

Figure 4.16: Triangulation of $\mathcal{K}^\star$

Since Π_i lies between the levels $3i\frac{n}{r} \mp \frac{3}{2} \cdot \frac{n}{r}$, $\Pi_1, \ldots, \Pi_{r/3}$ forms an $\frac{3n}{2r}$-approximate leveling of $\mathcal{A}(\mathcal{L})$. By applying Suri's algorithm [124] of computing a minimum link path in the simply connected region, lying between K_{3i-1} and K_{3i+1}, we obtain a path Π'_i, such that $|\Pi'_i| \le |\Pi_i|$. Note that Π'_i is also a $\frac{3n}{2r}$-approximate $\frac{3in}{r}$-level. Hence, we have

Corollary 4.34 *Given a set $\mathcal{L}$ of n lines and a parameter $1 < r < n$, we can compute a set of $\frac{3n}{2r}$-approximate levels of $\mathcal{A}(\mathcal{L})$, $\mathcal{K} = \{\Pi'_1, \dots, \Pi'_{r/3}\}$ in time $O(nr \log n \log^{\omega} r)$ with the property that $\sum_{i=1}^{r/3} |\Pi'_i| = O(r^2)$.*

□

Matoušek has also proved that

Lemma 4.35 (Matoušek [89]) *There are at most $O(nr)$ intersections between $\mathcal{K}$ and $\mathcal{L}$.*

Proof: We will only bound the number of intersections between lines of $\mathcal{L}$ and Π'_{2i} for $i \leq \lfloor r/6 \rfloor$, the intersections between the lines of $\mathcal{L}$ and Π'_{2i-1} can be counted similarly. Applying Lemma 4.2 with $\omega = n/r$ and $\mathcal{U}_i = [\frac{3n}{r}(2i-1) - \frac{3n}{2r}, \frac{3n}{r}(2i-1) + \frac{3n}{2r}]$, we can obtain levels $U_1, U_2, \dots, U_{\lfloor r/6 \rfloor}$ of $\mathcal{A}(\mathcal{L})$, with at most $O(nr)$ edges in total, such that Π'_{2i} lies between U_i and U_{i+1}.

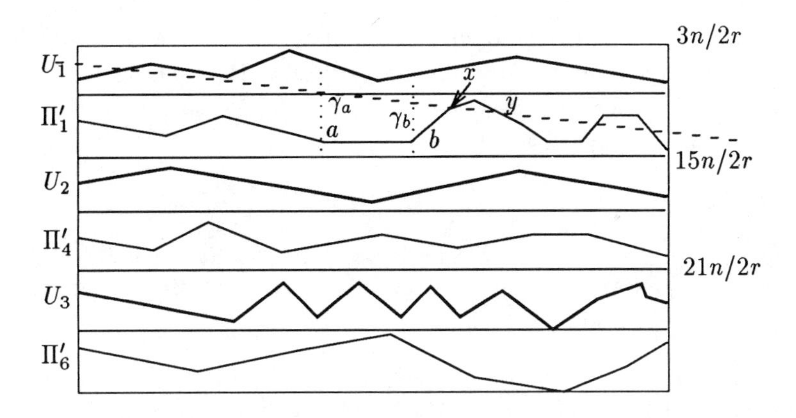

Figure 4.17: Levels U_i and paths Π'_{2i}; x: good intersection, y: bad intersection

Consider intersection points between the lines of $\mathcal{L}$ and Π'_{2i}. We refer to such an intersection as *good* if it is immediately preceded or followed (along ℓ) by an intersection of ℓ with U_i or U_{i+1}. The total number of good intersections over all i is easily seen to be $O(nr)$. As for the number of "bad" intersections, consider an edge ab of Π'_{2i}. Let γ_a (resp. γ_b) be the vertical segment passing through a (resp. b) with endpoints on U_i and U_{i+1}. If a line ℓ forms a bad intersection with ab, then ℓ intersects γ_a and γ_b. But U_i and U_{i+1} are $\frac{n}{r}$-approximate levels, therefore γ_a and γ_b intersect at most $\frac{9n}{r}$ lines of $\mathcal{L}$. Hence by Lemma 4.33, the total number of bad intersections is also $O(r^2 \times n/r) = O(nr)$.

□

In view of Lemma 4.35, $\mathcal{K}$ can be decomposed into $O(r^2)$ edges, each intersecting at most $O(\frac{n}{r})$ lines of $\mathcal{L}$. Next, construct the vertical decomposition $\mathcal{K}^\star$ of $\mathcal{K}$ by drawing a vertical line from every vertex of $\mathcal{K}$ in both directions until it meets an edge of $\mathcal{K}$ or $\mathbf{R}$ (see figure 4.16). Since every vertical edge added to $\mathcal{K} \cup \mathbf{R}$ lies within $\frac{3n}{r}$ levels, it intersects at most $\frac{3n}{r}$ lines of $\mathcal{L}$. Therefore, every trapezoid of $\mathcal{K}^\star$ intersects at most $O(\frac{n}{r})$ lines of $\mathcal{L}$. Finally, partition each trapezoidal cell of $\mathcal{K}^\star$ into two triangles. Hence, we can obtain the main and final result of the chapter.

Theorem 4.36 *Given a set $\mathcal{L}$ of n lines in the plane, and a parameter $1 < r < n$, we can decompose the enclosing rectangle* $\mathbf{R}$ *into $O(r^2)$ triangles in time, $O(nr \log n \log^\omega r)$, for some constant $\omega < 3.33$, so that no triangle meets more than $O(\frac{n}{r})$ lines of $\mathcal{L}$.*

□

Remark 4.37: Once $\mathbf{R}$ has been partitioned into $O(r^2)$ triangles, we can easily compute, for each triangle, the set of lines passing through its interior by spending $O(nr)$ additional time.

4.7 Coping with Degenerate Cases

In this section we show how to modify our partitioning algorithm so that it also works in degenerate cases, when more than two lines of $\mathcal{L}$ are concurrent, or more than one intersection points of lines in $\mathcal{L}$ lie on the same vertical line. To facilitate these modifications, we first need to redefine some of the terminology, introduced in Section 4.2 and Section 4.4.

(i) If $t \geq 2$ lines of $\mathcal{L}$ pass through a point p, then we consider p to be $\binom{t}{2}$ intersection points. Therefore, although the number of distinct intersection points in $\mathcal{L}$ can be less than $\binom{n}{2}$, the sum of their "weights", as just defined, is still $\binom{n}{2}$.

(ii) The level of a vertex of $\mathcal{A}(\mathcal{H})$ can no longer be uniquely defined, so we redefine the k-level of $\mathcal{A}(\mathcal{H})$ to be the polygonal path formed by the closure of the (open) edges of $\mathcal{A}(\mathcal{H})$ whose level is k. Let $p_0, \ldots, p_m$ be the vertices of a k-level of $\mathcal{A}(\mathcal{H})$. Let w_i denote the weight of p_i, and for any $\delta < m$, let p_{i_j} be the vertex of a k-level such that $\sum_{t=0}^{i_j-1} w_t < j\delta \leq \sum_{t=0}^{i_j} w_t$. We now define the *$\delta$-simplified k-level* to be the polygonal path connecting p_0 to p_{i_1}, p_{i_1} to $p_{i_2}, \ldots, p_{i_s}$ to p_m, for $s = \lfloor \frac{\sum_t w_t}{\delta} \rfloor$, concatenated with the left and right rays of the k-level incident to p_0 and p_m respectively.

(iii) For a convex quadrilateral $\mathcal{Q}$, we use $\mathcal{L}$ to denote the set of lines passing through the *interior* of $\mathcal{Q}$, and we let K denote the total weight of the intersection points contained in the interior of $\mathcal{Q}$.

(iv) The set A (resp. B) of left (resp. right) endpoints of the lines in $\mathcal{L}$, defined in Section 4.2, now becomes a multi-set, because many lines can have a common endpoint. If two lines ℓ_i and ℓ_j have a common left (resp. right) endpoint x and ℓ_i lies counterclockwise (resp. clockwise) from ℓ_j, when directed from x into $\mathcal{Q}$, then $\pi(i) < \pi(j)$ (resp. $\sigma(i) > \sigma(j)$). In other words, A and B are ordered in the way they should be, if we shrink $\mathcal{Q}$ slightly.

(v) Finally, the quantity δ_i defined in Section 4.4 denotes the number of pseudo edges whose relative interior intersects ℓ_i. Similarly we define the quantities δ_i^{xy}.

Next we briefly describe the modifications required to make our algorithm and in its analysis work in degenerate cases as well; we leave it for the reader to fill in the details.

Observe that in Section 4.3.1 we actually count the number of inversions to compute K_{rr}, K_{br} and K_{gr}, which gives the total weight of the intersection points (not the number of distinct intersection points) contained in $\mathcal{Q}$, so this part of the algorithm does not require any modification. However, now it is not always possible to find a vertical line that, for a given $k \leq K$, has exactly k intersection points in $\mathcal{Q}$ to its left (cf. Section 4.3.2). Instead, we find the rightmost vertical line having $\leq k$ intersections in $\mathcal{Q}$ to its left. If we interpret the order of the lines of $\mathcal{L}$ along a vertical line λ as the order that would result by moving λ slightly to the left, then it is easily checked that the procedure described in Section 4.3.2, with obvious and trivial modifications, would produce the desired line. We now apply this procedure, in the final step of our general partitioning algorithm (cf. Section 4.4.2), to partition a cell $\mathcal{F}_\alpha$ into subcells, each containing $\leq \zeta\sqrt{K}$ intersection point in its interior.

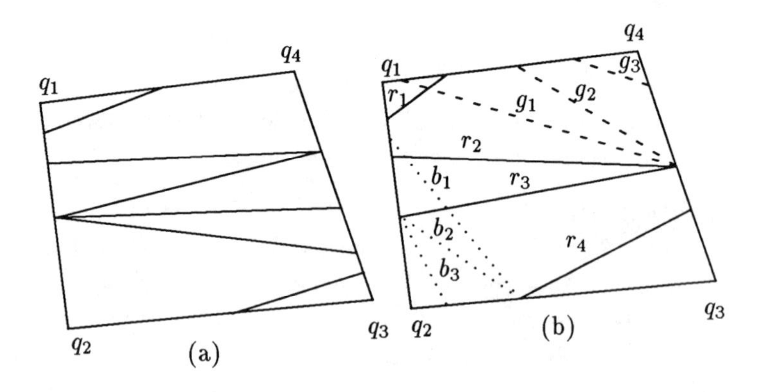

Figure 4.18: Cells in degenerate cases: (a) special case; (b) general case

The algorithm of Section 4.4.1 remains the same. Since many elements in A or B can have the same value, two or more pseudo edges can have a common endpoint (see figure 4.18a), or even completely overlap; however, we will regard overlapping pseudo edges as a single edge. It can be checked that our new conventions regarding A and B imply that a line $\ell_i \in \mathcal{L}$ intersects at least $|\pi(i) - \sigma(i)|$ lines inside $\mathcal{Q}$, that is, Lemma 4.8 still holds, which in conjunction with the new definitions of K and δ implies that Lemma 4.9 also continues to hold. Lemma 4.11 is trickier to adjust. We redefine μ_i as the number of lines passing through $\mathcal{Q}_i$ and not having an endpoint whose rank in A or in B is between $(i-1)\zeta+1$ and $i\zeta$. With some care, the proof can be modified to yield the same bound on μ_i.

Our general algorithm described in Section 4.4.2 also does not change except that in the final step we now use the modified algorithm mentioned above to partition the cells $\mathcal{F}_\alpha$ into $\lceil \frac{K_\alpha}{\zeta\sqrt{K}} \rceil$ cells, each containing at most $\zeta\sqrt{K}$ intersection points in its interior. Note that now we can have some triangular cells that do not have an edge lying along $\partial\mathcal{Q}$ but only have a vertex lying on $\partial\mathcal{Q}$; these cells are also considered boundary cells. We may even have some zero-area cells, caused by overlapping pseudo edges, but we discard these cells. In view of our convention regarding the weights of intersection points and the order of the lines along $\partial\mathcal{Q}$, it is easy to see that the charging scheme of Lemma 4.14 still works. (More specifically, we charge a red-green intersection to a red-green cell $\mathcal{F}_{ij}$ if the rank of the right endpoint of the red line is between $(j-1)\zeta+1$ and $j\zeta$ and the rank of the appropriate endpoint of the green line is between $(i-1)\zeta+1$ and $i\zeta$, and similarly for red-blue intersections.) Moreover, using the same argument as in Section 4.4.3 we can prove Lemma 4.17 and Lemma 4.18 in this degenerate setting because we are not counting those pseudo lines that intersect ℓ at their endpoints. Finally, it can be shown that Lemma 4.16, Lemma 4.19 and Lemma 4.20 also hold, because, although a line ℓ_i can meet the boundary of many cells, it meets the interior of exactly δ_i+1 cells. Lemma 4.21 can also be shown to hold, using our notational convention in degenerate cases.

As for the second phase of our algorithm, Lemma 4.29 and Lemma 4.30 are not affected by degeneracies in $\mathcal{L}$. However, computing the graphs $\mathcal{G}_k$ now becomes slightly more difficult because the level of a vertex v of $\mathcal{G}$ may be undefined (that is, when v is a "heavy-weighted" vertex of $\mathcal{A}(\mathcal{L})$ as well). But for each vertex v of $\mathcal{G}$, we can compute the lines of $\mathcal{L}$ passing through it in time $O(nr \log n \log^\omega r)$, because a line passing through a vertex v either lies in the interior of a triangle adjacent to v, or contains on of the edges adjacent to v. Moreover we can compute the levels of each edge incident to v in a sufficiently small neighborhood of v in time $O(nr \log n \log^\omega r)$. Therefore, we can still propagate the levels from left to right along the edges of $\mathcal{G}$, and can determine the levels crossed by each edge of $\mathcal{G}$. Thus, we can partition $\mathcal{G}$ into $\mathcal{G}_1, \ldots, \mathcal{G}_r$ and obtain an $\frac{n}{2r}$-approximate leveling of $\mathcal{A}(\mathcal{L})$. It can be checked that the technical results in [65] and [89] can be extended to the degenerate case, which in turn implies that Lemma 4.3, Lemma 4.33 still hold if we follow our new conventions regarding simplified levels and the weights of intersection points. It is also easy to check that the proof of Lemma 4.35 does not require the lines of $\mathcal{L}$ to be in general position.

Hence, we can conclude that Theorem 4.36 holds, with appropriate modifications as discussed above, even if the lines are not in general position.

4.8 Discussion and Open Problems

In this chapter we presented a deterministic algorithm that, given a set $\mathcal{L}$ of n lines and a parameter $1 < r < n$, partitions the plane into $O(r^2)$ triangles, each of which meets at most $O(\frac{n}{r})$ lines of $\mathcal{L}$. Recently Matoušek [91] improved the running time to $O(nr)$, which is optimal if we want to report the lines intersecting each triangle. He also gave a deterministic algorithm for higher dimensions, which, given a set $\mathcal{H}$ of n hyperplanes in $\mathbb{R}^d$, partitions the space, in time $O(nr^{d-1}\log^A n)$, into $O(r^d)$ simplices each of which intersects at most $\frac{n}{r}$ hyperplanes of $\mathcal{H}$, where $A > 0$ is some constant. It can be shown that the number of simplices produced by his algorithm is asymptotically tight, and the running time is almost optimal (within polylog factors). If S is a set of n points in $\mathbb{R}^d$, then, for the range space $(S, \{ S \cap \Delta \mid \Delta \text{ is a simplex} \})$, he can compute an $\frac{1}{r}$-net of size $O(r \log r)$ in time $O(nr^{d-1}\log^A n + r^B)$, where $A, B > 0$ are constants.

Although Matoušek's new algorithm settles several open questions in this area, there are still some interesting open problems:

(i) The $\Omega(nr)$ lower bound applies only if we want to report the lines passing through each triangle. No non-trivial lower bound is known if these crossing need not be reported. A challenging open problem is to establish a similar lower bound in this case too, or to come up with an algorithm whose running time is close to $O(n + r^2)$.

(ii) Suppose we have a collection of m blue lines and a collection of n red lines. For a given parameter r, how fast we can partition the plane into $O(r^2)$ triangles, so that each triangle meets $O(\frac{m}{r})$ blue lines and $O(\frac{n}{r})$ red lines? Note that the random sampling technique easily yields a similar partitioning, with each triangle meeting $O(\frac{m}{r}\log r)$ blue lines and $O(\frac{n}{r}\log r)$ red lines.

(iii) Suppose we have a collection $\mathcal{G}$ of n segments which intersect at K points. The random sampling technique shows that, for a given parameter r, we can partition the plane into $O(r+\frac{Kr^2}{n^2})$ triangles so that no triangle meets more than $O(\frac{n}{r}\log r)$ segments of $\mathcal{G}$. Can our algorithms be extended to yield a deterministic algorithm that produces that many triangles, each meeting $O(\frac{n}{r})$ segments, and runs in time close to $O(n+\frac{Kr}{n})$ (the lower bound on triangle-segment crossing in this case is $\Omega(n+\frac{Kr}{n})$)?

Chapter 5

Applications of the Partitioning Algorithm

5.1 Introduction

In the previous chapter we presented a deterministic algorithm that, given a collection $\mathcal{L}$ of n lines in the plane and a parameter $1 \leq r \leq n$, partitions the plane into $O(r^2)$ triangles, in time $O(nr \log n \log^{\omega} r)$, so that no triangle meets more than $O(\frac{n}{r})$ lines of $\mathcal{L}$ in its interior, where ω is some constant < 3.33. Such a partitioning algorithm is useful in obtaining divide and conquer algorithms for a variety of problems involving lines (or line segments) in the plane. Typically, an original problem involving the lines of $\mathcal{L}$ is split into $O(r^2)$ subproblems, one per triangle in the resulting partitioning, each involving only $O(\frac{n}{r})$ lines of $\mathcal{L}$ meeting the corresponding triangle. These subproblems are then solved either by recursive application of the partitioning technique, or, if the size of the subproblems is sufficiently small, by some different and direct method.

As mentioned in Chapter 4, a reasonable lower bound on the cost of the partitioning problem, in this divide and conquer context, is $\Omega(nr)$, because this is the worst-case total size of the input to the $O(r^2)$ subproblems, that is the total number of line-triangle crossings. The best previous technique for constructing such a partitioning, that of [89], is about an order of magnitude worse than this bound (in terms of r). As it turns out, this overhead is too expensive for most applications when r is large. Thus Matoušek's algorithm can be applied only with small, constant values of r. This has two disadvantages. One is that the algorithm becomes recursive and thus more complicated; the other disadvantage is that the resulting time complexity is larger, by a factor of the form $O(n^\delta)$, for any $\delta > 0$, than what could be obtained by a judicious choice of a large value of r.

The goal of this chapter is to obtain fast algorithms for a variety of problems involving lines or segments in the plane, using the partitioning algorithm of the previous chapter. The problems that benefit from our algorithm have the common property that they can be solved efficiently using the random sampling technique. Our algorithms for most of these problems have the same flavor. We divide the original problem into $O(r^2)$ subproblems, as explained above, then solve each subproblem directly by a simpler but slower algorithm, and finally merge the results of these problems. A considerable part of this chapter is devoted to the discussion of these simpler algorithms, and to details of the merging. In several applications the merging is trivial (e.g. in problems (i), (iv), (v) below), but in other applications it may require some extra nontrivial techniques.

The following list summarizes the results obtained in this chapter:

(i) *Computing incidences between lines and points* (Section 5.2): Given a set of n lines and a set of m points in the plane, compute how many lines pass through each given point. (Alternatively, compute the lines passing through each point, or just determine whether any line passes through any point.) Edelsbrunner et al. [59] have given a randomized algorithm for this problem whose expected running time is $O(m^{2/3-\delta}n^{2/3+2\delta} + (m+n)\log n)$, for any $\delta > 0$. A slightly improved, but still randomized, algorithm has been given in [55]. We present a deterministic algorithm with $O(m^{2/3}n^{2/3}\log^{2/3} n \log^{\omega/3} \frac{m}{\sqrt{n}} + (m + n)\log n)$ time complexity. Since the maximum number of incidences between m points and n lines is $\Theta(m^{2/3}n^{2/3} + m + n)$, our algorithm is close to optimal in the worst case.

(ii) *Computing many faces in an arrangement of lines* (Section 5.3): Given a set of n lines and a set of m points in the plane, compute the faces in the arrangement of the lines containing the given points. Edelsbrunner et al. [59] have given a randomized algorithm for this problem with expected running time $O(m^{2/3-\delta}n^{2/3+2\delta} + n\log n \log m)$, for any $\delta > 0$. As in the case of the incidence problem, a slightly better randomized algorithm has been given in [55]. We present a deterministic $O(m^{2/3}n^{2/3}\log^{5/3} n \log^{\omega/3} \frac{m}{\sqrt{n}} + (m+n)\log n)$ algorithm, again coming close to optimal in the worst case (see [38] for combinatorial bounds).

(iii) *Computing many faces in an arrangement of segments* (Section 5.4): This is the same problem as the previous one except that now we have a collection of segments instead of lines. The previous best solution is by Edelsbrunner et al. [59], which is randomized and has expected running time $O(m^{2/3-\delta}n^{2/3+2\delta} + n\alpha(n)\log^2 n \log m)$, for any $\delta > 0$. We present a deterministic algorithm with improved time complexity $O(m^{2/3}n^{2/3}\log n \log^{\omega/3+1} \frac{n}{\sqrt{m}} + n\log^3 n + m\log n)$.

(iv) *Counting segment intersections* (Section 5.5): We give a deterministic $O(n^{4/3}\log^{(\omega+2)/3} n)$ algorithm to count the number of intersections in a given collection of n segments; this is an improvement over Guibas et al.'s algorithm [73], which counts the intersections in $O(n^{4/3+\delta})$ randomized expected time, for any $\delta > 0$.

(v) *Counting and reporting red-blue intersections* (Section 5.6): Given a set Γ_r of n_r "red" segments and another set Γ_b of n_b "blue" segments in the plane, count the number of intersections between Γ_r and Γ_b, or report all of them. (In this problem, we need to ignore the potentially large number of intersections within Γ_r or within Γ_b.) In Chapter 3 we showed that all K red-blue intersections can be reported deterministically in $O((n_r\sqrt{n_b} + n_b\sqrt{n_r} + K)\log n)$ time, where $n = n_r + n_b$. Here we give a deterministic $O(n^{4/3}\log^{(\omega+2)/3} n)$ algorithm to count all red-blue intersections. It can report all K red-blue intersections in $O(n^{4/3}\log^{(\omega+2)/3} n + K)$ time.

(vi) *Implicit point location problem* (Section 5.7): Given a collection of m points and a collection of (possibly intersecting) n triangles in the plane, find which points lie in the union of the triangles. This turns out to be a special case of a general problem of implicit point location in planar maps formed by overlapping figures. We present a deterministic algorithm with $O(m^{2/3}n^{2/3}\log^{2/3} n \log^{\omega/3} \frac{n}{\sqrt{m}} + (m+n)\log n)$ time complexity.

(vii) *Approximate half-plane range query* (Section 5.8): Given a set S of n points in the plane and a parameter (not necessarily constant) $\epsilon > 0$, preprocess them so that for any query line ℓ, we can approximately count the number of points lying above ℓ with an error of at most $\pm\epsilon n$. We give an algorithm that preprocesses S, in time $O(\frac{n}{\epsilon}\log n \log^{\omega}\frac{1}{\epsilon})$, into a data structure of size $O(\frac{1}{\epsilon^2})$, so that a query can be answered in $O(\log n)$ time.

(viii) *Constructing spanning trees with low stabbing number* (Section 5.9): Given a set S of n points in the plane, we present an $O(n^{3/2}\log^{\omega+1} n)$ deterministic algorithm to construct a family of $k = O(\log n)$ spanning trees $\mathcal{T}_1, \ldots, \mathcal{T}_k$ of S with the property that, for any line ℓ there is tree $\mathcal{T}_i$, such that ℓ intersects at most $O(\sqrt{n})$ edges of $\mathcal{T}_i$. Moreover, with additional $O(n\log n)$ preprocessing and $O(n)$ space, the tree $\mathcal{T}_i$ corresponding to a query line ℓ can be determined in $O(\log n)$ time. The previously best known algorithm is by Matoušek [90], which runs in $O(n^{7/4}\log^2 n)$ time, and, moreover, produces a stabbing number $O(\sqrt{n}\log^2 n)$ instead of $O(\sqrt{n})$.

(ix) *Space query-time tradeoff in triangle range searching* (Section 5.10) Given a set S of n points in the plane, preprocess them so that for any query triangle, we can quickly compute the number of points contained in that triangle. We give an algorithm with $O(\frac{n}{\sqrt{m}}\log^{3/2} n)$ query time, using $O(m)$ space. The preprocessing time is bounded by $O(n\sqrt{m}\log^{\omega+1/2} n)$. Similar bounds have been obtained independently by Chazelle [22].

(x) *Overlapping planar maps* (Section 5.11): Given two planar maps P, Q, and a bivariate function $F_P(x,y)$, $F_Q(x,y)$ associated with each of them, such that over each face of P the function F_P has some simple structure (e.g. it is constant, linear, or convex over each face), and similarly for Q, determine a point that minimizes $F_P(x,y) - F_Q(x,y)$. We show that if the maps satisfy certain conditions, then an optimal

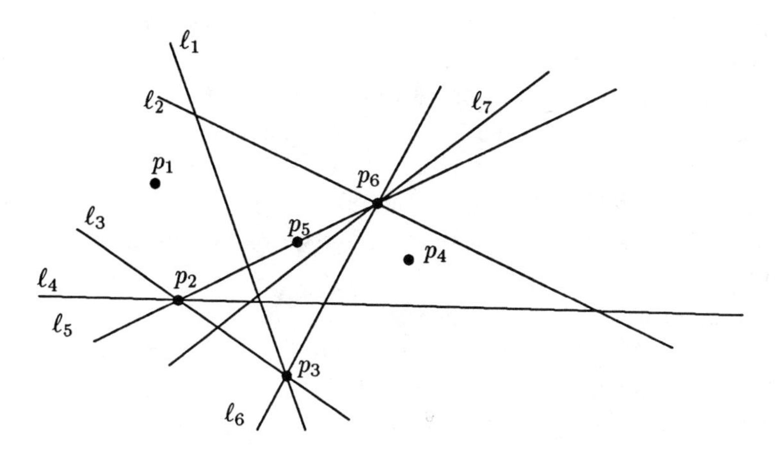

Figure 5.1: An instance of the incidence problem

point can be computed in $O(n^{4/3} \log^{(\omega+2)/3} n)$ time, where n is the total complexity of the two maps.

5.2 Computing or Detecting Incidences between Points and Lines

Consider the following problem:

> *Given a set $\mathcal{L} = \{\ell_1, \ldots, \ell_n\}$ of n lines and a set $P = \{p_1, \ldots, p_m\}$ of m points in the plane, for each point p_i compute the lines in $\mathcal{L}$ passing through it. This is an extension of Hopcroft's problem of determining whether there is a point in P lying on a line in $\mathcal{L}$.*

Szemerédi and Trotter [128] showed that the maximum number of incidences between n lines and m points is $\Theta(m^{2/3}n^{2/3} + m + n)$ (a much simpler proof, with a substantially smaller constant of proportionality, appears in [38]). Edelsbrunner et al. [59] have given a randomized algorithm for computing all incidences; its expected running time is $O(m^{2/3-\delta}n^{2/3+2\delta}+$

$(m+n)\log n)$, for any $\delta > 0$ (see also [45]). Like many other randomized algorithms of this kind, this algorithm can be made deterministic without any additional overhead, using Matoušek's algorithm [89]. A slightly faster randomized algorithm is given in [55] with $O(m^{2/3}n^{2/3}\log^4 n+(m+n^{3/2})\log^2 n)$ expected running time, which however is not known as yet to admit such "cheap" determinization. In this section we first present a very simple algorithm whose running time is roughly $m\sqrt{n}\log^{1/2} n$; this, combined with our partitioning algorithm, will yield a deterministic algorithm that is faster than the preceding ones.

We can assume that $m < n^2$, because otherwise we can compute all incidences in time $O(n^2+m\log n) = O(m\log n)$ by constructing the arrangement of $\mathcal{L}$ and locating in it each of the points.

Divide the set P into t disjoint subsets $P_1, \ldots, P_t$, each of size at most $\lceil \frac{m}{t} \rceil$. For each P_i, we compute the incidences between P_i and $\mathcal{L}$ as follows. Dualize the lines ℓ_j to points $\ell_j^\star$, and the points p_j to lines $p_j^\star$, so we have a set $P_i^\star$ of $\lceil \frac{m}{t} \rceil$ lines and a set $\mathcal{L}^\star$ of n points in the plane. Since duality preserves incidences, it suffices to determine the points of $\mathcal{L}^\star$ lying on each line $p_j^\star$; this can be done by constructing the arrangement $\mathcal{A}(P_i^\star)$, processing it for fast point location as in [60], and locating in it each of the points of $\mathcal{L}^\star$. The cost of all this is $O\left(\frac{m^2}{t^2} + n\log n\right)$ (cf. [62], [60]). Summing over all P_i's, the overall running time becomes

$$T(m,n) \quad = \quad O\left(t\left(\frac{m^2}{t^2} + n\log n\right)\right) \; = \; O\left(\frac{m^2}{t} + nt\log n\right).$$

For $t = \left\lceil \frac{m}{\sqrt{n\log n}} \right\rceil$, the total running time is

$$T(m,n) \;=\; O(m\sqrt{n}\log^{1/2} n + n\log n). \tag{5.1}$$

Next, we describe the main algorithm. First, we partition the plane into $M = O(r^2)$ triangles $\Delta_1, \ldots, \Delta_M$ so that no triangle meets more than $O(\frac{n}{r})$ lines of $\mathcal{L}$, for some r to be specified later. Let P_i (resp. $\mathcal{L}_i$) denote the set of points (resp. lines) lying inside (resp. meeting the interior of) the triangle Δ_i; let n_i (resp. m_i) be the size of $\mathcal{L}_i$ (resp. P_i). The sets $\mathcal{L}_i$ are computed by determining the triangles intersected by each line of $\mathcal{L}$, as described in Chapter 4, and the sets P_i are obtained, in time $O((r^2+m)\log r)$, by locating each point of P in the planar subdivision formed by the triangles Δ_i. The incidences between the lines and the points lying on the triangle boundaries can be easily computed in time $O((m+nr)\log n)$, once we have distributed the lines over the triangles. We then apply the above algorithm for each triangle Δ_i to determine the incidences between P_i and $\mathcal{L}_i$ within Δ_i. Since partitioning the plane takes $O(nr \log n \log^\omega r)$ time (cf. Theorem 4.36), the total time $T(m,n)$ spent in computing the incidences between n lines and m points is therefore at most

$$\begin{aligned} T(m,n) &\le \sum_{i=1}^{M} T(m_i,n_i) + O(r^2 \log r + m \log n + nr \log n \log^\omega r) \\ &= \sum_{i=1}^{M} O(m_i\sqrt{n_i}\log^{1/2} n_i + n_i \log n_i) + O((m + nr \log^\omega r) \log n) \qquad (5.2) \end{aligned}$$

Since $n_i = O(\frac{n}{r})$, (5.2) becomes

$$\begin{aligned} T(m,n) &= O\left(\sqrt{\frac{n}{r}} \log^{1/2} n \cdot \sum_{i=1}^{M} m_i\right) + O((m + nr \log^\omega r) \log n) \\ &= O\left(\frac{m\sqrt{n}}{\sqrt{r}} \log^{1/2} n + m \log n + nr \log^\omega r \log n\right) \qquad (5.3) \end{aligned}$$

because $\sum_{i=1}^{M} m_i = m$. Now choose $r = \max\left\{\frac{m^{2/3}}{n^{1/3}\log^{1/3} n \log^{2\omega/3} \frac{m}{\sqrt{n}}}, 2\right\}$; since $m < n^2$, we have $r \le n$ as required. Therefore (5.3) gives

$$T(m,n) = O\left(m^{2/3}n^{2/3}\log^{2/3} n \cdot \log^{\omega/3} \frac{m}{\sqrt{n}} + (m+n)\log n\right).$$

Hence, combining this with the case $m \ge n^2$, we have

Theorem 5.1 *Given a set of n lines and a set of m points in the plane, we can compute the lines passing through each point in time $O(m^{2/3}n^{2/3}\log^{2/3} n \cdot \log^{\omega/3} \frac{m}{\sqrt{n}} + (m+n)\log n)$. (In particular, we can determine whether any line passes through any point within the same amount of time.)*

□

5.3 Computing Many Faces in Arrangements of Lines

Next we consider the following problem:

> *Given a set $\mathcal{L} = \{\ell_1, \ldots, \ell_n\}$ of n lines and a set $P = \{p_1, \ldots, p_m\}$ of m points, compute the faces of $\mathcal{A}(\mathcal{L})$ containing one or more points of P.*

Clarkson et al. [38] have proved that the combinatorial complexity of m distinct faces in any arrangement of n lines in the plane is $O(m^{2/3}n^{2/3} + n)$ (see also [16]), and Edelsbrunner et al. [59] have given a randomized algorithm to compute m distinct faces, whose expected running time is $O(m^{2/3-\delta}n^{2/3+2\delta} + n \log n \log m)$, for any $\delta > 0$. This algorithm can be made deterministic, without substantially changing its time complexity, using the original technique of Matoušek [89]. As in the case of the incidence problem, a slightly faster randomized algorithm, for large values of m, is presented in [55] and has $O(n^{3/2}\log^2 n + m^{2/3}n^{2/3}\log^4 n)$ expected running time, but we do not know of any way to make it deterministic without substantially increasing its running time. We present a deterministic algorithm that computes these faces in time $O(m^{2/3}n^{2/3}\log^{5/3} n \log^{\omega/3} \frac{m}{\sqrt{n}} + n \log n)$.

Similar to the previous section, we first give a slower $O(m\sqrt{n}\log^2 n + n \log n)$ algorithm for this problem and then, using the same divide and conquer technique, we obtain an algorithm with the asserted time bound. Without loss of generality we can assume that $m \leq n^2$, for otherwise the faces can be computed in time $O(m \log n)$ by constructing the entire arrangement $\mathcal{A}(\mathcal{L})$. Our slower algorithm works as follows.

Partition the set P into t disjoint sets $P_1, \ldots, P_t$ so that P_i contains $m_i \leq \lceil \frac{m}{t} \rceil$ points. We show how to compute the faces of $\mathcal{A}(\mathcal{L})$ containing the points of P_i, and repeat this procedure for all $i \leq t$. Let $\mathcal{L}^\star$ denote the set of points dual to the lines $\mathcal{L}$, and let $P_i^\star$ denote the set of lines dual to the points in P_i. Let f be a face of $\mathcal{A}(\mathcal{L})$ containing some point p. For each line $\ell \in \mathcal{L}$ bounding f, its dual point $\ell^\star$ is such that the dual line $p^\star$ can be moved (actually rotated around some point) to touch $\ell^\star$, without crossing any other point of $\mathcal{L}^\star$ while rotating. In other words, the dual of the face f containing a point p corresponds to the portions of the convex

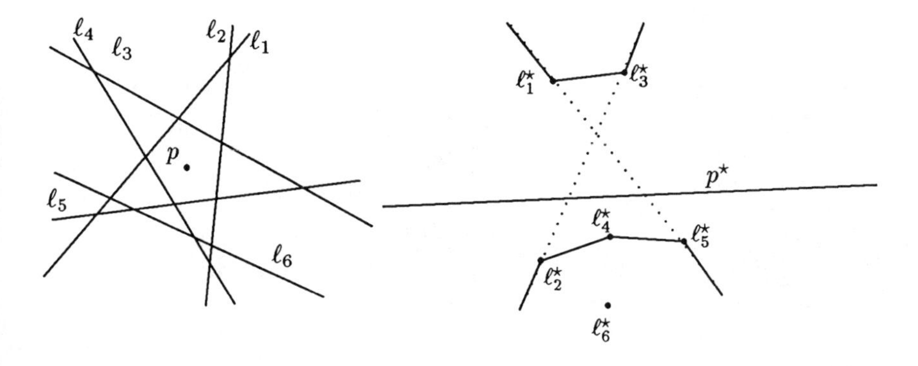

Figure 5.2: A face in an arrangement of lines, and its dual

hulls $CH(\mathcal{L}^\star \cap (p^\star)^+)$ and $CH(\mathcal{L}^\star \cap (p^\star)^-)$ between their common tangents, where $(p^\star)^+$, $(p^\star)^-$ denote the half planes lying respectively above and below $p^\star$, as shown in figure 5.2. Therefore, it suffices to describe how to compute the convex hull of the points in $\mathcal{L}^\star$ lying above or below the line $p^\star$, for each line $p^\star \in P_i^\star$.

First, compute the arrangement $\mathcal{A}(P_i^\star)$. Let $\mathcal{D}$ denote the dual of the planar graph formed by $\mathcal{A}(P_i^\star)$, i.e. the vertices of $\mathcal{D}$ correspond to the faces of $\mathcal{A}(P_i^\star)$, and there is an edge ϕ_{jk} between two vertices v_j, v_k of $\mathcal{D}$ if the corresponding faces f_j, f_k of $\mathcal{A}(P_i)$ share an edge e_{jk} in $\mathcal{A}(P_i^\star)$ (see figure 5.3). Let $\mathcal{L}_j^\star \subseteq \mathcal{L}^\star$ denote the set of points lying in the face $f_j \in \mathcal{A}(P_i^\star)$. For each $\mathcal{L}_j^\star$, compute its convex hull $CH(\mathcal{L}_j^\star)$. We associate $\mathcal{L}_j^\star$ and its hull with the node v_j of $\mathcal{D}$.

Let $\mathcal{T}$ denote any spanning tree of $\mathcal{D}$; it can be easily computed in time $O(m_i^2)$. If $\mathcal{T}$ contains a subtree of $\mathcal{T}$, all of whose nodes are associated with empty subsets of $\mathcal{L}^\star$, we remove that subtree from $\mathcal{T}$. It is easily seen that a line $p^\star \in P_i^\star$ intersects at most m_i edges of $\mathcal{T}$ (in the sense that the two faces of $\mathcal{A}(P_i^\star)$ connected by such an edge lie on different sides of $p^\star$). We perform a depth first search on $\mathcal{T}$ and connect the vertices of $\mathcal{T}$ in the order they are first visited by the depth first traversal; this gives a spanning path

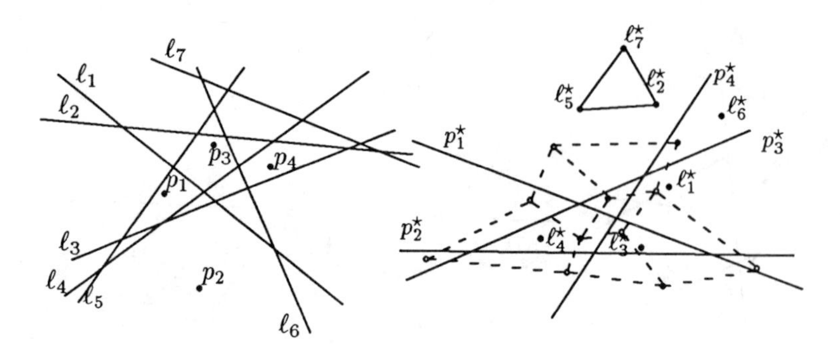

Figure 5.3: Arrangements $\mathcal{A}(\mathcal{L})$, $\mathcal{A}(P_i^\star)$ and the dual graph $\mathcal{D}$

Π with the property that a line $p^\star \in P_i^\star$ intersects at most $2m_i$ edges of Π (in the same sense as above cf. [32]), and that each edge of Π is intersected by exactly one line of $P_i^\star$. Next we construct a spanning path $\mathcal{C}$ of $\mathcal{L}^\star$ from Π by modifying each vertex v_j of Π, depending on the cardinality of $\text{CH}(\mathcal{L}_j^\star)$. There are three cases to consider:

(i) $|\text{CH}(\mathcal{L}_j^\star)| = 0$: remove the vertex v_j and the edges $\phi_{j-1,j}$, $\phi_{j,j+1}$ from Π, and add the edge $\phi_{j-1,j+1}$ to Π (figure 5.4b); this shortcuting may be repeated several times if needed, producing at the end a shortcut edge $\phi_{kk'}$.

(ii) $|\text{CH}(\mathcal{L}_j^\star)| \leq 1$: replace the vertex v_j by $\text{CH}(\mathcal{L}_j^\star)$ (figure 5.4c).

(iii) $|\text{CH}(\mathcal{L}_j^\star)| \geq 2$: let $\ell_x^\star$, $\ell_y^\star$ be two adjacent vertices of $\text{CH}(\mathcal{L}_j^\star)$. Replace v_j by $\text{CH}(\mathcal{L}_j^\star)$, make the edge $\phi_{j-1,j}$ (resp. $\phi_{j,j+1}$) incident to $\ell_x^\star$ (resp. $\ell_y^\star$) (figure 5.4d, e), and if $|\text{CH}(\mathcal{L}_j^\star)| > 2$, then remove the edge $\overline{\ell_x^\star \ell_y^\star}$ from $\text{CH}(\mathcal{L}_j^\star)$ (figure 5.4e).

It is easily seen that the resulting structure is a spanning path $\mathcal{C}$ of $\mathcal{L}^\star$ (see figure 5.5).

Lemma 5.2 *A line $p^\star \in P_i^\star$ intersects at most $2m_i$ edges of $\mathcal{C}$.*

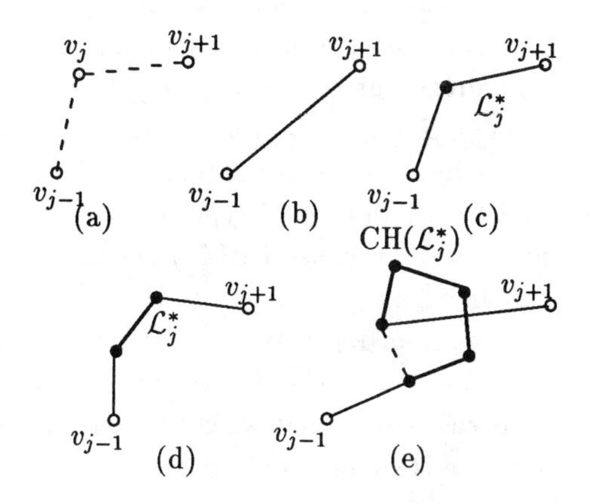

Figure 5.4: Modifying a vertex v_j of Π; a: vertex v_j of Π, b: v_j is deleted from Π, c, d, e: v_j is replaced by $\text{CH}(\mathcal{L}_j^*)$

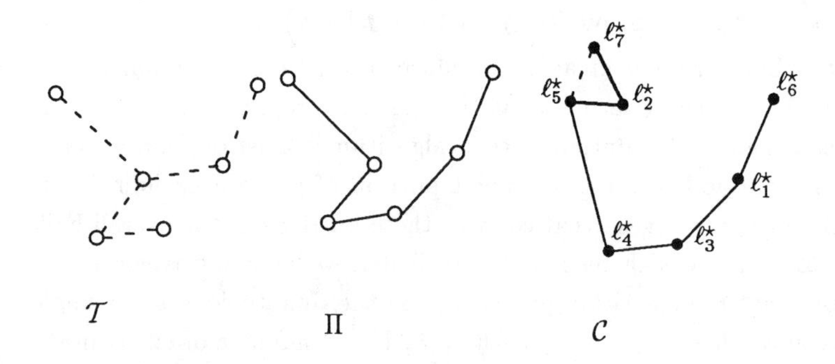

Figure 5.5: $\mathcal{T}$, Π and $\mathcal{C}$ for points of $\mathcal{L}^*$ shown in figure 5.3

Proof: Let $p^\star \in P_i^\star$ be a line intersecting s edges of Π. We prove that $p^\star$ intersects at most s edges of $\mathcal{C}$ by showing that each intersection between $p^\star$ and an edge of $\mathcal{C}$ can be charged to an edge ϕ of Π intersecting $p^\star$, in such a way that no edge of Π is charged more than once. There are three types of edges in $\mathcal{C}$: (i) edges that were already present in Π (e.g. $\overline{\ell_1^*\ell_6^*}$ in figure 5.5), (ii) edges of $\mathrm{CH}(\mathcal{L}_j^\star)$, for some $v_j \in \mathcal{T}$ (e.g. $\overline{\ell_2^*\ell_5^*}$ in figure 5.5), and (iii) edges that were introduced while removing a vertex of Π (e.g. $\overline{\ell_4^*\ell_5^*}$ in figure 5.5). We charge an intersection of $P^\star$ with an edge of type (i) to the edge itself. Edges of type (ii) do not intersect $p^\star$, because $\mathrm{CH}(\mathcal{L}_j^\star)$ lies inside a face of $\mathcal{A}(P_i^\star)$. Finally, if $p^\star$ intersects an edge $\phi_{k,k'}$ of type (iii) (i.e. a shortcut edge introduced while deleting vertices from Π), then $p^\star$ must intersect at least one edge $\phi_{j,j+1}$ of Π, for $j = k, k+1, \ldots, k'-1$. We can therefore charge this intersection to $\phi_{j,j+1}$. It is easily seen that we charge only those edges of Π that intersect $p^\star$ and no edge is charged twice. Hence $p^\star$ intersects at most $s \leq 2m_i$ edges of $\mathcal{T}$.

□

Edelsbrunner et al. [55] have shown that if T is a spanning path of a set S of k points in the plane, then T can be preprocessed in $O(k \log k)$ time so that, for any line ℓ intersecting at most s edges of $\mathcal{T}$, $\mathrm{CH}(S \cap \ell^+)$ can be computed in $O(s \log^3 k)$ time. Since in our case $k = n$ and $s \leq 2m_i$, $\mathrm{CH}(\mathcal{L}^\star \cap p^\star)$, for $p^\star \in P_i^\star$ can be computed in $O(m_i \log^3 m_i)$ time, which implies that the total time spent in computing the faces in $\mathcal{A}(\mathcal{L})$ containing the points of P_i is bounded by $O\left(\frac{m^2}{t^2} \log^3 n + n \log n\right)$.

However, Edelsbrunner et al.'s procedure returns only an implicit representation, which they referred to as the "necklace representation", of the desired faces. That is, the output of their algorithm is a list of pointers, each pointing to some node storing a disjoint portion of the convex hull, intermixed with "bridging edges" that connect these portions in the overall hull. If we want to compute each desired face explicitly, we have to traverse all the hull portions that the algorithm points to, and the time to compute a single face f_j becomes $O(m_i \log^3 n + k_j)$, where k_j is the number of edges in f_j. Therefore, the total time spent in computing the faces containing the points of P_i is at most $O\left(m_i^2 \log^3 n + n \log n + \sum_{p_j \in P_i} k_j\right)$. But in the worst case $\sum_{p_j \in P_i} k_j$ could be as large as $\Theta(m_i n_i)$, e.g. when all of the points lie in the same face, which happens to be bounded by all the lines of $\mathcal{L}$. This bound is too large for our purposes, which means that we cannot afford to output the same face too many times. We circumvent this problem by modifying the above algorithm as follows. Suppose we have already computed the faces containing $p_1, \ldots, p_j$ of P_i, and we are about to compute the face f_{j+1} containing p_{j+1}. Before computing this face we first check whether p_{j+1} lies in any of the faces computed so far; we compute f_{j+1}, as described above, only if it is indeed a new face. Since each face of $\mathcal{A}(\mathcal{L})$ is a convex polygon, we can easily test p_{j+1} for containment in each of the already computed faces of $\mathcal{A}(\mathcal{L})$ in $O(\log n)$ time, so the total time needed to decide whether f_{j+1} should be computed is at most $O(j \log n)$. Thus, the total time required to compute the collection S of the desired faces is

$$\begin{aligned} T(m_i, n) &= \sum_{j=1}^{m_i} O(m_i \log^3 n + j \log n) + O(\sum_{f_j \in S} |f_j|) + O(n \log n) \\ &= O(m_i^2 \log^3 n + n \log n) + O(\sum_{f_j \in S} |f_j|). \end{aligned}$$

Edelsbrunner and Welzl [66] (see also [16]) have proved that the complexity of m distinct faces in an arrangement of n lines is at most $O(m\sqrt{n})$. Therefore

$$T(m_i, n) = O(m_i^2 \log^3 n + n \log n) + O(m_i \sqrt{n}).$$

Since $m_i \leq \lceil \frac{m}{t} \rceil$, summing over all P_i's we obtain

$$\begin{aligned} T(m, n) &= \sum_{i=1}^{t} O\left(\frac{m^2}{t^2} \log^3 n + n \log n + \frac{m}{t}\sqrt{n}\right) \\ &= O\left(\frac{m^2}{t} \log^3 n + nt \log n + m\sqrt{n}\right). \end{aligned}$$

Choosing $t = \left\lceil \frac{m \log n}{\sqrt{n}} \right\rceil$, we obtain $T(m, n) = O\left(m\sqrt{n} \log^2 n + n \log n\right)$.

Remark 5.3: We believe that using, in the above procedure, the algorithm of [55] of merging the convex hulls to obtain the explicit face representation is an overkill, and a simpler, more naive solution should exist. But at present we do not know how to simplify the algorithm.

Now we describe the main algorithm. As in the previous section, we partition the plane into $M = O(r^2)$ triangles $\Delta_1, \ldots, \Delta_M$ each of which

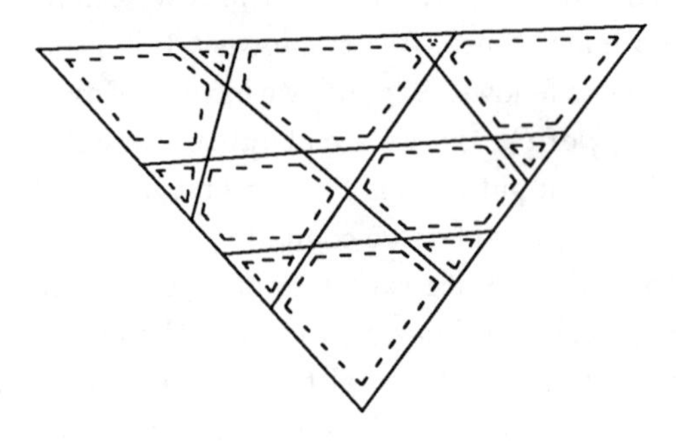

Figure 5.6: Zone of a triangle $\triangle_i$

meets at most $O(\frac{n}{r})$ lines of $\mathcal{L}$. Let P_i (resp. $\mathcal{L}_i$) denote the set of points of P (resp. lines of $\mathcal{L}$) contained in (resp. meeting) $\triangle_i$, and let $f_i(p)$ denote the face of $\mathcal{A}(\mathcal{L}_i)$ containing a point p. The *zone* of $\triangle_i$ in $\mathcal{A}(\mathcal{L}_i)$ is defined as the collection of the face portions $f \cap \triangle_i$, for all faces $f \in \mathcal{A}(\mathcal{L}_i)$, that intersect the boundary of $\triangle_i$ (see figure 5.6). Clarkson et al. [38] have observed that the total number of edges in the zone of $\triangle_i$ is $O(\frac{n}{r})$ (see also [30] and [62], where a zone is defined with respect to a half plane). If a face $f_i(p)$ is fully contained in the interior of $\triangle_i$, then $f_i(p) = f(p)$. Otherwise if $f_i(p)$ intersects the boundary of $\triangle_i$, then it is a face of the zone of $\triangle_i$. Moreover, if a face $f \in \mathcal{A}(\mathcal{L})$ does not lie in the interior of a triangle $\triangle_i$, it is split into two or more pieces, each being a face in the zone of some triangle. Also, such a face f intersects a triangle $\triangle_i$ if and only if f is a face in the zone of $\triangle_i$. Thus, all the faces in $\mathcal{A}(\mathcal{L})$ containing the points of P can be obtained by computing, for every $\triangle_i$, (i) the faces of $\mathcal{A}(\mathcal{L}_i)$ that contain the points of P_i, and (ii) the zone of $\triangle_i$. The faces of $\mathcal{A}(\mathcal{L})$ (containing points of P) that are split among the zones, can be easily glued together by matching their edges that lie on triangle edges.

Edelsbrunner and Guibas [54] have given an $O(n \log n)$ algorithm to compute a zone with respect to a half-plane in an arrangement of n lines. The same algorithm can be applied to calculate the zone of each $\triangle_i$. As for computing the faces that lies in the interior of $\triangle_i$, we use the simplified algorithm, given as above. Thus, the total time spent in processing $\triangle_i$ is at most $O(m_i\sqrt{n_i}\log^2 n_i + n_i \log n_i)$. Finally, the total time spent in merging the zones is at most $O(nr\log n)$ because zones of two different triangles do not intersect, and each zone has at most $O(\frac{n}{r})$ edges. Hence the total time $T(m,n)$ spent in computing m distinct faces in an arrangement of n lines in the plane is (provided $m \le n^2$)

$$\begin{aligned} T(m,n) &= \sum_{i=1}^{M} O\left(m_i\sqrt{n_i}\log^2 n_i + n_i\log n_i\right) + O(nr\log n) + \\ &\quad O(nr\log n\log^{\omega} r) \\ &= O\left(\sqrt{\frac{n}{r}}\log^2 n\sum_{i=1}^{M} m_i + nr\log n\right) + O(nr\log n\log^{\omega} r) \\ &\qquad \left(\text{because } n_i \le \frac{n}{r} \text{ and } M = O(r^2)\right) \\ &= O\left(\frac{m\sqrt{n}}{\sqrt{r}}\log^2 n + nr\log n\log^{\omega} r\right) \end{aligned}$$

because $\sum_{i=1}^{M} m_i = m$. For $r = \max\left\{\frac{m^{2/3}\log^{2/3} n}{n^{1/3}\log^{2\omega/3}\frac{m}{\sqrt{n}}}, 2\right\}$, the above bound becomes

$$T(m,n) = O\left(m^{2/3}n^{2/3}\log^{5/3} n\log^{\omega/3}\frac{m}{\sqrt{n}} + n\log n\right).$$

Combining this with the trivial bound $O(m\log n)$, for $m > n^2$, we obtain

Theorem 5.4 *Given a set of n lines in the plane, we can compute the faces of its arrangement that contain m given points in time*

$$O(m^{2/3}n^{2/3}\log^{5/3} n\log^{\omega/3}\frac{m}{\sqrt{n}} + (m+n)\log n).$$

□

5.4 Computing Many Faces in Arrangements of Segments

Consider the following problem:

Given a set $\mathcal{G} = \{e_1, \ldots, e_n\}$ of n segments and a set $P = \{p_1, \ldots, p_m\}$ of m points, compute the faces of $\mathcal{A}(\mathcal{G})$ containing the points of P.

Aronov et al. [4] have shown that the combinatorial complexity of m distinct faces in an arrangement of n segments is bounded by

$$O\left(m^{2/3}n^{2/3} + n\alpha(n) + n\log m\right).$$

Edelsbrunner et al. [59] have given a randomized algorithm to compute m distinct faces in an arrangement of n segments whose expected running time is $O(m^{2/3-\delta}n^{2/3+2\delta} + n\alpha(n)\log m \log^2 n)$, for any $\delta > 0$. Our algorithm for computing many faces in an arrangement of lines cannot be easily extended to the case of segments, so we present an alternative technique that proceeds by applying the partitioning algorithm in the dual plane rather than in the primal. Our algorithm is closely related to the proof of the combinatorial bound given in [4]. Again we assume that $m \le n^2$ for otherwise we can compute the faces in $O(m \log n)$ time by constructing the entire arrangement $\mathcal{A}(\mathcal{G})$.

Let ℓ denote the line containing the segment e of $\mathcal{G}$. We dualize each line ℓ to a point $\ell^\star$, and each point p of P to a line $p^\star$; this yields a set $P^\star$ of m lines, and a set $\mathcal{L}^\star$ of n points in the dual plane. We partition the dual plane into $t = O(r^2)$ triangles $\Delta'_1, \ldots, \Delta'_t$ so that no triangle meets more than $O(\frac{m}{r})$ lines of $P^\star$. By Theorem 4.36, this can be done in $O(mr \log m \log^\omega r)$ time. If a triangle Δ'_i contains $n_i > \frac{n}{r^2}$ points of $\mathcal{L}^\star$, we split it further into $\lceil \frac{n_i r^2}{n} \rceil$ triangles, none of which contains more than $\frac{n}{r^2}$ points. Clearly, the distribution of the points of $\mathcal{L}^\star$ among the triangles, and the further partitioning of the triangles can be done in $O(n \log n)$ time. Let $\Delta_1, \ldots, \Delta_M$ denote the set of resulting triangles; we still have $M = O(r^2)$. Let $\mathcal{L}^\star_i$ denote the set of points contained in Δ_i, and $P^\star_i$ the set of lines meeting Δ_i. Let $\mathcal{G}_i$ denote the set of segments corresponding to the points $\mathcal{L}^\star_i$. If a line $p^\star_j$ does not meet Δ_i, then p_j lies either above all lines containing the segments of $\mathcal{G}_i$ or below all such lines, which implies that p_j lies in the unbounded face of $\mathcal{A}(\mathcal{G}_i)$. Hence, for each subcollection $\mathcal{G}_i$, it suffices to compute the unbounded face of $\mathcal{A}(\mathcal{G}_i)$ and the faces that contain the points of P_i. As a matter of fact, we compute the entire arrangement $\mathcal{A}(\mathcal{G}_i)$ in time $O(\frac{n^2}{r^4})$, and select the desired faces from it. Let $f_i(p)$ denote the face of $\mathcal{A}(\mathcal{G}_i)$ containing the point p. Note that the face $f(p)$ of $\mathcal{A}(\mathcal{G})$ containing p is the connected component of $\bigcap_{i=1}^{M} f_i(p)$ containing p. Therefore for each $p \in P_i$, we have to "merge", i.e. compute the connected component containing p of the intersection of, all M corresponding faces.

Recall that our partitioning algorithm first computes r approximate levels, which are disjoint polygonal chains with a total of $O(r^2)$ vertices, and then triangulates each "corridor" lying between two adjacent polygonal chains. We construct a binary tree $\mathcal{T}$ of height $H = O(\log r)$ whose leaves correspond to these triangles and whose root corresponds to the enclosing rectangle $\mathbf{R}$ (see [4]). We first construct a binary tree $\mathcal{T}_C$, as described in [4], for each corridor C on the set of triangles lying in C so that the preorder traversal of $\mathcal{T}_C$ visits the leaves (i.e. the triangles in C) in the order in which they appear along C from left to right (see figure 5.7). $\mathcal{T}$ is then constructed with the trees $\mathcal{T}_C$ as its leaves, in a similar manner.

Each node v of $\mathcal{T}$ is associated with a simply connected region $\mathcal{P}_v$, which is the union of the regions associated with the leaves of the subtree $\mathcal{T}_v$ of $\mathcal{T}$ rooted at v (the construction of $\mathcal{T}$ implies that each $\mathcal{P}_v$ is simply connected). For each node v of $\mathcal{T}$, let $\mathcal{G}_v = \bigcup_{\Delta_i \subseteq \mathcal{P}_v} \mathcal{G}_i$ and $P_v = \bigcup_{\Delta_i \subseteq \mathcal{P}_v} P_i$. Let $n_v = |\mathcal{G}_v|$ and $m_v = |P_v|$. Observe that any point $p \in P - P_v$ lies either above all the lines containing the segments of $\mathcal{G}_v$ or below all these lines, and therefore all these points lie in the unbounded face of $\mathcal{A}(\mathcal{G}_v)$. Let w and z denote the children of the interior node v. It is easily seen that $P_v = P_w \cup P_z$. For every node v of T, we compute the unbounded face of $\mathcal{A}(\mathcal{G}_v)$ and the faces containing the points of P_v. Let F_v denote the set of these faces and R_v denote the total number of edges in the faces of F_v. Note that the face $f_v(p)$ of $\mathcal{A}(\mathcal{G}_v)$ is the connected component of $f_w(p) \cap f_z(p)$ that contains the point p, where $f_w(p)$ (resp. $f_z(p)$) is the face of $\mathcal{A}(\mathcal{G}_w)$ (resp. $\mathcal{A}(\mathcal{G}_z)$) containing the point p. Thus if we have already computed F_w and F_z, then F_v can be computed by sweeping a vertical line first from left to right and then from right to left as described in [59]. Let $\mathcal{M}_v$ denote the time spent in merging F_w and F_z. It follows from the analysis of [59] that

$$\mathcal{M}_v = O((R_v + m_v + n_v\alpha(n_v)) \log n_v). \tag{5.4}$$

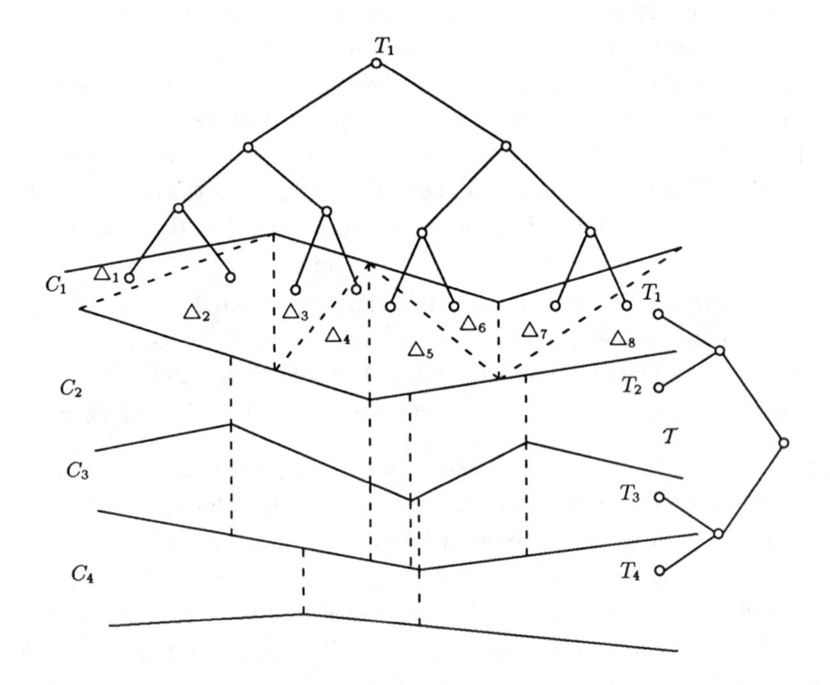

Figure 5.7: Binary tree formed on the triangles; $\mathcal{T}_{C_1} = T_1$ and $\mathcal{T}$

Therefore, the total time $\mathcal{M}(m,n)$ spent in merging the faces is

$$\begin{aligned}\mathcal{M}(m,n) &= \sum_{v\in\mathcal{T}} O((R_v + m_v + n_v\alpha(n_v))\log n_v) \\ &= \sum_{i=1}^{H}\sum_{h(v)=i} O((R_v + m_v + n_v\alpha(n_v))\log n_v) \qquad (5.5)\end{aligned}$$

where $h(v)$ is the height of v. But it has been proved in [59] that

$$R_v \le R_w + R_z + 4m_v + 6n_v. \qquad (5.6)$$

Let $\mathcal{U}_v$ (resp. $\mathcal{Z}_v$) denote the set of leaves (resp. interior nodes) in the subtree $\mathcal{T}_v$. If $h(v) = i$, then by (5.6)

$$R_v \le \sum_{u\in\mathcal{U}_v} R_u + 4\sum_{z\in\mathcal{Z}_v} m_z + O(n_v \cdot i)$$

where the last term follows from the fact that $\sum_x n_x$, over all nodes at the same level of $\mathcal{T}_v$, is n_v, and the height of v is i. Let $k_v = |\mathcal{U}_v|$ denote the number of leaves of $\mathcal{T}_v$. As shown in [4], the special way in which $\mathcal{T}$ was constructed guarantees that

$$m_v \le \frac{ck_v m}{r} + 1, \qquad (5.7)$$

where c is some constant > 0. Moreover, for each leaf u of $\mathcal{T}$, $|n_u| = O(\frac{n}{r^2})$. Therefore $R_u = O(\frac{n^2}{r^4})$, and

$$R_v = O\left(O\left(k_v\frac{n^2}{r^4}\right) + \sum_{z\in\mathcal{Z}_v}\left(\frac{k_z m}{r} + 1\right) + n_v \cdot i\right),$$

which implies that

$$\mathcal{M}(m,n) = \sum_{i=1}^{H}\sum_{h(v)=i} O\left(\left(\frac{k_v n^2}{r^4} + \sum_{z\in\mathcal{Z}_v}\frac{k_z m}{r} + n_v(\alpha(n) + i)\right)\log n\right).$$

It can be easily proved that

$$\sum_{h(v)=i} k_v = O(r^2), \quad \sum_{h(v)=i} n_v = n \quad \text{and} \quad \sum_{h(v)=i}\sum_{z\in\mathcal{Z}_v} k_z = O(ir^2).$$

Therefore

$$\begin{aligned}\mathcal{M}(m,n) &= \sum_{i=1}^{H} O\Big(\Big(\frac{n^2}{r^2} + imr + n(\alpha(n)+i)\Big)\log n\Big) \\ &= O\Big(\Big(\frac{n^2}{r^2}\log r + n\alpha(n)\log r + (n+mr)\log^2 r\Big)\log n\Big)\end{aligned}$$

because $H = O(\log r)$.

Now going back to the original problem, we spent $O(mr\log m\log^{\omega} r)$ time in partitioning the plane into M triangles, and $O(\frac{n^2}{r^4})$ time in constructing $\mathcal{A}(\mathcal{G}_i)$ for each Δ_i (cf. [62]). Thus, the total time $T(m,n)$ spent in computing m distinct faces of an arrangement of n segments in the plane is at most

$$\begin{aligned}T(m,n) &= \sum_{i=1}^{M} O\left(\frac{n^2}{r^4}\right) + O(mr\log m\log^{\omega} r) + \\ &\quad O\left(\left(\frac{n^2}{r^2}\log r + n\alpha(n)\log r + (n+mr)\log^2 r\right)\log n\right) \\ &= O\left(\left(\frac{n^2}{r^2} + mr\log^{\omega-1} r + n\alpha(n) + n\log r\right)\log n\log r\right).\end{aligned}$$

Hence, by choosing $r = \max\left\{\frac{n^{2/3}}{m^{1/3}\log^{(\omega-1)/3}\frac{n}{\sqrt{m}}}, 2\right\}$, we obtain

$$\begin{aligned}T(m,n) &= O\left((m^{2/3}n^{2/3}\log^{(2\omega+1)/3}\frac{n}{\sqrt{m}} + n\log^2\frac{n}{\sqrt{m}} + m)\log n\right) \\ &= O\left(m^{2/3}n^{2/3}\log n\log^{(2\omega+1)/3}\frac{n}{\sqrt{m}} + n\log^3 n + m\log n\right).\end{aligned}$$

Theorem 5.5 *One can compute the faces of an arrangement of n line segments, which contain m given points, in time*

$$O\left(m^{2/3}n^{2/3}\log n\log^{(2\omega+1)/3}\frac{n}{\sqrt{m}} + n\log^3 n + m\log n\right).$$

□

Remark 5.6: If we partition Δ_i' into $\left\lceil\frac{n_i}{\sqrt{m/r}\log^{1/2} r}\right\rceil$ triangles (instead of $\left\lceil\frac{n_i r^2}{n}\right\rceil$), each containing at most $\sqrt{\frac{m}{r}}\log^{1/2} r$ points of $\mathcal{L}^*$, and choose $r =$

$\max\left\{\frac{n^{2/3}}{m^{1/3}}\log^{2\omega/3-1}\frac{n}{\sqrt{m}}, 2\right\}$, then the running time of the algorithm can be improved slightly to

$$O\left(m^{2/3}n^{2/3}\log n \log^{\omega/3+1}\frac{n}{\sqrt{m}} + n\log^3 n + m\log n\right).$$

5.5 Counting Segment Intersections

In this section we consider the following problem:

> *Given a set $\mathcal{G} = \{e_1, \ldots, e_n\}$ of n line segments in the plane, we wish to count the number of intersection points between them.*

This is a variant of one of the most widely studied problems in computational geometry, namely that of reporting all intersections (cf. [11], [15], [23]). The recent algorithm of Chazelle and Edelsbrunner [23] reports all k intersections in time $O(n\log n + k)$, using $O(n+k)$ space. It requires quadratic working storage in the worst case. Guibas et al. [74] gave an $O(n^{4/3+\delta} + k)$ randomized algorithm, for any $\delta > 0$, using only $O(n)$ working storage (see also [37], [98]). The only algorithms known for counting the intersection points in time that does not depend on k, are by Chazelle [19] and by Guibas et al. [74]. The latter algorithm is faster but randomized, and has expected running time $O(n^{4/3+\delta})$ for any $\delta > 0$. We modify Guibas et al.'s algorithm to give a slightly faster and deterministic algorithm, although the space requirement goes up roughly to $n^{4/3}$. Their algorithm relies on a procedure that, for a given triangle Δ, counts the number of intersection points contained in Δ in $O((m^2+n)\log n)$ time, where n is the number of segments meeting Δ, and $m \le n$ is the number of segments having at least one of their endpoints inside Δ. For the sake of completeness, we briefly overview this procedure because we will also make use of it.

Partition the segments of $\mathcal{G}$ meeting Δ into two subsets:

(i) $\mathcal{G}_l$: "long" segments of $\mathcal{G}$ whose endpoints do not lie inside Δ.

(ii) $\mathcal{G}_s$: "short" segments of $\mathcal{G}$ having at least one endpoint inside Δ.

There are three types of intersections to be counted:

- "Short-short" intersections: intersections between the segments of $\mathcal{G}_s$.
- "Long-long" intersections: intersections between the segments of $\mathcal{G}_l$.
- "Long-short" intersections: intersections between a segment of $\mathcal{G}_l$ and another segment of $\mathcal{G}_s$.

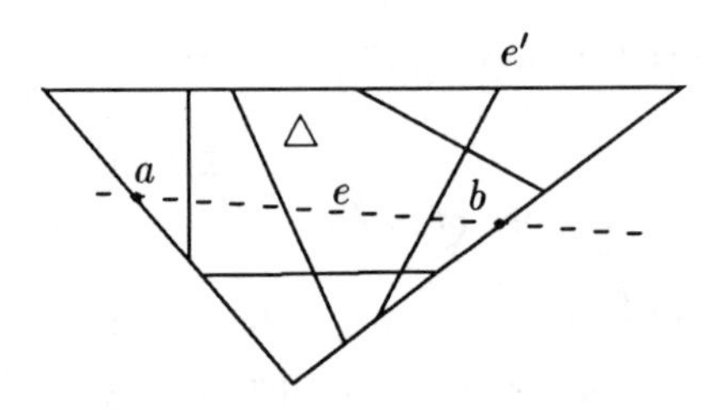

Figure 5.8: Long segment e with its endpoints a, b

Counting short-short intersections: The short-short intersections can be counted in $O(m^2)$ time by testing all pairs of segments of $\mathcal{G}_s$.

Counting long-long intersections: For a segment $e \in \mathcal{G}_l$, we refer to the intersection points of $\partial\Delta$ and $\mathcal{G}_l$ as *endpoints* of e. Let S denote the sequence of endpoints of segments in $\mathcal{G}_l$ sorted along $\partial\Delta$ in counter-clockwise direction, starting from one of its vertices. Let $a, b \in S$ be the endpoints of a segment $e \in \mathcal{G}_l$. It is easily seen that another long segment e' intersects e inside Δ if and only if exactly one endpoint of e' lies between a and b in S (see figure 5.8). Therefore, to count all "long-long" intersections we scan S once and do the following operations at each point of S. If we encounter a segment e for the first time, we insert e on top of a stack, maintained as a binary tree $\mathcal{B}$, and if we encounter e for the second time, we remove it from $\mathcal{B}$, but before doing so we count the number of segments in the tree that were inserted after e. It can be easily verified that no intersection is counted more than once. Since each operation takes at most $O(\log n)$ time, the total time spent in counting the number of long-long intersections is $O(n \log n)$.

Counting long-short intersections: For every segment $e \in \mathcal{G}_s$, let $\tilde{e}$ denote $e \cap \Delta$; and $\tilde{\mathcal{G}}_s = \{\tilde{e} | e \in \mathcal{G}_s\}$. Let $\overline{\mathcal{G}}_l$ denote the lines containing the segments of $\mathcal{G}_l$. It clearly suffices to count, for each $\ell \in \overline{\mathcal{G}}_l$, the number of intersections between ℓ and $\tilde{\mathcal{G}}_s$.

We dualize each segment $\tilde{e} \in \tilde{\mathcal{G}}_s$ to a double wedge $e^\star$, and construct the arrangement $\mathcal{H}$ of these double wedges. For any double wedge $e^\star$, each face f of $\mathcal{H}$ is either contained in $e^\star$ or does not intersect $e^\star$. The *weight* of a face f is the number of double wedges containing f; the weights of all faces of $\mathcal{H}$ can be easily determined while constructing $\mathcal{H}$.

A line $\ell \in \overline{\mathcal{G}}_l$ intersects a segment $\tilde{e} \in \tilde{\mathcal{G}}_s$ if and only if the point $\ell^\star$ lies in the double wedge $e^\star$. Thus, for every segment e in $\mathcal{G}_l$, the number of segments in $\tilde{\mathcal{G}}_s$ intersecting e is equal to the weight of the face in $\mathcal{H}$ containing the point $\ell^\star$. Therefore, we determine the number of segments intersecting ℓ by locating $\ell^\star$ in $\mathcal{H}$. Computing $\mathcal{H}$ and preprocessing it for fast point location queries can be done in time $O(m^2)$ ([62], [60]), so all long-short intersections can be computed in time $O(m^2 + n \log n)$.

The above discussion implies that we can count all intersection points of $\mathcal{G}$ contained in Δ in $O(m^2 + n \log n)$ time. The time complexity of the above procedure can be improved to $O(m\sqrt{n \log n} + n \log n)$ by partitioning $\mathcal{G}_s$ into $\lceil \frac{m}{\sqrt{n \log n}} \rceil$ subsets of size at most $\sqrt{n \log n}$ each, and counting the number of intersection points between each of the subsets and $\mathcal{G}_l$.

Next we describe the main algorithm. We partition the plane into $M = O(r^2)$ triangles $\Delta_1, \ldots, \Delta_M$, each meeting at most $O(\frac{n}{r})$ lines containing the segments of $\mathcal{G}$. Using the algorithm described above, we count the number of intersections contained in each Δ_i, for $i \leq M$, and add up the results. If m_i denotes the number of endpoints lying inside Δ_i, the time spent in counting intersections within Δ_i is $O(m_i\sqrt{\frac{n}{r}} \log^{1/2} n + \frac{n}{r} \log n)$. Using the same analysis as in previous sections, the total time of the algorithm is

$$\sum_{i=1}^{M} O\left(m_i\sqrt{\frac{n}{r}} \cdot \log^{1/2} n\right) + O\left(nr \log n \log^{\omega} r\right)$$
$$= O\left(\frac{n^{3/2}}{r^{1/2}} \log^{1/2} n + nr \log n \log^{\omega} r\right)$$

because $\sum_{i=1}^{M} m_i \leq 2n$. Hence by choosing $r = \dfrac{n^{1/3}}{\log^{(2\omega+1)/3} n}$, we obtain

Theorem 5.7 *Given a set of n line segments, their intersection points can be counted in time* $O(n^{4/3} \log^{(\omega+2)/3} n)$ *and* $O\left(\dfrac{n^{4/3}}{\log^{(2\omega+1)/3} n}\right)$ *space.*

□

Remark 5.8: We can combine this algorithm with the algorithm of [23] that computes the number of intersections k in time $O(n \log n + k)$. That is, we first run the algorithm of [23] and stop it as soon as the number of intersections exceeds $n^{4/3} \log^{(\omega+2)/3} n$. Then we use our algorithm. We thus have

Corollary 5.9 *One can count the number of k intersections between n line segments in time*

$$O(\min\{n \log n + k, n^{4/3} \log^{(\omega+2)/3} n\})$$

and space $O\left(\min\left\{n + k, \dfrac{n^{4/3}}{\log^{(2\omega+1)/3} n}\right\}\right)$.

□

5.6 Counting and Reporting Red-blue Intersections

Next, we consider a variant of the segment intersection problem:

> *Given a set* Γ_r *of* n_r *"red" line segments and another set* Γ_b *of* n_b *"blue" line segments, count or report all intersections between* Γ_r *and* Γ_b.

Let $n = n_r + n_b$. Mairson and Stolfi [88] gave an $O(n \log n + K)$ algorithm to report all K red-blue intersections when red-red and blue-blue intersections are not present. The algorithm of Chazelle and Edelsbrunner [23] for reporting segment intersections can also be applied to report all red-blue intersections in this special case. However, in the general case these algorithms cannot avoid encountering red-red and blue-blue intersections. For the general case we have presented an $O((n_r\sqrt{n_b} + n_b\sqrt{n_r} + K)\log n)$ algorithm to report all K red-blue intersections (see Section 3.4). We also showed that a restricted version of this problem, in which we only want to detect a red-blue intersection, can be solved in $O(n^{4/3+\delta})$ (randomized expected) time, for any $\delta > 0$, by reducing it to the problem of computing at most $2n$ faces in $\mathcal{A}(\Gamma_r)$ and in $\mathcal{A}(\Gamma_b)$. As for the counting problem, in the absence of monochromatic intersections, Chazelle et al. [24] have developed an $O(n \log n)$ algorithm to count all red-blue intersections. In this section we present an $O(n^{4/3} \log^{(\omega+2)/3} n)$ algorithm to count all red-blue intersections in the general case, using roughly $n^{4/3}$ space. Our algorithm actually computes, for every red segment e, the number of blue segments intersecting e. The algorithm can be modified to report all K red-blue intersections in time $O(n^{4/3} \log^{(\omega+2)/3} n + K)$.

As in the previous section, we first consider a restricted version of the problem. Let Γ_r and Γ_b be two sets of segments as defined above, all meeting the interior of a triangle Δ, such that m of these segments contain at least one endpoint inside Δ; we wish to count the number of red-blue intersections lying inside Δ. We describe an $O((m^2 + n)\log n)$ algorithm that, for every red segment e, counts the number of blue segments intersecting e, and can be modified to report all red-blue intersections with $O(1)$ overhead per intersection. The algorithm proceeds as follows:

Partition the segments of Γ_r and Γ_b into four subsets:

(i) A: "long" segments in Γ_r whose endpoints do not lie inside Δ; let $|A| = a$.

(ii) B: "short" segments in Γ_r having at least one endpoint inside Δ; let $|B| = b$.

(iii) C: "long" segments in Γ_b whose endpoints do not lie inside Δ; let $|C| = c$.

(iv) D: "short" segments in Γ_b having at least one endpoint inside Δ; let $|D| = d$.

Note that $a + c = n - m$ and $b + d = m$. We have to count (or report) four types of red-blue intersections:

- intersections between A and C,
- intersections between A and D,
- intersections between B and C, and
- intersections between B and D.

Our approach is similar to the one used by Guibas et al. [74] for counting segment intersections, as described in the previous section.

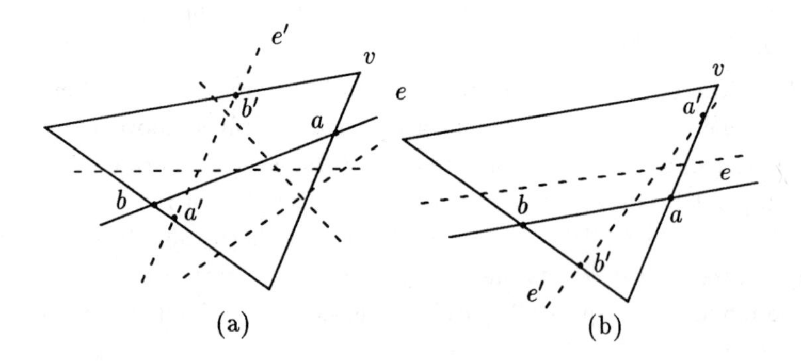

Figure 5.9: Intersections between long segments

Intersections between A and C: For a segment $e \in A \cup C$, its intersection points with $\partial\Delta$ are called the *endpoints* of e. Let S denote the set of endpoints of segments in $A \cup C$ sorted along $\partial\Delta$ in clockwise direction, starting from one of its vertices v. Let $a, b \in S$ be the endpoints of a red segment e with a appearing before b in S. Similarly, let a', b' be the endpoints of a blue segment e'. It is easily seen that e intersects e' if a, b, a' and b' appear in one of the following two orders:

(i) a, a', b and b' (see figure 5.9a), or

(ii) a', a, b' and b (see figure 5.9b).

For each red segment e, we show how to count red-blue intersections along e. We scan the boundary of $\triangle$ in clockwise direction. When we encounter a blue segment for the first time, we insert it on top of a stack, maintained as a binary tree $\mathcal{B}$, and when it is encountered for the second time, we delete it from $\mathcal{B}$. On the other hand, when we encounter a red segment e for the first time, we do nothing, but when we encounter it for the second time, we count the number of (blue) segments in the stack that were inserted before encountering the first endpoint of e. This gives the number of type (i) intersections between e and C. Type (ii) red-blue intersections can be counted in a symmetric way by scanning $\partial\triangle$ in counter-clockwise direction.

We leave it for the reader to verify that this algorithm can be extended to report all red-blue intersections between A and C.

For each segment $e \in A$, we spend $O(\log n)$ time, therefore the total time spent in counting (resp. reporting all K_{AC}) such red-blue intersections is at most $O((a+c)\log(a+c)) = O(n \log n)$ (resp. $O(n \log n + K_{AC})$).

Intersections between A and D: For every $e \in D$, let $\tilde{e}$ denote $e \cap \triangle$; and let $\tilde{D} = \{\tilde{e} | e \in D\}$. Let $\overline{A}$ denote the set of lines containing the segments of A, and let A^* denote the set of points dual to the lines of $\overline{A}$. Let $\mathcal{L}$ denote the set of lines dual to the endpoints of the segments in $\tilde{D}$. Construct the arrangement $\mathcal{A}(\mathcal{L})$. For each $\ell \in \overline{A}$, we count the number of intersections between ℓ and $\tilde{D}$ by locating the point ℓ^* in $\mathcal{A}(\mathcal{L})$, as in Section 5.5. The total time spent in counting these intersections is easily seen to be $O(m^2 + n \log n)$.

As for reporting the intersections between A and D contained in $\triangle$, let $\psi(f)$ denote the set of double wedges dual to the segments of $\tilde{D}$ containing a face f of $\mathcal{A}(\mathcal{L})$. If two faces f_1 and f_2 share an edge γ, contained in a line $\ell \in \mathcal{L}$, then $\delta_\gamma = \psi(f_1) \oplus \psi(f_2)$ is the set of segments having dual of ℓ as an endpoint[1]. Therefore by first constructing the arrangement $\mathcal{A}(\mathcal{L})$, and then locating each point of A^* in $\mathcal{A}(\mathcal{L})$, we can report all K_{AD} intersections between A and D contained in $\triangle$, in time $(m^2 + n \log n + \sum_\gamma |\delta_\gamma| + K_{AD})$. Thus, it suffices to bound $\sum_{e \in \mathcal{A}(\mathcal{L})} |\delta_e|$. Suppose the segments of $\tilde{D}$ have $t \leq 2d$ distinct endpoints and ν_i segments are incident to the ith endpoint. Obviously, $\sum_{i=1}^{t} \nu_i = 2d$ and, for each line $\ell \in \mathcal{L}$, there are t edges of $\mathcal{A}(\mathcal{L})$

[1] We use $A \oplus B$ to denote the symmetric difference of sets A and B.

contained in ℓ, therefore

$$\sum_{e \in \mathcal{A}(\mathcal{L})} |\delta_e| = \sum_{i=1}^{t} t\nu_i \leq 2d^2. \tag{5.8}$$

Hence, the total time spent is $O(n \log n + m^2 + K_{AD})$.

Intersections between B and C: If we just want to report or count the total number of intersections between B and C contained in $\triangle$, we can use the same procedure as in the previous case. But if we want to count the number of red-blue intersections for each red segment separately, we need a different technique.

Let $\tilde{B} = \{e \cap \triangle | e \in B\}$, and let $B^\star$ denote the set of double wedges dual to the segments in $\tilde{B}$. Let $\overline{C}$ denote the set of lines containing the segments of C, and $C^\star$ the set of points dual to the lines in $\overline{C}$. The number of intersections between a segment $e \in \tilde{B}$ and C is equal to the number of points of $C^\star$ in the double wedge $e^\star$. Therefore, for every double wedge, we want to find the number of points of $C^\star$ lying in it. This can be done in time $O(b^2 + c \log(b + c)) = O(m^2 + n \log n)$, using the algorithm described in Edelsbrunner et al. [55].

Intersections between B and D: For every segment $e \in B$, we can determine the segments of D intersecting it by testing all such pairs of segments. This takes $O(m^2)$ time.

Thus for every segment in Γ_r, we can count the number of blue segments intersecting it inside $\triangle$ in time $O(m^2 + n \log n)$, and we can report all red-blue intersections inside $\triangle$ within the same time plus $O(1)$ overhead per intersection. The running time can be improved to $O(m\sqrt{n} \log^{1/2} n + n \log n)$ by partitioning the collection of short segments (that is $B \cup D$) into $\lceil \frac{m}{\sqrt{n \log n}} \rceil$ subsets of size $\sqrt{n \log n}$ each, and then repeating the above procedure for each subset and the entire $A \cup C$.

Going back to the original problem, we partition the plane into $O(r^2)$ triangles, each meeting at most $O(\frac{n}{r})$ lines containing the segments of $\Gamma_r \cup \Gamma_b$. Using the algorithm described above, we count (or more generally report all K_i) red-blue intersections within the ith triangle in $O(m_i \sqrt{\frac{n}{r}} \log^{1/2} \frac{n}{r} + \frac{n}{r} \log \frac{n}{r})$ (resp. $O(m_i \sqrt{\frac{n}{r}} \log^{1/2} \frac{n}{r} + \frac{n}{r} \log \frac{n}{r} + K_i))$ time, where m_i is the number of segment endpoints falling inside the ith triangle. Following the same analysis as in Section 5.5, we obtain

Theorem 5.10 *Given a set of n_r "red" line segments and another set of n_b "blue" line segments, we can count, for each red segment, the number of blue segments intersecting it in overall time $O(n^{4/3}\log^{(\omega+2)/3} n)$ using $O\left(\frac{n^{4/3}}{\log^{(2\omega+1)/3}}\right)$ space, where $n = n_r + n_b$. Moreover, we can report all K red-blue intersections in time $O(n^{4/3}\log^{(\omega+2)/3} n + K)$.*

□

Remark 5.11:

(i) Our algorithm uses roughly $n^{4/3}$ space only for partitioning the plane into $O(r^2)$ triangles; all other stages of the algorithm require $O(n)$ space. If we choose $r = O(1)$ and solve the problem recursively as in [74], we can reduce the space complexity to $O(n)$, but the running time increases to $O(n^{4/3+\delta})$, for any $\delta > 0$ (which can be made as small as we wish by choosing r sufficiently large).

(ii) If we allow randomization, then using the random sampling technique of [36] or of [78], we can count all red-blue intersections in $O(n^{4/3}\log n)$ expected time, and can report all K red-blue intersections in expected time $O(n^{4/3}\log n + K)$. We leave it for the reader to fill in the missing details.

(iii) Note that if Γ_r is a set of lines, then we have to consider only the first two cases, because $B = \emptyset$.

5.7 Batched Implicit Point Location

The planar point location problem is a well studied problem in computational geometry [82], [60], [111]. In this problem we want to preprocess a given planar subdivision so that, for a query point, the face containing p can be computed quickly. Guibas et al. [73] have considered a generalization of this problem, in which the map is defined as the arrangement (i.e. overlay) of n polygonal objects of some simple shape, and we want to compute certain information for the query points related to their arrangement (for example, to determine which query points lie in the union of these polygons). For simplicity we break the given polygonal objects into a collection of line segments, and consider the following formal statement of the problem:

> *We are given a collection* $\mathcal{G} = \{e_1, \ldots, e_n\}$ *of* n *segments, and with each segment* e *we associate a function* ψ_e *defined on the entire plane, which assumes values in some associative and commutative semigroup* S *(denote its operation by* $+$*), and let* $\Psi(x) = \sum_{e \in \mathcal{G}} \psi_e(x)$*. Given a set* $P = \{p_1, \ldots, p_m\}$ *of* m *points, compute* $\Psi(p_1), \ldots, \Psi(p_m)$ *efficiently.*

We assume that ψ_e and Ψ satisfy the following conditions:

(i) The function ψ_e has constant complexity, that is, we can partition the plane into $O(1)$ convex regions so that within each region ψ_e is constant. This also implies that, for any given point x, $\psi_e(x)$ can be computed in $O(1)$ time.

(ii) Any two values in S can be added in $O(1)$ time.

(iii) Given a set $\mathcal{G}$ of n segments in the plane, we can preprocess $\mathcal{G}$ in time $O(n \log^k n)$, for some $k \geq 0$, into a linear size data structure so that, for a query point x lying either above all the lines containing the segments of $\mathcal{G}$, or below all these lines, $\Psi(x)$ can be calculated in $O(\log n)$ time.

We will see that several natural problems, including the containment problem mentioned above, can be cast into this abstract framework. Note that we consider here the *batched* version of the problem, in which all query points are known in advance. In the next chapter we present an algorithm for the preprocessing-and-query version of the problem and solve it using different techniques based on spanning trees with low stabbing number.

A naive approach to solving this problem is to construct the arrangement $\mathcal{A}(\mathcal{G})$ (more precisely, the arrangement obtained by overlapping all the convex subdivisions associated with each of the functions ψ_e), so that the value of Ψ is constant within each resulting face. Now $\Psi(p_1), \ldots, \Psi(p_m)$ can be easily computed in $O(m \log n)$ time by locating the points of P in the above planar map. If $m \geq n^2$ then this is the method of choice, and it runs in overall $O(m \log n)$ time, but if $m < n^2$ this procedure takes $\Omega(n^2)$ time in the worst case, so the goal is to come up with a subquadratic algorithm. Guibas et al. [73] have indeed given a randomized algorithm whose expected running time is $O(m^{2/3-\delta}n^{2/3+\delta} + m \log n + n \log^{k+1} n)$, for any $\delta > 0$. Our (deterministic) algorithm improves their result and works as follows.

Let $\mathcal{L}$ denote the set of lines containing the segments of $\mathcal{G}$. Let $\mathcal{L}^*$ (resp. P^*) denote the set of points (resp. lines) dual to the lines (resp. points) of $\mathcal{L}$ (resp. P). Partition the dual plane, in time $O(mr \log m \log^{\omega} r)$, into $t = O(r^2)$ triangles $\Delta'_1, \ldots, \Delta'_t$, each meeting at most $O(\frac{m}{r})$ lines of P^*. If a triangle contains $n_i > \frac{n}{r^2}$ points of $\mathcal{L}^*$, then partition it further, in time $O(n \log n)$, into $\lceil \frac{n_i r^2}{n} \rceil$ triangles, none of which contains more than $\frac{n}{r^2}$ points. Let $\Delta_1, \ldots, \Delta_M$ denote the resulting triangles; we have $M = O(r^2)$. Let P_i^* denote the set of lines passing through Δ_i, and let $\mathcal{L}_i^*$ denote the set of points contained in Δ_i; thus $|P_i^*| = O(\frac{m}{r})$, $|\mathcal{L}_i^*| \leq \frac{n}{r^2}$. Let $\Psi_i = \sum_{e \in \mathcal{G}_i} \psi_e$. For each $p \in P_i$, we compute $\Psi_i(p)$ by constructing the entire arrangement $\mathcal{A}(\mathcal{G}_i)$, as discussed above (see also [73]). The total time spent in computing $\Psi_i(p)$, for all $p \in P_i$, is $O(\frac{n^2}{r^4} + \frac{m}{r} \log n)$ [62], [60].

Next, we show how to add the values computed within each triangle to calculate $\Psi(p) = \sum_i \Psi_i(p)$. We use a procedure similar to the one used in Section 5.4 for computing many faces in an arrangement of segments. In particular we construct a binary tree $\mathcal{T}$ with the properties defined in Section 5.4. For each node v of $\mathcal{T}$, let $\mathcal{G}_v$, P_v be as defined in Section 5.4, let $m_v = |P_v|$, $n_v = |\mathcal{G}_v|$, and let $\Psi_v = \sum_{e \in \mathcal{G}_v} \psi_e$. At each node v of $\mathcal{T}$, the goal is to compute Ψ_v for all $p \in P_v$. At the end of this process, we will have obtained, at the root u of $\mathcal{T}$, the value of $\Psi_u = \Psi$ for all $p \in P_u = P$. We calculate Ψ_v bottom-up, starting at the leaves of $\mathcal{T}$, as described above.

Let v be an internal node of $\mathcal{T}$ having children w and z. We preprocess $\mathcal{G}_w$, $\mathcal{G}_z$ to obtain data-structures $\mathcal{D}_w$, $\mathcal{D}_z$ of linear size so that, for any point lying either above all the lines containing the segments of $\mathcal{G}_w$ (resp. $\mathcal{G}_z$) , or below all these lines, Ψ_w (resp. Ψ_z) can be computed in logarithmic time. Now for each $p \in P_v$, $\Psi_v(p) = \Psi_w(p) + \Psi_z(p)$. If $p \in P_w$, we already have computed $\Psi_w(p)$ at w. Otherwise p lies either above all lines containing the segments of $\mathcal{G}_w$, or below all these lines, so we can use $\mathcal{D}_w$ to compute $\Psi_w(p)$

in $O(\log n_w)$ time. Similar actions are taken to compute $\Psi_z(p)$. Thus we can obtain Ψ_v for all points in P_v in time $O(m_v \log n_v)$. By the third property of ψ_e, $\mathcal{D}_w$ can be constructed in $O(n_w \log^k n_w)$ time, and similarly for $\mathcal{D}_z$. Hence, the total time spent in computing Ψ_v over all nodes v of $\mathcal{T}$, including the initial partitioning of the dual plane, is

$$\begin{aligned} T(m,n) &= \sum_{v \in \mathcal{T}} O(m_v \log n_v + n_v \log^k n_v) + O(mr \log m \log^\omega r) \\ &\quad + O\left(r^2\left(\frac{n^2}{r^4} + \frac{m}{r}\log n\right)\right) \\ &= \sum_{i=1}^{H} \sum_{h(v)=i} O(m_v \log n_v + n_v \log^k n_v) + O\left(\frac{n^2}{r^2} + mr \log n \log^\omega r\right) \end{aligned}$$

where $H = O(\log r)$ is the height of $\mathcal{T}$ and $h(v)$ is the height of a node v of $\mathcal{T}$. As mentioned in Section 5.4, it was shown in [4] that

$$m_v \leq \frac{ck_v m}{r} + 1,$$

where c is some constant > 0 and k_v is the number of leaves in the subtree of $\mathcal{T}$ rooted at v. Moreover, we argued in Section 5.4 that

$$\sum_{h(v)=i} k_v = O(r^2) \quad \text{and} \quad \sum_{h(v)=i} n_v = n.$$

Therefore

$$\begin{aligned} T(m,n) &= \sum_{i=1}^{H} O\left(r^2 \cdot \frac{m}{r} \log n + n \log^k n\right) + O\left(\frac{n^2}{r^2} + mr \log n \log^\omega r\right) \\ &= O\left(mr \log n \log^\omega r + n \log^k n \log r + \frac{n^2}{r^2}\right). \end{aligned}$$

By choosing $r = \max\left\{\dfrac{n^{2/3}}{m^{1/3}\log^{1/3} m \log^{\omega/3} \frac{n}{\sqrt{m}}}, 2\right\}$, we obtain

Theorem 5.12 *Given a collection $\mathcal{G}$ of n segments, a function ψ_e associated with each $e \in \mathcal{G}$ with the properties listed above, and a set P of m points, we can compute $\sum_{e \in \mathcal{G}} \psi_e(p)$, for each $p \in P$, in time*

$$O(m^{2/3} n^{2/3} \log^{2/3} m \log^{2\omega/3} \frac{n}{\sqrt{m}} + n \log^k n \log \frac{n}{\sqrt{m}} + m \log n). \square$$

Remark 5.13:

(i) In several special cases it is possible to obtain $\mathcal{D}_v$, in $O(n_v)$ time, by merging $\mathcal{D}_w$ and $\mathcal{D}_z$. In such cases the second term of the above bound reduces to $O(n \log^k n)$.

(ii) As in Section 5.4 the running time can be improved to

$$O(m^{2/3}n^{2/3} \log^{2/3} n \log^{\omega/3} \frac{n}{\sqrt{m}} + n \log^k n \log \frac{n}{\sqrt{m}} + m \log n)$$

by partitioning Δ_i' into $\left\lceil \frac{n_i}{\sqrt{m/r} \log^{1/2} n} \right\rceil$ triangles none of which contains more than $\sqrt{\frac{m}{r}} \log^{1/2} r$ points of $\mathcal{L}^*$, and by choosing

$$r = \max\left\{ \frac{n^{2/3}}{m^{1/3} \log^{1/3} n \log^{2/3} \frac{n}{\sqrt{m}}}, 2 \right\}.$$

Various applications of the batched implicit point location problem have been discussed in [73]. The running time of all these applications can be improved by using the algorithm provided in Theorem 5.12. We briefly describe a couple of these application, and refer to [73] for more details.

5.7.1 Polygon containment problem — batched version

Consider the following problem:

> *Given a set* $\mathbf{T}$ *of* n *(possibly intersecting) triangles and a set* P *of* m *points, for every point* p *of* P*, count the number of triangles in* $\mathbf{T}$ *containing* p*, or more generally for each point* p*, report all triangles containing* p*.*

We review the solution technique of [73]. Let $\mathcal{G}$ be the set of the edges of triangles in $\mathbf{T}$, and let $\mathcal{L}$ be the set of lines containing the segments of $\mathcal{G}$. For each edge e of a triangle Δ, let $B(e)$ denote the semi-infinite trapezoidal strip lying directly below e. Define a function ψ_e in the plane so that $\psi_e(p) = 0$ if p lies outside $B(e)$, $\psi_e(p) = 1$ if p is in $B(e)$ and Δ lies below the line containing e, otherwise $\psi_e(p) = -1$.

It is easily seen that, for any point p, $\Psi(p)$ gives the number of triangles containing p, and ψ_e satisfies the properties (i)–(ii). As for property (iii), if a point p lies above all lines of $\mathcal{L}$, then $\Psi(p) = 0$, by definition. If p lies below all lines of $\mathcal{L}$, then we do the following. Let $\hat{e}$ denote the x-projection of an edge e of a triangle. It is easily checked that

$$\Psi(p) = \sum_{\{\epsilon_j : p \in \hat{e}_j\}} \epsilon_j,$$

where ϵ_j is the non-zero value of ψ_{e_j} at p. Note that the sum of the right hand side does not change between two consecutive endpoints of the projected segments, and that the value of Ψ over each interval can be computed in overall $O(n \log n)$ time, by scanning these projected segments from left to right. Hence, we can preprocess $\mathbf{T}$, in time $O(n \log n)$, into a data structure $\mathcal{D}$ so that, for a point p lying below all lines of $\mathcal{L}$, $\Psi(p)$ can be computed in $O(\log n)$ time. Moreover, for a node v in $\mathcal{T}$, $\mathcal{D}_v$ can be obtained in $O(n_v)$ time by merging $\mathcal{D}_w$ and $\mathcal{D}_z$, where w, z are the children of v. Hence, Theorem 5.12 and the remark following it imply that

Corollary 5.14 *Given a set* $\mathbf{T}$ *of* n *triangles and a set* P *of* m *points, we can compute, for each point* $p \in P$*, the number of triangles in* $\mathbf{T}$ *containing* p *in time*

$$O(m^{2/3} n^{2/3} \log^{2/3} m \log^{2\omega/3} \frac{n}{\sqrt{m}} + (m+n) \log n).$$

□

5.7.2 Implicit hidden surface removal — batched version

The next application of the implicit point location problem is the following version of hidden surface removal problem:

> *Given a collection of objects in 3-dimensional space, and a viewing point* a*, calculate the scene obtained by viewing these objects from* a.

The hidden surface removal problem has been extensively studied by many researchers (see e.g. [49], [94]), because of its applications in graphics and other areas. For the sake of simplicity let us restrict our attention to polyhedral objects, whose boundary $\mathbf{T}$ is a collection $\{\Delta_1, \ldots, \Delta_n\}$ of n non-intersecting triangles. In the case of *implicit* hidden surface removal problem, we do not want to compute the scene explicitly; instead we wish to determine the objects seen at given pixels [44], [73]. In this subsection, we consider the following special case of the implicit hidden surface removal problem. Let $\mathbf{T} = \{\Delta_1, \ldots, \Delta_n\}$ be a collection of n non-intersecting horizontal triangles in $\mathbb{R}^3$ such that Δ_i lies in the plane $z = c_i$, where $c_1 \leq c_2 \leq \cdots \leq c_n$ are some fixed heights. Let $P = \{p_1, \ldots, p_m\}$ be a set of m points lying in a horizontal plane below all triangles of $\mathbf{T}$. The problem is to determine, for each point $p \in P$, the lowest triangle Δ_i hit by the vertical line passing through p.

We review the techniques used by Guibas et al. [73]. A point $p \in P$ is said to be *blocked* by $\mathbf{T}$, if the vertical line from p intersects at least one triangle $\Delta_i \in \mathbf{T}$. First consider the following problem: Given a set $\mathbf{T}$ of n triangles and a set P of m points, determine which points of P are blocked by Δ. This problem can be solved by applying our implicit point location algorithm to P and the xy-projection of the triangles in $\mathbf{T}$. Hence, we can compute the blocked points in $O(m^{2/3}n^{2/3}\log^{2/3} m \log^{2\omega/3} \frac{n}{\sqrt{m}} + (m+n)\log n)$ time.

Going back to the original problem, if the number of the points or the number of the triangles is ≤ 1, then we solve the problem directly; otherwise we split $\mathbf{T}$ into two subsets $\mathbf{T}_1$, $\mathbf{T}_2$, so that $\mathbf{T}_1$ contains the lower half of the triangles $\Delta_1, \ldots, \Delta_{n/2}$ and $\mathbf{T}_2$ contains the upper half of the triangles $\Delta_{n/2+1}, \ldots, \Delta_n$. We apply the blocking algorithm to P and $\mathbf{T}_1$. Let $P_1 \subset P$ be the subset of points blocked by $\mathbf{T}_1$, and let $P_2 = P - P_1$. We recursively compute the lowest triangle in $\mathbf{T}_1$ (resp. in $\mathbf{T}_2$) above each of the points in P_1 (resp. P_2). Using the same analysis as in [73], we can show that the total running time is

$$O\left(m^{2/3}n^{2/3}\log^{2/3} m \log^{2\omega/3} \frac{n}{\sqrt{m}} + m\log n + n\log^2 n\right).$$

Hence, we can conclude

Theorem 5.15 *Given an ordered collection* $\mathbf{T}$ *of* n *triangles in* $\mathbb{R}^3$ *and a set* P *of* m *points lying below all of them, the triangle seen from each point* *of P in upward vertical direction can be determined in time*

$$O(m^{2/3}n^{2/3}\log^{2/3} m \log^{2\omega/3} \frac{n}{\sqrt{m}} + m\log n + n\log^2 n).$$

□

Remark 5.16:

(i) In fact this algorithm works for a more general case where triangles in $\mathbf{T}$ have the property that they can be linearly ordered so that if a vertical line hits two triangles Δ_i and Δ_j with Δ_i lying below Δ_j, then $\Delta_i < \Delta_j$.

(ii) We can extend the above algorithm to the case, where the points of P do not lie below all of the triangles in $\mathbf{T}$. Now at each level of recursion, for each point p of P_1, we also find the highest triangle Δ_p of $\mathbf{T}_1$ whose projection contains p. If Δ_p lies below p, then we remove p from P_1 and add it to P_2. Using the above algorithm we can find Δ_p, for each $p \in P_1$, in time

$$O(m^{2/3}n^{2/3}\log^{2/3} m \log^{2\omega/3} \frac{n}{\sqrt{m}} + m\log n + n\log^2 n).$$

Therefore the overall running time is

$$O(m^{2/3}n^{2/3}\log^{2/3} m \log^{2\omega/3} \frac{n}{\sqrt{m}} + m\log n + n\log^3 n).$$

5.8 Approximate Half-plane Range Queries

The half-plane range query problem is defined as: Given a set S of n points in the plane, preprocess it so that for any query line ℓ, we can quickly count the number of points in S lying above ℓ. In the dual setting, S becomes a set $S^\star$ of n lines, ℓ becomes a point $\ell^\star$, and the number of points lying above ℓ is same as the level of $\ell^\star$ in $\mathcal{A}(S^\star)$. Therefore, if we allow $O(n^2)$ space, the query can be obviously answered in time $O(\log n)$ by precomputing $\mathcal{A}(S^\star)$ and locating $\ell^\star$ in it. Chazelle and Welzl [32] recently gave an algorithm that answers a query in time $O(\sqrt{n}\log n)$ using only $O(n)$ space. A result of Chazelle [20] shows that if we restrict the space to be linear, the query takes at least $\Omega(\sqrt{n})$ time in the semigroup model (in particular subtraction is not allowed, see [20] for details), which implies that we cannot hope for a much better algorithm if we want to count the exact number of points. However, in several applications it suffices to count the number of points approximately (one such example is described by Matoušek [90]). Therefore, in the dual setting, the approximate half-plane range query problem is: Given a set $S^\star$ of n lines and a parameter (not necessarily a constant) $\epsilon > 0$, preprocess it so that for any query point, we can quickly compute an approximate level for it in $\mathcal{A}(S^\star)$, namely a level that lies within $\pm\epsilon n$ from the true level. It is easily seen that the problem can be reduced to an instance of point location problem in an $\frac{\epsilon n}{4}$-approximate leveling of $\mathcal{A}(S^\star)$ (see also [65], [89]). Hence by Theorem 4.36, we obtain

Theorem 5.17 *Given a set of n points in the plane and a positive real number $\epsilon < 1$, we can preprocess it in time $O(\frac{n}{\epsilon}\log n \log^\omega \frac{1}{\epsilon})$, into a data structure of size $O(\frac{1}{\epsilon^2})$, so that for any query line ℓ we can obtain, in $O(\log n)$ time, an approximate count of the number of points in S lying above ℓ, which deviates from the true number by at most $\pm\epsilon n$.*

□

5.9 Computing Spanning Trees with Low Stabbing Number

Let S be a set of n points in $\mathbb{R}^d$ and $\mathcal{T}$ a spanning tree on S whose edges are line segments. The *stabbing number* $\sigma(\mathcal{T})$ of $\mathcal{T}$ is the maximum number of edges of $\mathcal{T}$ that can be crossed by a hyperplane h. Chazelle and Welzl [32] (see also [132]) have proved that, for any set of n points in $\mathbb{R}^d$, there exists a spanning tree with stabbing number $O(n^{1-1/d})$, and that this bound is tight in the worst case. For a family $\mathbf{T}$ of trees, the stabbing number $\sigma(\mathbf{T})$ is s if for each hyperplane h there is a tree $\mathcal{T} \in \mathbf{T}$ such that h intersects at most s edges of $\mathcal{T}$.

Edelsbrunner et al. [55] gave a randomized algorithm with expected running time $O(n^{3/2}\log^2 n)$ for computing a family $\mathbf{T} = \{\mathcal{T}_1, \ldots, \mathcal{T}_k\}$ of $k = O(\log n)$ spanning trees with the property that, for any line ℓ, there exists at least one tree $\mathcal{T}_i$ such that ℓ intersects at most $O(\sqrt{n}\log^2 n)$ edges of $\mathcal{T}_i$. They also showed that a spanning tree on S with stabbing number $O(\sqrt{n})$ can be deterministically constructed in time $O(n^3 \log n)$. Recently Matoušek [90] has improved the running time of these algorithms. He has given a randomized algorithm with expected running time $O(n^{4/3}\log^2 n)$ to construct a family of $O(\log n)$ spanning trees with the above property; this algorithm can be converted into a deterministic one with $O(n^{7/4}\log^2 n)$ running time. He has also given an $O(n^{5/2}\log n)$ deterministic algorithm (or a randomized algorithm with expected running time $O(n^{1.4+\delta})$, for any $\delta > 0$) to construct a single spanning tree with stabbing number $O(\sqrt{n}\log n)$. His algorithms actually compute spanning paths of S.

In this section we describe a deterministic algorithm for constructing a family $\mathbf{T}$ of $O(\log n)$ spanning trees with $\sigma(\mathbf{T}) = O(\sqrt{n})$. The crux of Matoušek's algorithms lies in the following lemma.

Lemma 5.18 (Matoušek [90]) *Given a set S of n points in the plane, we can find a set $\mathcal{L}$ of $O(n)$ lines with the property that, for any spanning path $\mathcal{T}$ on S, and for every line ℓ, there is a line $\ell' \in \mathcal{L}$ such that if ℓ' intersects s edges of $\mathcal{T}$, then ℓ intersects at most $s + O(\sqrt{n}\log n)$ edges of $\mathcal{T}$.*

□

Matoušek describes an $O(n^{7/4}\log^2 n)$ deterministic algorithm for computing this set of lines. Using Theorem 4.36 we can strengthen Lemma 5.18 as follows:

Lemma 5.19 *Given a set S of n points in the plane, we can deterministically construct a set $\mathcal{L}$ of $O(n)$ lines in time $O(n^{3/2}\log^{\omega+1} n)$ with the property that, for any spanning path $\mathcal{T}$ on S, and for every line ℓ, there is a line $\ell' \in \mathcal{L}$ such that if ℓ' intersects s edges of $\mathcal{T}$, then ℓ intersects at most $s + O(\sqrt{n})$ edges of $\mathcal{T}$.*

Proof: We dualize the points of S to obtain a set $S^\star$ of n lines. By Theorem 4.36, choosing $r = \sqrt{n}$, we can partition the plane into $O(n)$ triangles in time $O(n^{3/2}\log^{\omega+1} n)$, so that no triangle meets more than $O(\sqrt{n})$ lines of S. We pick up a point $\ell^\star$ from each triangle, let $\mathcal{L}^\star$ denote the set of these points, and let $\mathcal{L}$ be the set of their dual lines in the primal plane.

Arguing as in [90], let ζ be an arbitrary line in the primal plane. By construction, there exists a line $\ell \in \mathcal{L}$ such that the segment $e = \overline{\zeta^\star \ell^\star}$ does not cross more than $O(\sqrt{n})$ lines of $S^\star$. Going back to the primal plane, if an edge g of $\mathcal{T}$ intersects ζ but not ℓ, then one endpoint of g must lie in the double wedge $e^\star$ dual to e, but our construction implies that $e^\star$ contains at most $O(\sqrt{n})$ points of S. Thus, there are at most $O(\sqrt{n})$ edges of $\mathcal{T}$ that intersect ζ but not ℓ, and the lemma follows.

□

We construct a family of $O(\log n)$ spanning paths with low stabbing number only for the lines in $\mathcal{L}$. Although the basic approach is the same as in [90] or [55], we need some additional techniques to improve the running time. Here we briefly sketch the main idea, and refer the reader to [90] or [55] for more details.

Suppose we have constructed $\mathcal{T}_1, \ldots, \mathcal{T}_{i-1}$, and have obtained a set $\mathcal{L}_i \subset \mathcal{L}$ such that $m_i = |\mathcal{L}_i| \leq \frac{m}{2^{i-1}}$ (where $m = |\mathcal{L}| = O(n)$) and $\mathcal{L}_i$ is "bad" for all paths constructed so far, that is, a line in $\mathcal{L}_i$ intersects every tree at more than $C\sqrt{n}$ edges, for some constant C to be specified below. We now show how to construct $\mathcal{T}_i$ and $\mathcal{L}_{i+1}$. Initially $\mathcal{L}_1 = \mathcal{L}$.

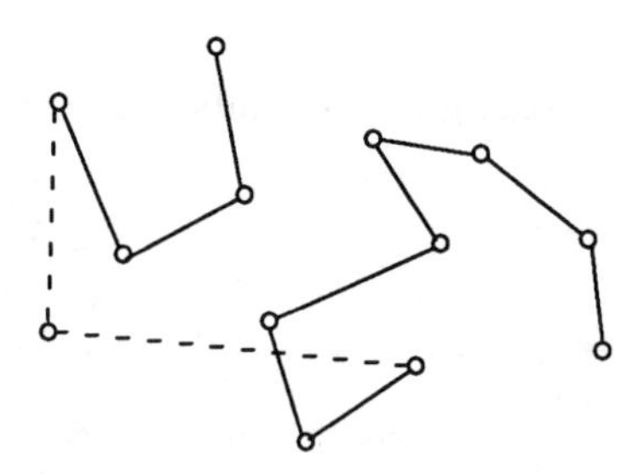

Figure 5.10: Spanning path $\mathcal{T}_i$; connecting the endpoints of Π_j

The spanning path $\mathcal{T}_i$ is constructed in $O(\log n)$ phases. In the beginning of the jth phase we have a current collection Π_j of n_j vertex-disjoint paths on S (in the beginning of the first phase the collection Π_1 consists of all singleton paths on the points of S). Our algorithm ensures that $n_j \leq n \cdot \left(\frac{2}{3}\right)^{j-1}$. If

$n_j \leq \sqrt{n}$, we connect the endpoints of the paths in Π_j to form a single spanning path on S, and we are done (see figure 5.10). Otherwise, if $n_j > \sqrt{n}$, we proceed as follows. Choose $r = c_1\sqrt{n_j}$ and partition the plane into $n_j/3$ triangles so that no triangle meets more than $c_2\frac{m_i}{\sqrt{n_j}}$ lines of $\mathcal{L}_i$, for appropriate constants c_1, c_2, which exist by Theorem 4.36. If a triangle contains endpoints of several paths in Π_j, we obtain a maximal matching of these endpoints and connect each pair of matched points by an edge (see figure 5.11), thereby combining two paths in Π_j into a new path. To avoid creating cycles, we only choose one endpoint of each path of Π_j. The resulting paths form the set Π_{j+1}. It can be easily proved that we add at least $\frac{n_j}{3}$ new edges to the current set of paths, which implies that $n_{j+1} \leq \frac{2n_j}{3}$.

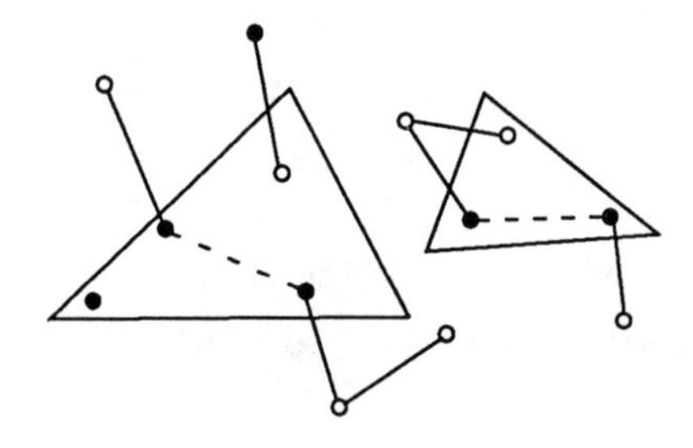

Figure 5.11: Maximal matching of endpoints within Δ_i; bullets denote the selected endpoints of Π_j

Lemma 5.20 *There are at least $\frac{m_i}{2}$ lines of $\mathcal{L}_i$ that intersect T_i in $\leq C\sqrt{n}$ edges, for some constant $C > 0$.*

Proof: We bound the total number of intersection points between edges of T_i and $\mathcal{L}_i$. In the jth phase we add at least $n_j - n_{j+1}$ edges, and each edge intersects at most $c_2\frac{m_i}{\sqrt{n_j}}$ lines of $\mathcal{L}_i$. In the final phase we add at most $\sqrt{n}$ edges, each crossed by at most m_i lines. Therefore the total number of intersections I between T_i and $\mathcal{L}_i$ is at most

$$I \leq \sum_{j=1}^{O(\log n)} (n_j - n_{j+1})c_2\frac{m_i}{\sqrt{n_j}} + \sqrt{n}\cdot m_i$$

$$\begin{aligned}
&\le c_2 m_i \sum_{j=1}^{O(\log n)} \sqrt{n_j} + \sqrt{n} m_i \\
&\le c_2 m_i \sqrt{n} \sum_{j=1}^{O(\log n)} \left(\frac{2}{3}\right)^{\frac{j-1}{2}} + \sqrt{n} m_i \\
&\qquad (\text{because } n_j \le n \cdot \left(\frac{2}{3}\right)^{j-1}) \\
&\le \frac{c_2 m_i \sqrt{n}}{1 - \sqrt{2/3}} + \sqrt{n} m_i \\
&= ((3+\sqrt{6})c_2 + 1) \cdot m_i \sqrt{n}.
\end{aligned}$$

Now it follows immediately that at least half of the lines in $\mathcal{L}_i$ intersect $\mathcal{T}_i$ in at most $C\sqrt{n}$ edges, for $C = (6 + 2\sqrt{6})c_2 + 2$.

□

Lemma 5.20 implies that at most half of the lines are "bad". For every line $\ell \in \mathcal{L}_i$, we count the number of intersections between ℓ and $\mathcal{T}_i$, using our red-blue intersection algorithm given in Section 5.6. We pick up those lines of $\mathcal{L}_i$ that intersect $\mathcal{T}_i$ at more than $C\sqrt{n}$ points. The resulting set is $\mathcal{L}_{i+1}$.

Next we analyze the running time of our algorithm. We first bound the time to compute $\mathcal{L}_i$, for $i \le k$. Since $m_i \le n$ and there are only n edges in $\mathcal{T}_i$, it follows from Theorem 5.10 that we can compute $\mathcal{L}_i$ in $O(n^{4/3} \log^{(\omega+2)/3} n)$ time. Moreover $k = O(\log n)$, so the total time spent in computing the incidences between $\mathcal{T}_i$ and $\mathcal{L}_i$, over all k phases, is at most $O(n^{4/3} \log^{(\omega+5)/3} n)$.

As for the time spent in computing $\mathcal{T}_i$, we choose $r = c_1\sqrt{n_j}$ in the jth phase, therefore it requires

$$O(m_i \sqrt{n_j} \log m_i \log^\omega n_j + n_j \log n_j)$$

time. (It is easily checked that this also bounds the time needed to distribute the path endpoints among the triangles, and to match them to obtain the new set of paths.) Hence the total time spent in computing $\mathcal{T}_i$ is at most

$$\sum_{j=1}^{O(\log n)} O\left(m_i \sqrt{n} \cdot \left(\frac{2}{3}\right)^{(j-1)/2} \log m_i \log^\omega n + n_j \log n_j\right)$$

$$
\begin{aligned}
&= O\left(m_i\sqrt{n}\cdot\left(\frac{2}{3}\right)^{(j-1)/2}\log m_i \log^\omega n + \left(\frac{2}{3}\right)^{j-1} n\log n\right) \\
&= O(m_i\sqrt{n}\log m_i \log^\omega n + n\log n).
\end{aligned}
$$

Summing over all i, we obtain

$$
\begin{aligned}
& \sum_{i=1}^{k} O\left(m_i\sqrt{n}\log m_i \log^\omega n + n\log n\right) \\
&= \sum_{i=1}^{k} O\left(m_i\sqrt{n}\log^{\omega+1} n + n\log n\right) \\
&= O\left(\sum_{i=1}^{k} \frac{n}{2^{i-1}}\sqrt{n}\log^{\omega+1} n\right) \\
&\quad (\text{because } m_i \le \frac{m}{2^{i-1}} \text{ and } m = O(n)) \\
&= O\left(n^{3/2}\log^{\omega+1} n\right).
\end{aligned}
$$

Hence, we have

Theorem 5.21 *Given a set S of n points in the plane, we can deterministically construct, in time $O(n^{3/2}\log^{\omega+1} n)$, a family $\mathbf{T}$ of $k = O(\log n)$ spanning paths on S with the property that, for any line ℓ, there exists a path $T \in \mathbf{T}$, such that ℓ intersects at most $O(\sqrt{n})$ edges of T.*

□

Moreover, we have

Lemma 5.22 *The set of $O(\log n)$ spanning paths computed by the above algorithm have the property that, for any query line ℓ, we can determine, in $O(\log n)$ time, a path that ℓ intersects in at most $O(\sqrt{n})$ edges. This requires an additional linear preprocessing time and storage.*

Proof: Let $\mathbf{\Delta}$ be the set of triangles computed in the proof of Lemma 5.19. Suppose the dual of $\ell^\star$ lies in $\Delta_k \in \mathbf{\Delta}$, and let $\ell_k^\star$ be the point selected from Δ_k. Then ℓ_k is a good line for at least one path $\mathcal{T}_i$, i.e. it meets at most $O(\sqrt{n})$ edges of $\mathcal{T}_i$. By Lemma 5.19, ℓ also meets only $O(\sqrt{n})$ edges of that path. Moreover, for any given ℓ, we can find ℓ_k (and thus the corresponding path $\mathcal{T}_i$) in $O(\log n)$ time, using an efficient point location algorithm; since the map formed by Δ has only $O(n)$ faces, linear preprocessing time and storage suffices [60]). Hence, the lemma follows.

□

Remark 5.23:

(i) Note that the best known deterministic algorithm for constructing a single spanning path with $O(\sqrt{n}\log n)$ stabbing number has time complexity of $O(n^{5/2}\log^2 n)$. Therefore it follows from Theorem 5.21 and Lemma 5.22 that the multi-tree structure is better than the single path structure for all purposes except that the storage requirement is worse by a factor of $O(\log n)$. In some applications, however, it may not be possible to use a multi-tree structure (e.g. the reporting version of the simplex range searching [32] and also the counting version if subtraction is not allowed).

(ii) The spanning path obtained by our algorithm may have intersecting edges. However, if the application requires the paths to be non self-intersecting, we can apply a technique of [55] that converts a polygonal path $\mathcal{T}$ with n edges into another, non self-intersecting path $\mathcal{T}'$, in time $O(n\log n)$, with the property that a line intersects $\mathcal{T}'$ in at most twice as many edges as it intersects $\mathcal{T}$.

(iii) If we use the randomized version of our red-blue intersection algorithm to count the intersections between the edges of $\mathcal{T}_i$ and $\mathcal{L}$, in Matoušek's randomized algorithm [90] for constructing $\mathbf{T}$, then $\sigma(\mathbf{T})$ can be improved to $O(\sqrt{n}\log n)$ without increasing the time complexity of his algorithm.

Chazelle and Welzl [32] have shown that spanning trees with low stabbing number can be used to develop an almost optimal algorithm for simplex range queries. Other applications of spanning trees with low stabbing number include computing a face in an arrangement of lines (cf. [55]), ray shooting in non-simple polygons (cf. Chapter 6) and implicit point location (cf. Chapter 6). Our algorithm improves the preprocessing time as well as query time of most of these applications. For example, Edelsbrunner et al. [55] have shown that given a set $\mathcal{L}$ of n lines, it can be preprocessed in $O(n^{3/2}\log^2 n)$ (randomized expected) time, into a data structure of size $O(n\log n)$, using a family $\mathbf{T}$ of $O(\log n)$ spanning trees with $\sigma(T) = s$, so that for a query point p, the face in $\mathcal{A}(\mathcal{L})$ containing p can be computed in time $O(s\log^2 n + K)$, where K is the number of edges bounding the desired face. The result of Matoušek [90] implies that the preprocessing can be done deterministically in $O(n^{7/4}\log^2 n)$ time. However, if we use our algorithm for constructing the spanning trees, we obtain

Corollary 5.24 *Given a set $\mathcal{L}$ of n lines, we can preprocess it deterministically in $O(n^{3/2}\log^{\omega+1} n)$ time into a data structure of size $O(n\log n)$ so that, for a query point p, we can compute the face in $\mathcal{A}(\mathcal{L})$ containing the point p in $O(\sqrt{n}\log^3 n + K)$ time, where K is the number of edges bounding the desired face.*

□

Another result of [55], combined with our algorithm, implies that

Corollary 5.25 *Given a set $\mathcal{L}$ of n lines, we can preprocess it deterministically, in $O(n^{3/2}\log^{\omega+1} n)$ time, into a data structure of size $O(n\log^2 n)$ so that, for any ray ρ emanating from a point p in direction d, we can compute, in time $O(\sqrt{n}\log n)$, the intersection point between ρ and the lines of $\mathcal{L}$, that lies nearest to p.*

□

Similarly using the result of [32][2], we obtain

Corollary 5.26 *Given a set S of n points in the plane, we can preprocess it deterministically, in $O(n^{3/2}\log^{\omega+1} n)$ time into a data structure of size $O(n\log n)$ so that, for a query line ℓ, we can compute the number of points of S lying above ℓ in $O(\sqrt{n}\log n)$ time.*

□

Remark 5.27: Recently Matoušek and Welzl [93] gave an alternative deterministic algorithm to perform such half-plane range queries. Their algorithm has the same storage and query time bounds, and its preprocessing time is only $O(n^{3/2}\log n)$.

[2]The half-plane range searching algorithm of Chazelle and Welzl uses a single spanning tree, but it works even if we use a family of $O(\log n)$ spanning trees instead of a single tree structure, though the space complexity rises to $O(n\log n)$.

5.10 Space Query-time Tradeoff in Triangle Range Search

Finally we consider the following problem:

> *Given a set S of n points in the plane, preprocess S so that for a query triangle Δ, we can quickly count the number of points of S lying in Δ.*

As just noted, the problem has been solved by Chazelle and Welzl [32], using a spanning tree with low stabbing number, in $O(n)$ space and $O(\sqrt{n}\log n)$ query time. In this section we study the issue of tradeoff between the allowed space and query time. Chazelle [20] has proved that if we allow $O(m)$ space, then the query time is at least $\Omega(\frac{n}{\sqrt{m}})$. (However, this lower bound applies to an arithmetic model involving operations in a semigroup; in particular no subtractions are allowed.) For $m = n^2$, a query can be easily answered in $O(\log n)$ time, so the interesting case is when $n < m < n^2$. In this section we show that our partitioning algorithm in conjunction with Chazelle and Welzl's technique yields an algorithm that counts the number of points lying in a query triangle in $O(\frac{n}{\sqrt{m}}\log^{3/2} n)$ time using $O(m)$ space, where $n^{1+\epsilon_0} \le m \le n^{2-\epsilon_1}$, for some constants $\epsilon_0, \epsilon_1 > 0$. The preprocessing time of our algorithm is bounded by $O(n\sqrt{m}\log^{\omega+1/2} n)$, which is faster than that of any previously known algorithm.

We first establish this tradeoff for the half-plane range search problem: "Given a set S of n points, preprocess S so that, for any query line ℓ, we can quickly count the number of points lying below ℓ."

We dualize S to a set of n lines, S^*, and partition the plane into $M = O(r^2)$ triangles $\Delta_1, \ldots, \Delta_M$ so that no triangle meets more than $O(\frac{n}{r})$ lines of S^*. The dual of a triangle Δ is a 3-corridor, namely the region lying between the upper and lower envelopes of the three lines dual to the vertices of Δ (see [78] and figure 5.12). Let Δ_i^* denote the dual of Δ_i. A line ℓ fully lies in Δ_i^* if and only if ℓ^* lies in Δ_i, and a point p is in Δ_i^* if and only if p^* meets Δ_i. Let $S_i \subset S$ denote the points of S contained in Δ_i^*; by construction $|S_i^*| = O(\frac{n}{r})$. For each Δ_i^*, we construct a family $\mathbf{T}^i$ of $O(\log \frac{n}{r})$ spanning paths on the set S_i with the property that, for every line ℓ there exists a path $T_j^i \in \mathbf{T}^i$, such that ℓ intersects at most $O(\sqrt{\frac{n}{r}})$ edges of T_j^i

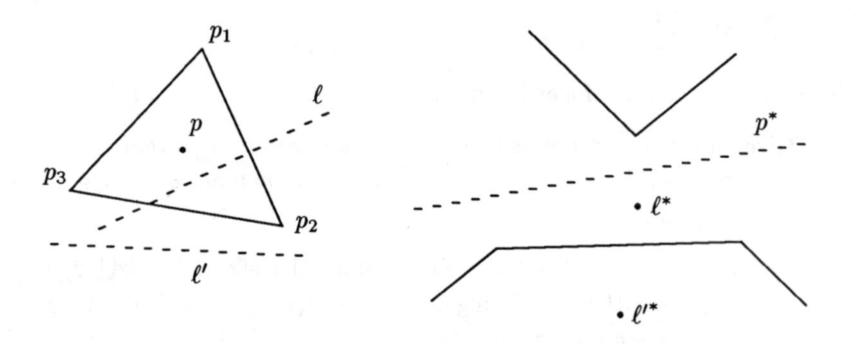

Figure 5.12: A triangle $\triangle$ and its dual $\triangle^*$

(see Section 5.9). We preprocess every $\mathcal{T}_j^i \in \mathbf{T}^i$ into a data structure of size $O(\frac{n}{r})$ for half plane range searching, as described in [32], so that a query can be answered in $O(\sqrt{\frac{n}{r}} \log \frac{n}{r})$ time.

To answer a query, we first find the 3-corridor $\triangle_i^*$ containing the query line ℓ. That is, we locate the triangle $\triangle_i$ containing the dual point ℓ^*. Let Φ_i denote the number of points in $S - S_i$ lying below ℓ, which we will have precomputed for each i. We thus only need to count the number of points of S_i lying below ℓ. By Lemma 5.22, we can find, in $O(\log \frac{n}{r})$ time, a path $\mathcal{T}_j^i \in \mathbf{T}^i$ that intersects ℓ in at most $O(\sqrt{\frac{n}{r}})$ edges. Moreover, the number of points of S_i lying below ℓ can be counted in $O(\sqrt{\frac{n}{r}} \log \frac{n}{r})$ time using $\mathcal{T}_j^i$, as in Corollary 5.26. Hence, the total query time is bounded by $O(\sqrt{\frac{n}{r}} \log \frac{n}{r})$. Since each $\mathbf{T}^i$ requires $O(\frac{n}{r} \log \frac{n}{r})$ space, the total space used is at most $O(nr \log \frac{n}{r})$. We choose $r = \dfrac{m}{n \log \frac{n}{\sqrt{m}}}$ to achieve $O(m)$ space, and the query time is therefore $O(\frac{n}{\sqrt{m}} \log^{3/2} \frac{n}{\sqrt{m}} + \log n)$.

As for the preprocessing time, we spend $O(nr \log n \log^{\omega} r)$ time in partitioning the plane into M triangles. Let $W_i \subset S^*$ denote the set of lines lying below the triangle $\triangle_i$, so $\Phi_i = |W_i|$. It is easily seen that for two adjacent triangles $\triangle_i, \triangle_j$, $W_i - W_j \subset S_i^* \cup S_j^*$. Therefore, Φ_i for each $\triangle_i$, can be computed in time $O(nr)$, spending $O(n)$ time at the first triangle plus $O(\frac{n}{r})$ time at each of the remaining triangle. We can compute Φ_i, for each triangle Δ_i, in $O(nr \log n)$ time, and, by Theorem 5.21, we can construct $\mathbf{T}^i$ in $O\left((\frac{n}{r})^{3/2} \log^{\omega+1} \frac{n}{r}\right)$ time. It follows from [32] that $\mathcal{T}_j^i$ can be preprocessed in $O(\frac{n}{r} \log \frac{n}{r})$ time for answering half-plane range searching. Therefore the total time spent in preprocessing is at most

$$\begin{aligned}
P(n) &= O(nr \log n \log^{\omega} r) + O\left(\left(\frac{n}{r}\right)^{3/2} \cdot r^2 \log^{\omega+1} \frac{n}{r}\right) \\
&= O\left(n \frac{m}{n \log \frac{n}{\sqrt{m}}} \log n \log^{\omega} r + n^{3/2} \sqrt{\frac{m}{n \log \frac{n}{\sqrt{m}}}} \log^{\omega+1} \frac{n}{\sqrt{m}}\right) \\
&= O\left(m \log^{\omega+1} n + n\sqrt{m} \log^{\omega+1/2} \frac{n}{\sqrt{m}}\right).
\end{aligned}$$

By Chazelle's lower bound mentioned above, we obtain

Theorem 5.28 *Given a set S of n points in the plane and $n \log n \le m \le n^2$ storage, we can preprocess S, in $O(m \log^{\omega+1} n + n\sqrt{m} \log^{\omega+1/2} \frac{n}{\sqrt{m}})$ time, so that for any query line ℓ, we can count the number of points of S lying below ℓ in time $O(\frac{n}{\sqrt{m}} \log^{3/2} \frac{n}{\sqrt{m}} + \log n)$, using $O(m)$ space. This is optimal up to a polylog factor.*

□

Remark 5.29:

(i) Matoušek's original algorithm [89] can also be used to obtain the same tradeoff. However, since we use large values of r, our preprocessing is faster than that obtainable by Matoušek's algorithm. We have recently learnt that Chazelle [21] has also independently obtained a similar result.

(ii) We can reduce a $\sqrt{\log \frac{n}{\sqrt{m}}}$ factor in the query time, if we compute a single spanning path instead of $O(\log n)$ paths. But then the (deterministic) time complexity of computing one such path rises to $O(\frac{n^3}{r^3} \log \frac{n}{r})$.

(iii) Notice that the counting version of the half-plane range query problem is more difficult than the reporting version; for the latter version,

Chazelle et al. [30] have given an $O(\log n + K)$ algorithm to report all K points lying below the query line, using only $O(n)$ space and $O(n \log n)$ preprocessing.

Next, we extend the above algorithm to obtain a similar tradeoff for the *slanted range search* problem: "Given a set S of n points, preprocess S so that, for a query segment e, we can count efficiently the number of points that lie in the semi-infinite trapezoidal strip lying directly below e." Let us denote the number of such points by $\Psi(e)$ (see figure 5.13).

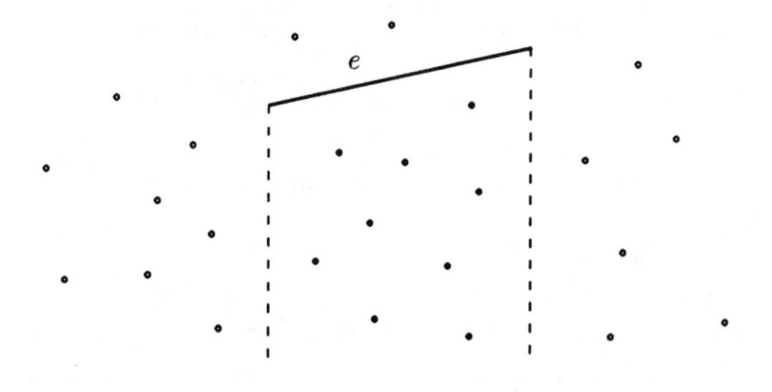

Figure 5.13: Instance of a slanted range searching; $\Psi(e) = 9$

Chazelle and Guibas [29] have given an optimal algorithm for the reporting version of the slanted range query problem, which reports all K such points in $O(\log n + K)$ time using $O(n)$ space and $O(n \log n)$ preprocessing. Since the half plane range search problem is a special case of the slanted range search problem, the lower bound on the query time for the slanted range search problem, with $O(m)$ storage, is also $\Omega(\frac{n}{\sqrt{m}})$. Our tradeoff is obtained as follows.

Construct a binary tree $\mathcal{B}$ on the x-projections of the points in S as follows. Sort the points of S in increasing x order. Decompose the sorted set into $\frac{n}{c}$ blocks, each containing at most c points, for some fixed constant $c > 0$, and associate each block with a leaf of $\mathcal{B}$. Each node v of $\mathcal{B}$ is thus associated with the set $S_v \subseteq S$ of points stored in the leaves of the subtree of $\mathcal{B}$ rooted at v. For each node v of $\mathcal{B}$ we preprocess the points in S_v for answering half-plane range queries, using the above algorithm, with $r = r_i$, where r_i is a parameter depending on the level i of v in $\mathcal{B}$. A segment e is called a *canonical segment* if there is a node $v \in \mathcal{B}$ such that the x-projection of e covers the x-projections of all the points in S_v, and of no other point in $S - S_v$. Observe that, for a canonical segment e, $\Psi(e)$ can be computed by solving a half-plane range query at the corresponding node. In general, a query segment e can be decomposed into $k \leq 2\log n$ canonical subsegments $e_1, \ldots, e_k$, such that at most two of them correspond to nodes at the same level of $\mathcal{B}$ (cf. [107]). Thus $\Psi(e) = \sum_{i=1}^{k} \Psi(e_i)$, which implies that $\Psi(e)$ can be computed by answering at most $2\log n$ half-plane range queries.

Since the nodes of the same level are associated with pairwise disjoint sets of points, and we are choosing the same value of r for all nodes of the same level, the space $s(n)$ used by our algorithm is

$$s(n) = O(\sum_{i=1}^{\log n} nr_i \log n).$$

Let $m = n^\gamma$, where $1 + \epsilon_0 \leq \gamma \leq 2 - \epsilon_1$, for some constants $\epsilon_0, \epsilon_1 > 0$. If we choose $r_i = \dfrac{n_i^{\gamma-1}}{\log n}$, where $n_i = \dfrac{n}{2^{i-1}}$ is the size of each set S_v at level i, we have

$$\begin{aligned} s(n) &= O\left(n \log n \sum_{i=1}^{\log n} \frac{n_i^{\gamma-1}}{\log n}\right) \\ &= O\left(n \sum_{i=1}^{\log n} \left(\frac{n}{2^{i-1}}\right)^{\gamma-1}\right) \\ &= O(n^\gamma) \qquad\qquad (\text{because } \gamma > 1 + \epsilon_0) \\ &= O(m). \end{aligned}$$

Next, the total time spent in answering a query is

$$\begin{aligned} Q(n) &= O\left(\sum_{i=1}^{\log n} \sqrt{\frac{n_i}{r_i}} \log n\right) \\ &= O\left(\sum_{i=1}^{\log n} \sqrt{n_i^{2-\gamma}} \log^{3/2} n\right) \end{aligned}$$

$$\begin{aligned} &= O\left(n^{1-\gamma/2}\log^{3/2} n\right) \quad \text{(because } \gamma \le 2-\epsilon_1) \\ &= O\left(\frac{n}{\sqrt{m}}\log^{3/2} n\right). \end{aligned}$$

As for the preprocessing time, we spend $O((n_i r_i + n_i^{3/2}\sqrt{r_i})\log^{\omega+1} n)$ at a node of the ith level. Since there are 2^i nodes at level i, the overall preprocessing time is bounded by

$$\begin{aligned} P(n) &= \sum_{i=0}^{\log n} O\left(2^i(n_i r_i + n_i^{3/2}\sqrt{r_i}) \cdot \log^{\omega+1} n\right) \\ &= \sum_{i=1}^{\log n} O\left((n \cdot r_i + \frac{n^{3/2}}{2^{i/2}} \cdot \sqrt{r_i}) \cdot \log^{\omega+1} n\right) \\ &= \sum_{i=1}^{\log n} O\left(\left(n\frac{n_i^{\gamma-1}}{\log n} + \frac{n^{3/2}}{2^{i/2}}\sqrt{\frac{n_i^{\gamma-1}}{\log n}}\right)\log^{\omega+1} n\right) \\ &= O\left(\left(\frac{n^\gamma}{\log n} + \frac{n^{1+\gamma/2}}{\sqrt{\log n}}\right)\log^{\omega+1} n\right) \\ &= O(n\sqrt{m}\log^{\omega+1/2} n) \qquad \text{(because } m \le n^2). \end{aligned}$$

Therefore, we have

Theorem 5.30 *Given a set S of n points in the plane, and $n^{1+\epsilon_0} \le m \le n^{2-\epsilon_1}$, for some $\epsilon_0, \epsilon_1 > 0$, we can preprocess S, in $O(n\sqrt{m}\log^{\omega+1/2} n)$ time, into a data structure of size $O(m)$ so that, for a query segment e, $\Psi(e)$ can be computed in $O(\frac{n}{\sqrt{m}}\log^{3/2} n)$ time. This is optimal up to a polylog factor.* □

Remark 5.31:

(i) The remarks following Theorem 5.28 apply here as well.

(ii) If $n\log^2 n \le m < n^{1+\epsilon}$, for all $\epsilon > 0$, then $Q(n) = O(\frac{n}{\sqrt{m}}\log^2 n)$. Similarly, if $m > n^{2-\epsilon}$ for all $\epsilon > 0$, then a more careful analysis shows that $Q(n) = O(\frac{n}{\sqrt{m}}\log^{5/2}\frac{n}{\sqrt{m}} + \log n)$.

Finally, we show how to solve the simplex range query problem, using Theorem 5.30. Let $\triangle$ denote a triangle with vertices p_1, p_2 and p_3. Assume

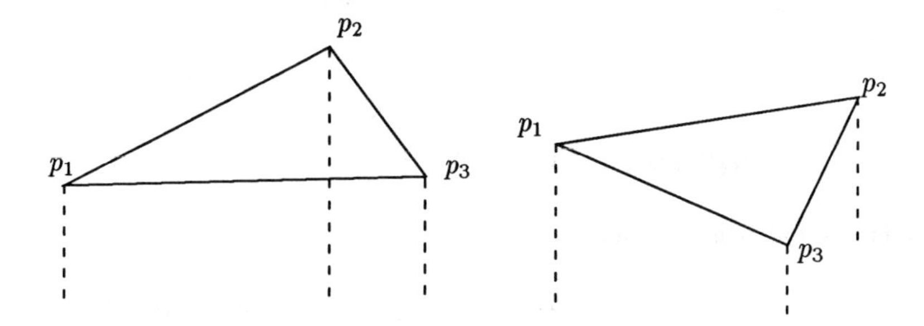

Figure 5.14: Two types of triangles

that p_1 is the leftmost vertex and $\overline{p_1p_2}$ lies above $\overline{p_1p_3}$ (see figure 5.14). If $x(p_2) \leq x(p_3)$, then the number of points in Δ is

$$\Psi(\overline{p_1p_2}) + \Psi(\overline{p_2p_3}) - \Psi(\overline{p_1p_3}),$$

and if $x(p_2) > x(p_3)$, then the number is

$$\Psi(\overline{p_1p_2}) - \Psi(\overline{p_1p_3}) - \Psi(\overline{p_2p_3})).$$

It thus follows from Theorem 5.30 that

Theorem 5.32 *Given a set S of n points in the plane and $n^{1+\epsilon_0} \leq m \leq n^{2-\epsilon_1}$, for some constants $\epsilon_0, \epsilon_1 > 0$, we can preprocess S, in time $O(n\sqrt{m} \cdot \log^{\omega+1/2} n)$, into a data structure of size $O(m)$ so that, for a query triangle Δ, we can count the number of points contained in Δ in $O(\frac{n}{\sqrt{m}} \log^{3/2} n)$ time.*

Remark 5.33:

(i) If $n \log^2 n \leq m < n^{1+\epsilon}$, for any $\epsilon > 0$, then the query time becomes $O(\frac{n}{\sqrt{m}} \log^2 n)$. Similarly, if $m > n^{2-\epsilon}$ for all $\epsilon > 0$, then $Q(n) = O(\frac{n}{\sqrt{m}} \log^{5/2} \frac{n}{\sqrt{m}} + \log n)$.

(ii) Notice that we use subtraction to count the number of points lying inside a triangle. It is not known whether Chazelle's lower bound [20] can be extended to the case where we use subtraction, that is to the group model. Therefore, we do not know how sharp our bounds are in that model.

5.11 Overlapping Planar Maps

Consider the following problem:

Let P and Q be two piecewise linear planar maps. Let $F_P(x,y)$ (resp. $F_Q(x,y)$) be a bivariate function associated with P (resp. Q). We wish to compute

$$\min_{x,y}\{F_P(x,y) - F_Q(x,y)\}.$$

Let $G(x,y) = F_P(x,y) - F_Q(x,y)$. In this section we give an efficient algorithm to compute an optimal point in a special case, where F_P and F_Q satisfy the following conditions:

(i) In the ith face of P (resp. Q), F_P (resp. F_Q) is given by a function f_P^i (resp. f_Q^i), which can be computed in $O(1)$ time for any given point. We also assume that F_P and F_Q are continuous.

(ii) There exists a vertex ν of $P \cap Q$, such that

$$G(\nu) = \min_{x,y} G(x,y).$$

A vertex of $P \cap Q$ is either a vertex of P or Q, or an intersection point between an edge of P and an edge of Q.

(iii) For any edge γ of P, the function $G(x,y)$ is unimodal along γ. (The function F along an edge of P is achieved by a fixed function f_P^i.)

Remark 5.34: Notice that we do not make a similar unimodality assumption along the edges of Q. In the more special case, where $G(x,y)$ satisfies the above properties, and it is also unimodal along each edge of Q, then an optimal point can be computed in $O(n \log n)$ time, using the approach of [24] (see also [25]).

Let n denote the total number of edges in P and Q. Let v_P (resp. v_Q) denote a vertex of P (resp. Q) that minimizes $G(x,y)$ over all vertices of P (resp. Q). Such vertices can be easily computed in $O(n \log n)$ time, using efficient planar point location algorithms (see [60], [111]). Therefore it suffices to show how to find quickly an intersection point between an edge of P and an edge of Q, which minimizes $G(x,y)$ over all such intersection points. This problem can be reformulated as follows. Consider the edges of P (resp. Q) as a collection Γ_r (resp. Γ_b) of red (resp. blue) segments. Let S denote the set of all red-blue intersection points induced between Γ_r and Γ_b. The goal is to compute a point Ψ such that

$$G(\Psi) = \min_{\xi \in S} G(\xi).$$

A naive approach is to compute the set S using the reporting version of our red-blue intersection algorithm (cf. Section 5.6), and to choose the point that minimizes $G(x,y)$ over the entire set S. But $|S| = \Omega(n^2)$ in the worst case, so the aim is to come up with a subquadratic algorithm for this problem. In this section we show that a variant of our red-blue intersection algorithm (cf. Section 5.6) yields an $O(n^{4/3} \log^{(\omega+2)/3} n)$ algorithm for this problem. We first consider a restricted version of the problem. Let Γ_r and Γ_b be the two sets of segments, all meeting the interior of a triangle $\triangle$, such that m of these segments contain at least one endpoint inside $\triangle$; we wish to compute a point $\overline{\Psi}$ such that

$$G(\overline{\Psi}) = \min_{\xi \in S \cap \triangle} G(\xi).$$

For any $X \subseteq \Gamma_r$, $Y \subseteq \Gamma_b$, we use $S_{XY} \subseteq S$ to denote the set of red-blue intersections between X and Y, and let $G(\Psi_{X,Y}) = \min_{\xi \in S_{XY}} G(\xi)$; similarly define $\overline{\Psi}_{XY}$. If $X = \{\gamma\}$, then we denote Ψ_{XY} (resp. $\overline{\Psi}_{XY}$) by $\Psi_{\gamma,Y}$ (resp. $\overline{\Psi}_{\gamma,Y}$). The point Ψ_{γ,Γ_b} is called the *minimum* of γ. Our algorithm proceeds as follows.

Partition the segments of Γ_r, Γ_b into four subsets A, B, C and D as in Section 5.6, where $a = |A|$, $b = |B|$, $c = |C|$ and $d = |D|$. We compute the four points $\overline{\Psi}_{A,C}$, $\overline{\Psi}_{A,D}$, $\overline{\Psi}_{B,C}$ and $\overline{\Psi}_{B,D}$; obviously $\overline{\Psi}$ is one of these four points. The intersection points of a segment γ and $\partial\triangle$ are called the *endpoints* of γ.

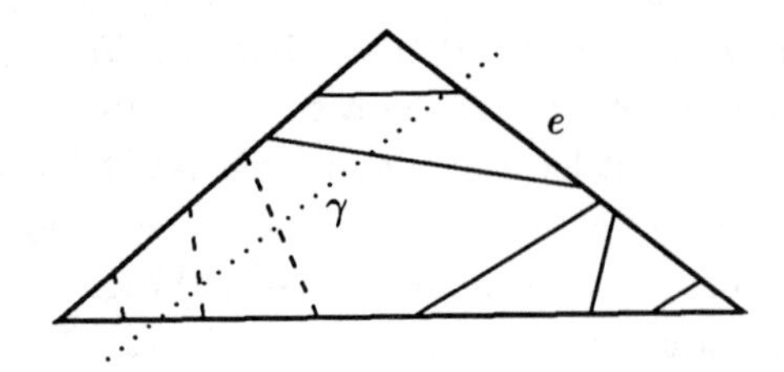

Figure 5.15: Long blue segments: solid lines denote C_e and dashed lines denote $C_{\overline{e}}$; $\gamma \in A_e$

Computing $\overline{\Psi}_{\mathbf{A,C}}$: Let e be a fixed edge of $\triangle$. Let A_e (resp. $A_{\overline{e}}$) denote the set of segments of A intersecting (resp. not intersecting) e; similarly define C_e and $C_{\overline{e}}$ (see figure 5.15). $\overline{\Psi}_{A_e,C_e}$ is computed as follows. Sort the intersection points $e \cap (\cup C_e)$ along the edge e. Let γ be a segment of A_e. First, check whether the minimum of γ lies in the interior of $\triangle$; this can be done in $O(\log n)$ time by locating the intersection points $\gamma \cap \partial\triangle$ in the map Q. If the minimum lies outside $\triangle$, then by property (iii), $\overline{\Psi}_{\gamma,C_e}$ is either the first or the last intersection point of γ nad C_e, and therefore can be easily computed in $O(\log n)$ time. On the other hand if the minimum does lie in $\triangle$, we can determine it, in $O(\log n)$ time, by a binary search, because the segments in C_e are non-intersecting, and $G(x, y)$ is unimodal along γ. Repeating the same process for all segments in A_e, we can compute $\overline{\Psi}_{A_e,C_e}$ in $O(a \log c)$ time. Similarly, we compute $\overline{\Psi}_{A_{\overline{e}},C_e}$ and $\overline{\Psi}_{A,C_{\overline{e}}}$, and then choose the one that minimizes $G(x, y)$.

It is easy to see that the total time spent in computing $\overline{\Psi}_{A,C}$ is $O((a + c)\log(a + c)) = O(n \log n)$.

Computing $\overline{\Psi}_{\mathbf{B,C}}$: Use the same method as above for computing $\overline{\Psi}_{B,C}$; the total time spent in computing $\overline{\Psi}_{B,C}$ is $O(n \log n)$.

Computing $\overline{\Psi}_{\mathbf{A,D}}$: $\overline{\Psi}_{A,D}$ is computed by determining $\overline{\Psi}_{\gamma,D}$, for each $\gamma \in A$. To compute $\overline{\Psi}_{\gamma,D}$, we first check in $O(\log n)$ time whether the minimum of γ lies in the interior of $\triangle$. If it does, then we compute all intersection points $\gamma \cap D$ lying inside $\triangle$, and choose the one that minimizes

$G(x, y)$; this can be easily done in $O(d)$ time. If Ψ_{γ,Γ_b} lies outside Δ, then $\overline{\Psi}_{\gamma,D}$ is either the first or the last intersection point of γ and D. Let $\phi_\gamma(p)$ denote the first intersection point of γ and a segment of D, as we walk along γ from one of its endpoints p towards the interior of Δ. To compute $\phi_\gamma(p)$, we do the following preprocessing. Sort the intersection points of $(\cup D) \cap \partial\Delta$ along the boundary of Δ so that, for any point $p \in \partial\Delta$, we can find, in time $O(\log d)$, the face of the planar map H_D formed by $\{\gamma \cap \Delta | \gamma \in D\} \cup \partial\Delta$, containing p (see figure 5.16). We preprocess each face f of H_D that touches $\partial\Delta$ for ray shooting queries, namely for finding the first edge of f hit by a ray emanating from a point p in a direction d. If f is convex, then such a query can be answered by a binary search on the edges of f, otherwise we can use the algorithm of Chazelle and Guibas [27]. The latter algorithm preprocesses f, in time $O(|f| \log |f|)$, into a data structure of size $O(|f|)$ so that a ray shooting query can be answered in $O(\log |f|)$ time. Now it is easy to see that $\phi(\gamma, p)$ and therefore, $\overline{\Psi}_{\gamma,D}$ can be computed in $O(\log d)$ time.

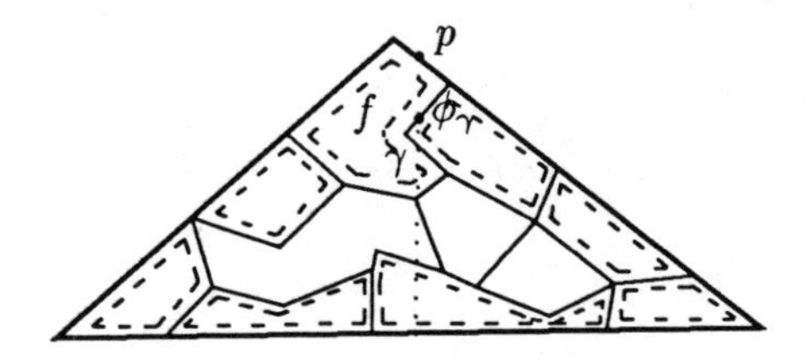

Figure 5.16: Planar map $(D \cap \Delta) \cup \partial\Delta$

Let K denote the number of segments of A having their minimum in the interior of Δ, then the total time spent in computing $\overline{\Psi}_{A,D}$ is $O(n \log n + (K + 1)m)$.

Remark 5.35: We actually do not need the algorithm of [27], because we can compute, in advance, the set of segments of A without their minimum inside Δ. Since the segments in A and D do not intersect among themselves, and the endpoints of the segments of A lie on $\partial\Delta$, $\phi(\gamma, p)$, for each segment of A can be computed in $O(n \log n)$ time by using a sweep line method [115].

Computing $\overline{\Psi}_{\mathbf{B,D}}$: $\overline{\Psi}_{B,D}$ can be determined by calculating all points of $S_{B,D}$. The total time spent is obviously $O(m^2)$.

Thus the total running time spent in computing $\overline{\Psi}$ is $O(n \log n + m^2 + K \cdot m)$, where K is the number of segments in Γ_r having minimum in the interior of Δ. The running time of the above algorithm can be further improved to $O((m + K)\sqrt{n} \log^{1/2} n + n \log n)$ by partitioning Δ into $t = \lceil \frac{m}{\sqrt{n \log n}} \rceil$ triangles, each of which contains at most $\sqrt{n} \log^{1/2} n$ short segments, and then applying the above procedure in each triangle separately.

Now going back to the problem of computing Ψ, using our partitioning algorithm, we partition the plane, in time $O(nr \log n \cdot \log^{\omega} r)$, into $M = O(r^2)$ triangles $\Delta_1, \ldots, \Delta_M$, each meeting at most $O(\frac{n}{r})$ lines containing the segments of Γ_r and Γ_b. In each triangle Δ_i, we compute the minimum using the algorithm described above, and then choose the overall minimum. Let K_i be the number of segments of Γ_r with minimum in the interior of Δ_i. Let m_i (resp. n_i) be the number of segments having an endpoint in (resp. meeting) Δ_i. Following the same analysis as in Section 5.6, we obtain

$$\begin{aligned} T(n) &\leq \sum_{i=1}^{M} O((m_i + K_i)\sqrt{n_i} \log^{1/2} n_i + n_i \log n_i) + O(nr \log n \log^{\omega} r) \\ &= O\left(\frac{n^{3/2}}{\sqrt{r}} \log^{1/2} n\right) + O\left(\sqrt{\frac{n}{r}} \cdot \sum_{i=1}^{M} K_i \log^{1/2} n_i\right) + \\ &\quad O(nr \log n \log^{\omega} r) \text{ (because } \sum_{i=1}^{M} m_i \leq 2m). \end{aligned}$$

Since the minimum of a red segment lies in the interior of at most one triangle, $\sum K_i \geq n$, therefore

$$T(n) = O\left(n\left(\sqrt{\frac{n}{r}} \log^{1/2} n + r \log n \log^{\omega} r\right)\right).$$

By choosing $r = \dfrac{n^{1/3}}{\log^{(2\omega+1)/3} n}$, we obtain $T(n) = O(n^{4/3} \log^{(\omega+2)/3})$. Hence, we can conclude

Theorem 5.36 *Given two planar maps P, Q having n edges in total, and the functions F_P, F_Q associated with them satisfying the conditions (i)–(iii), we can compute a point, in time $O(n^{4/3}\log^{(\omega+2)/3} n)$, that minimizes $F_P(x,y) - F_Q(x,y)$.*

□

5.12 Discussion and Open Problems

In this chapter we presented various applications of our partitioning algorithm described in Chapter 4. Most of the algorithms described in this chapter have a similar flavor. In particular, we first gave a simple but slower algorithm with running time roughly $m\sqrt{n}$ or $n\sqrt{m}$, and then combined it with our partitioning algorithm to obtain a faster algorithm. As mentioned in the introduction of this chapter, we do not need the second phase of our partitioning algorithm in several applications, because the number of triangles produced in the first phase is sufficiently small to imply the asserted running time. For example, consider the problem of computing incidences between a given set of m points and a set of n lines in the plane, and suppose we perform only the first phase of our partitioning algorithm. Equation (5.2) implies that the running time of the algorithm is

$$T(m,n) \le \sum_{i=1}^{M} O(m_i\sqrt{n_i}\log^{1/2} n_i + n_i \log n_i) + O((m + nr\log^{\omega} r)\log n),$$

where $\sum_{i=1}^{M} m_i = m$, $n_i = O(\frac{n}{r})$ and $M = O(r^2 \log^{\omega} r)$. Therefore,

$$\begin{aligned} T(m,n) &= O\left(\sqrt{\frac{n}{r}}\log^{1/2}\frac{n}{r}\sum_{i=1}^{M} m_i\right) + O\left((nr\log^{\omega} r + m)\log n\right) \\ &= O\left(\frac{m\sqrt{n}}{\sqrt{r}}\log^{1/2} n + nr\log^{\omega} r \log n\right) + O(m\log n). \end{aligned}$$

Again, if we choose $r = \max\left\{\dfrac{m^{2/3}}{n^{1/3}\log^{1/3} n \log^{2\omega/3} \frac{m}{\sqrt{n}}}, 2\right\}$, we get

$$T(m,n) = O\left(m^{2/3}n^{2/3}\log^{2/3} n \log^{\omega/3}\frac{m}{\sqrt{n}} + (m+n)\log n\right).$$

Similarly, we can show that we do not need the second phase of the partitioning procedure for the algorithms presented in Sections 5.3–5.7. However, we do need it for approximate half plane range searching, constructing spanning trees with low stabbing number and simplex range searching.

Although this chapter gives efficient algorithms for several problems, which improve previous, often randomized, techniques there is no reason to believe that all the algorithms presented here are close to optimal. Some of these problems that deserve further attention are Hopcroft's problem, counting segment intersections, red-blue intersection, and constructing spanning trees with low stabbing number. One of the most intriguing open problems is whether there exists an $O(n \log n)$ algorithm (or, for that matter any algorithm faster than those given above) for counting segment intersections, or for counting (or just detecting) red-blue segment intersections. The "red-blue" version of such an algorithm would also be able to detect an incidence between points and lines (Hopcroft's problem) in the same time. Another interesting open problem is to obtain a faster algorithm for constructing a spanning tree (or family of spanning trees) with low stabbing number, because that will improve the preprocessing time of various other problems (as in [55], or Chapter 6).

Chapter 6

Spanning Trees with Low Stabbing Number

6.1 Introduction

In the previous chapter (see Section 5.9) we presented an efficient deterministic algorithm for constructing a family of spanning trees with low stabbing number for a given set of points in the plane. This chapter describes several new applications of spanning trees with low stabbing number. The algorithms presented here are faster than the previously best known algorithms for these problems. One of the main goals of this chapter is to demonstrate that such a spanning tree is a versatile tool that can be applied to obtain efficient algorithms for a large class of problems, much beyond the simplex range searching problem for which they were originally introduced. We also show that by combining the spanning tree data structure with the the partitioning algorithm described in Chapter 4 we can reduce the query time if we allow more space. Similar tradeoffs between space and query time have been obtained earlier [55], [22] and also in Section 5.10.

The first and perhaps the most interesting application that we consider is *ray-shooting* in arrangements of segments. There are two versions of this problem, one for segments that are non-intersecting, and one for an arbitrary collection of segments. Formally, these problems can be stated as follows:

(a) Given a collection $\mathcal{G} = \{e_1, \ldots, e_n\}$ *of* n *non-intersecting line*

Figure 6.1: Ray shooting in an arrangement of (a): non-intersecting, (b): arbitrary segments

segments in the plane, preprocess it so that, given a query ray ρ emanating from a point p in direction d, we can quickly compute the intersection point $\Phi(\mathcal{G}, \rho)$ of ρ with the segment of $\mathcal{G}$ that lies nearest to p (see figure 6.1a).

(b) Same problem, except that the segments in $\mathcal{G}$ can intersect arbitrarily (see figure 6.1b).

If the segments in $\mathcal{G}$ form the boundary of a simply connected region, then the algorithm of Chazelle and Guibas [27] preprocesses $\mathcal{G}$ into a data structure of linear size so that, for any ray ρ, $\Phi(\mathcal{G}, \rho)$ can be computed in $O(\log n)$ time. The preprocessing time of their algorithm has been reduced to $O(n \log\log n)$ by Guibas et al. [72]. We are not aware of any ray-shooting algorithm for non-simple polygons (or for an arrangement of segments) that answers a query in $O(\log^{O(1)} n)$ time, using roughly linear space. If we allow quadratic space, then a query is easy to answer in time $O(\log n)$ (see Section 6.4.1). Our goal in this chapter is to obtain efficient solutions that use roughly linear space, and to establish a trade-off between space and query time.

For a special case, where $\mathcal{G}$ is a set of lines, a result of Edelsbrunner et al. [55] implies that one can construct, in randomized expected time $O(n^{3/2} \log^2 n)$, a data structure of size $O(n \log^2 n)$, so that a ray shooting query in $\mathcal{A}(\mathcal{L})$ can be answered in $O(\sqrt{n}\log^3 n)$ time. (In Section 5.9 we determinize the preprocessing and improve the query time to $O(\sqrt{n}\log n)$.) Unfortunately, this algorithm does not apply to segments. An algorithm with a sublinear query time for the case of segments can be developed using the "recursive space-cutting tree" of Dobkin and Edelsbrunner [50] (see also [67]). The best known algorithm for computing $\Phi(\mathcal{G}, \rho)$ is by Guibas et al. [73], which constructs a data structure of size $O(n)$, so that a query can be answered in $O(n^{2/3+\delta})$ time, for any $\delta > 0$. Their algorithm is based on the random sampling technique of [36] and [78], and constructs a multi-level partition tree. The preprocessing of their algorithm is randomized with $O(n \log n)$ expected running time. However, the preprocessing can be made deterministic without any additional overhead using the partitioning algorithm described in Chapter 4.

We show that ray shooting can be performed in roughly (that is, up to poly-log factors) $\sqrt{n}$ time, while still using only roughly linear space and employing deterministic, rather than randomized, preprocessing techniques. We first give an algorithm for the case of non-intersecting segments. This algorithm constructs, in time $O(n^{3/2}\log^{\tau} n)$, a data structure of size $O(n\log^3 n)$ so that, for a given ray ρ, $\Phi(\mathcal{G}, \rho)$ can be computed in $O(\sqrt{n}\log^2 n)$ time, where $\tau = \omega + 1$ is a constant < 4.33. Our algorithm is simpler than that of [73] because it maintains only a two-level data structure. We then extend the above algorithm to general arrangements of segments. Although the basic idea remains the same, we need several new techniques, and the algorithm is more complex. In this case a query can be answered in $O(\sqrt{n\alpha(n)}\log^2 n)$ time, using $O(n\alpha(n)\log^4 n)$ space, after $O(n^{3/2}\log^{\tau} n)$ preprocessing. Another major difference between the two cases is that in the first case we can report all K intersections between a query ray ρ and $\mathcal{G}$ in $O(\sqrt{n}\log^2 n + K\log n)$ time, while we still do not know how to report these intersections in a comparably efficient manner in the general case. One disadvantage of our algorithms over those of [73], [50] is that our preprocessing time is roughly $n^{3/2}$ instead of roughly linear. This is the price we have to pay to achieve deterministic preprocessing and to reduce the query time.

The second problem for which we give an efficient algorithm using the spanning tree data structure is the preprocessing version of the *implicit point location*, defined in Section 5.7. A formal description is given in Section 6.7.

Guibas et al. [73] have presented an algorithm with $O(n^{2/3+\delta})$ query time, for any $\delta > 0$, using the random sampling technique. We improve the query time to $O(\sqrt{n}\log^2 n)$ and use deterministic preprocessing. The algorithm of [73] uses $O(n)$ space, while ours requires $O(n\log^2 n)$ space.

Guibas et al. [73] have described several applications of the implicit point location problem, such as polygon containment, implicit hidden surface removal, polygon placement, etc. We show that our implicit point location algorithm improves the query time of these algorithms as well.

This chapter is organized as follows. In Section 6.2 we discuss spanning trees with low stabbing number. Section 6.3 describes our ray shooting algorithm for arrangements of non-intersecting segments. In Section 6.4 we show that ray shooting queries can be performed faster, if we are allowed to use more space. Section 6.5 extends the algorithms of Sections 6.3 and 6.4 to report all intersections between $\mathcal{G}$ and a query ray ρ at logarithmic cost per intersection. In Section 6.6 we generalize our ray shooting algorithms to arrangements of arbitrary (possibly intersecting) segments. Section 6.7 describes the algorithm for implicit point location and Section 6.8 discusses other applications of the spanning tree data structure. We conclude in Section 6.9 with some final remarks.

6.2 Spanning Trees of Low Stabbing Number

Let S be a set of n points in $\mathbb{R}^2$. As mentioned in Section 5.9, Chazelle and Welzl [32] (see also [132]) have proved that, for any set of n points in $\mathbb{R}^2$, there exists a spanning tree with stabbing number $O(\sqrt{n})$, and that this bound is tight in the worst case. In the previous chapter, we have described a deterministic $O(n^{3/2}\log^{\tau} n)$ algorithhm for constructing a family $\mathbf{T}$ of $O(\log n)$ spanning trees of S, such that the stabbing number $\sigma(\mathbf{T}) = O(\sqrt{n})$ (cf. Theorem 5.21).

Let $\mathcal{C}$ be a spanning path on S. For our applications we need to convert $\mathcal{C}$ into a balanced binary tree $\mathcal{B}$ whose leaves store the points of S in their order along $\mathcal{C}$. Each node v of $\mathcal{B}$ is associated with the subpath $\mathcal{C}_v$ of $\mathcal{C}$ connecting the points stored at the leaves of the subtree rooted at v; let us denote by S_v the subset of S consisting of these points (see figure 6.2).

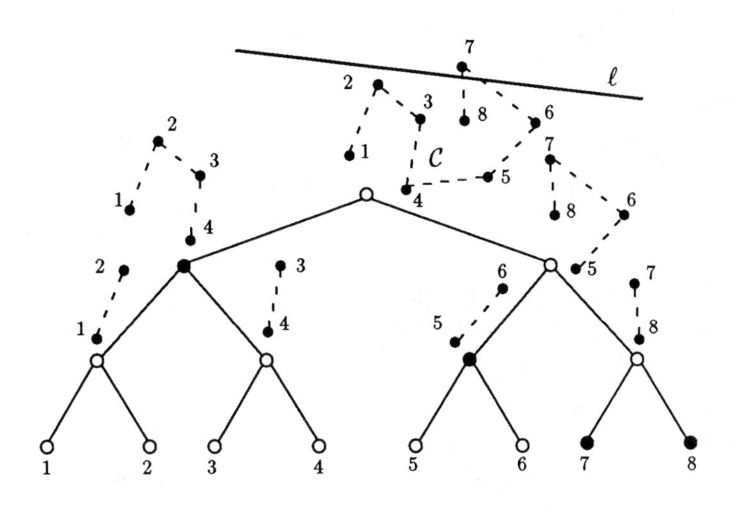

Figure 6.2: $\mathcal{C}$ and $\mathcal{B}(\mathcal{C})$: black nodes of $\mathcal{B}$ denote $V_{\mathcal{B}}(\ell)$

A line ℓ *stabs* a node v of $\mathcal{B}$ if ℓ intersects C_v. Let $V_{\mathcal{B}}(\ell)$ denote the set of nodes v of $\mathcal{B}$ such that v is not stabbed by ℓ but its parent (if one exists) is stabbed. It is easily seen that $\{S_v : v \in V_{\mathcal{B}}(\ell)\}$ is a disjoint partitioning of S. Moreover, we can easily verify

Lemma 6.1 *If a line ℓ intersects s edges of $\mathcal{C}$, then $|V_{\mathcal{B}}(\ell)| \leq 2(s+1)\log n$ and the nodes of $V_{\mathcal{B}}(\ell)$ lie on at most $2(s+1)$ paths of $\mathcal{B}$.*

□

Another simple but key observation is that

Lemma 6.2 *A line ℓ intersects a polygonal path C if and only if ℓ intersects the convex hull of the vertices of C.*

□

Lemma 6.2 implies that ℓ stabs a node v if and only if ℓ intersects the convex hull of S_v. Since an intersection between a line and a convex n-gon can be detected in $O(\log n)$ time, it follows that $V_{\mathcal{B}}(\ell)$ and the paths containing its nodes can be computed in $O(|V_{\mathcal{B}}(\ell)| \log n)$ time if we store the convex hull of the subpath C_v at each node v of $\mathcal{B}$. The running time of this computation can actually be improved to $O(|V_{\mathcal{B}}(\ell)| + \log n)$, using fractional cascading (cf. [29]).

6.3 Ray Shooting among Non-intersecting Segments

In this section we present an algorithm that preprocesses a given set $\mathcal{G}$ of n non-intersecting segments so that, given a query ray ρ emanating from a point p in direction d, $\Phi(\mathcal{G}, \rho)$ can be computed quickly. (For technical reasons we consider ρ as an open ray, i.e. the point p does not belong to ρ.) We will also use $\Phi(\mathcal{G}, \rho)$ to denote the distance of that point from p; if no such intersection exists, we put $\Phi(\mathcal{G}, \rho) = +\infty$. Without loss of generality, we restrict our attention to rightward-directed rays; leftward-directed rays can be handled in a symmetric way. Denote the set of endpoints of the segments of $\mathcal{G}$ as $S = \{p_1, \ldots, p_m\}$, where $m \leq 2n$. For simplicity of exposition, we assume that no segment is vertical. Let $\mathbf{C} = \{C_1, \ldots, C_k\}$ denote a family of $k = O(\log n)$ spanning paths on S, with $\sigma(\mathbf{C}) = O(\sqrt{n})$.

We show how to preprocess a single path $C \in \mathbf{C}$. First, construct $\mathcal{B} = \mathcal{B}(C)$. A segment e is associated with a node v of $\mathcal{B}$ if one of its endpoints is in S_v. Let $\mathcal{G}_v^l$ (resp. $\mathcal{G}_v^r$) denote the set of those segments of $\mathcal{G}$ whose left (resp. right) endpoint is in S_v (see figure 6.3; note that these two sets are not necessarily disjoint). Let ℓ denote the line containing the query ray ρ. Since $\bigcup_{v \in V_{\mathcal{B}}(\ell)} (\mathcal{G}_v^l \cup \mathcal{G}_v^r) = \mathcal{G}$, we have

$$\Phi(\mathcal{G}, \rho) = \min_{v \in V_{\mathcal{B}}(\ell)} \{ \min\{\Phi(\mathcal{G}_v^l, \rho), \Phi(\mathcal{G}_v^r, \rho)\}\}. \tag{6.1}$$

Note that C_v is a connected path, so either all points in S_v, for a node $v \in V_{\mathcal{B}}(\ell)$, lie above ℓ or all of them lie below ℓ. In what follows we assume that all points of S_v lie above ℓ. We will show below that $\Phi(\mathcal{G}_v^l, \rho)$ and $\Phi(\mathcal{G}_v^r, \rho)$, for $v \in V_{\mathcal{B}}(\ell)$, can be computed in $O(\log n)$ time. First, a few notations: Let ℓ^- (resp. ℓ^+) denote the half plane lying below (resp. above) the line ℓ. We distinguish between the two sides of a segment e, the top (resp.

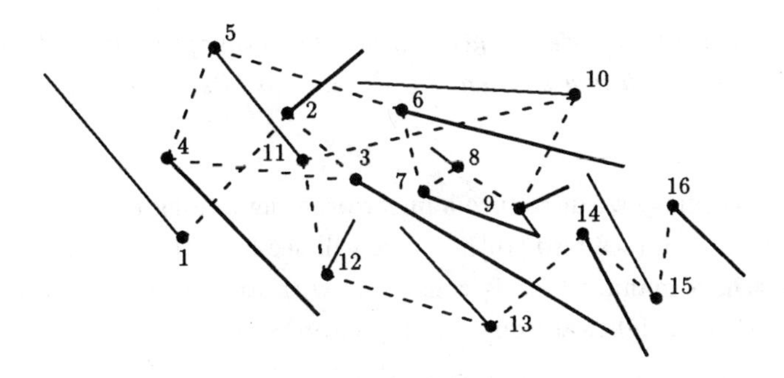

Figure 6.3: $\mathcal{G}_v^l$ and $\mathcal{G}_v^r$: dashed path denotes $\mathcal{C}_v$; bold lines denote $\mathcal{G}_v^l$; bullets denote S_v

bottom) side of e is denoted by e^+ (resp. e^-). We say that a ray ρ hits e from *above* (resp. *below*) if slightly to the left of their intersection, ρ lies above (resp. below) e. If we think of e as expanded into a very thin rectangle and of e^+, e^- as denoting the top and bottom sides of that rectangle, respectively, then ρ hits e from above if, when traversed from left to right, ρ first intersects e^+ and then e^-, and symmetrically for rays that hit e from below (see figure 6.4). If ρ hits e from above (resp. below), then we also say that it hits e^+ (resp. e^-).

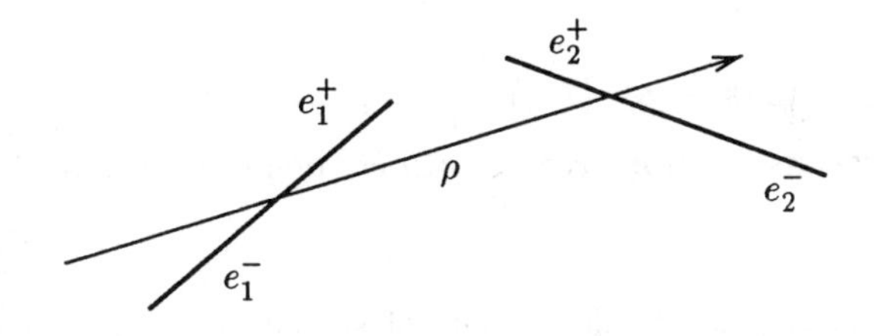

Figure 6.4: Two sided segments: ρ hits e_1 from above and e_2 from below

Since the segments of $\mathcal{G}_v^l$, for $v \in V_{\mathcal{B}}(\ell)$, have the property that their left endpoints lie in the half plane ℓ^+, it can be shown that

Lemma 6.3 *Let v be a node of $V_{\mathcal{B}}(\ell)$. Under the assumption that all points of S_v lie above ℓ, if ρ hits a segment $e \in \mathcal{G}_v^l$, then it hits e from below.*

□

Before proceeding, we introduce a linear ordering among the segments of $\mathcal{G}_v^l$, as defined in [73] (see also [76]). As we will see later, this ordering sorts the segments in a manner that is consistent with any order in which they can be crossed by a rightward-directed ray (from below).

Definition 6.4 [Guibas et al. GOSa]: For a given set $\mathcal{G} = \{e_1, \ldots, e_n\}$ of segments,

(i) $e_i \underset{s}{<} e_j$ if there exists a (non-vertical) line ℓ hitting both e_i^- and e_j^- such that its intersection with e_i lies to the left of its intersection with e_j, and such that ℓ does not hit any e_k^+, for $k \neq i, j$, at a point between e_i and e_j.

(ii) $e_i \underset{v}{<} e_j$ if there exists a vertical line intersecting both e_i and e_j such that its intersection with e_i lies below its intersection with e_j.

(iii) $e_i \underset{x}{<} e_j$ if e_i and e_j have non-overlapping x-projections and the projection of e_i lies to the left of that of e_j.

(iv) $e_i \underset{v^\star/x}{<} e_j$ (also denoted simply as $e_i < e_j$) if either e_i precedes e_j in the transitive closure $\underset{v^\star}{<}$ of $\underset{v}{<}$, or e_i and e_j are not related by $\underset{v^\star}{<}$ and $e_i \underset{x}{<} e_j$.

Theorem 6.5 (Guibas et al. [73]) *$\underset{s}{<}$ is a partial order, and $\underset{v^\star/x}{<}$ is a linear order that extends $\underset{s}{<}$. Moreover $\underset{v^\star/x}{<}$ can be computed in time $O(n \log n)$.*

□

Remark 6.6: It is possible, for a pair of segments e_1, e_2 that $e_2 \underset{v^\star/x}{<} e_1$ within some set $\mathcal{G}$, but $e_1 \underset{v^\star/x}{<} e_2$ relative to a subset $\mathcal{G}' \subset \mathcal{G}$ (see figure 6.5). Therefore it is important to mention the set relative to which we are ordering the segments.

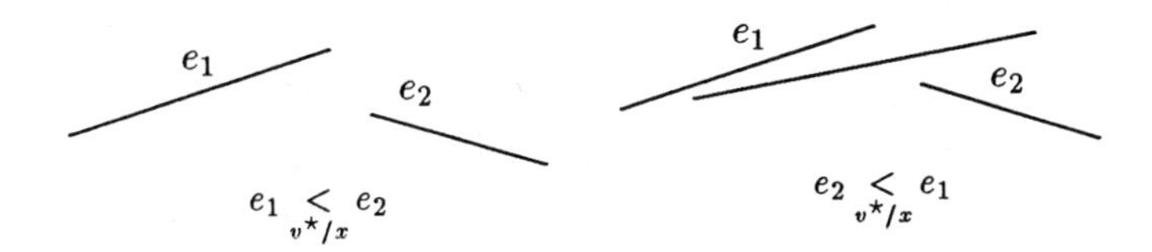

Figure 6.5: Ordering of a pair of segments is relative to a set

Next we prove a technical lemma about $\underset{v^*/x}{<}$ that we will need later. Let l_e (resp. r_e) denote the left (resp. right) endpoint of a nonvertical segment e.

Lemma 6.7 *For all segments $a, b \in \mathcal{G}_v^l$, if r_a lies below ℓ and $x(r_a) < x(l_b)$, then $a \underset{v^*/x}{<} b$ (relative to $\mathcal{G}_v^l$).*

Proof: Suppose, to the contrary that, there is a pair of segments $a, b \in \mathcal{G}_v^l$ such that r_a lies below ℓ and $x(r_a) < x(l_b)$, but $b \underset{v^*/x}{<} a$. Since the x–projection of b lies to the right of that of a, the only way b can precede a in $\underset{v^*/x}{<}$-ordering is by the transitive relation $\underset{v^*}{<}$. Thus there exists a sequence of segments in $\mathcal{G}_v^l$ such that $b = e_1 \underset{v}{<} e_2 \underset{v}{<} \cdots \underset{v}{<} e_k = a$. Let $\pi_{b,a}$ denote a shortest sequence among all such sequences, and let $d_{b,a}$ denote the length of $\pi_{b,a}$. We obtain a contradiction by showing that for every $k > 0$, there is no sequence $\pi_{b,a}$ such that $d_{b,a} = k$.

Obviously $d_{b,a} > 2$, because $x(r_a) < x(l_b)$. If $d_{b,a} = 3$, then there is a segment $c \in \mathcal{G}_v^l$ such that $b \underset{v}{<} c \underset{v}{<} a$. This implies that $x(r_c) \geq x(l_b) > x(r_a) \geq x(l_c)$. Let q be the intersection point of c and the vertical line $x = x(r_a)$. Note that a and c satisfy the following properties: (i) $c \underset{v}{<} a$, (ii) $x(r_c) > x(r_a)$, (iii) a does not intersect c, and (iv) the point l_c lies above ℓ (because $c \in \mathcal{G}_v^l$). Using these properties it can be easily proved that the segment $\overline{qr_c}$ lies below ℓ (see figure 6.6), which contradicts the assumption that $b \underset{v}{<} c$ (because $x(r_c) \geq x(l_b)$ and c lies below b at $x = x(l_b)$). Hence $d_{b,a} > 3$.

Now assume that, for all segments $a, b \in \mathcal{G}_v^l$ satisfying the conditions of

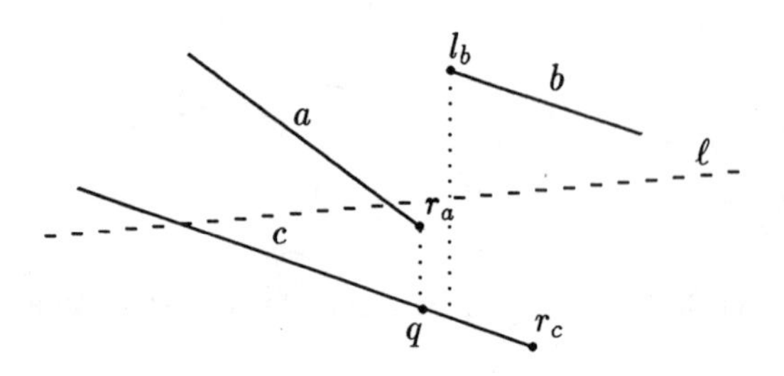

Figure 6.6: Illustration for Lemma 6.7; $d_{ba} = 3$

the lemma, either $a \underset{v^*/x}{<} b$ or $d_{b,a} \geq k$. Suppose there exists a pair a, b such that $b \underset{v^*}{<} a$ and $d_{b,a} = k$. Let $b = e_1 \underset{v}{<} \cdots \underset{v}{<} e_{k-1} \underset{v}{<} e_k = a$ be a corresponding shortest sequence $\pi_{b,a}$, and let $c = e_{k-1}$. Since $\pi_{b,a}$ is a shortest sequence, it is easily seen that $x(r_c) > x(r_a)$. Indeed, let e_j be the first segment in this sequence whose x-projection overlaps that of a. Then e_j must lie below a, for otherwise we would have obtained a cycle in $\underset{v^*}{<}$. Hence $e_j \underset{v}{<} a$ and we can shortcut the sequence after e_j. Clearly, $x(r_j) > x(r_a)$. Let q be the intersection of c with $x = x(r_a)$, as above. Again we can argue that $\overline{qr_c}$ lies below ℓ. If $x(r_c) < x(l_b)$, then c and b satisfy the property of the lemma, and thus contradict the inductive hypothesis because $d_{b,c} < k$. On the other hand if $x(r_c) \geq x(l_b)$, then we have $c \underset{v}{<} b$ (because c lies below b at $x = x(l_b)$), contradicting the assumption that $b \underset{v^*/x}{<} c$.

Hence, we can conclude that $a \underset{v^*}{<} b$.

□

Using Lemma 6.3 and Theorem 6.5 we obtain

Lemma 6.8 *Let $\mathcal{E} = (e_1, \ldots, e_m)$ denote the segments of $\mathcal{G}_v^l$ ordered with respect to $\underset{v^*/x}{<}$ (relative to $\mathcal{G}_v^l$), and suppose $\Phi(\mathcal{G}_v^l, \rho) = \rho \cap e_f$, for some $1 \leq f \leq m$. Then for all $i < f$, e_i does not intersect ρ.*

Proof: If e_i intersects ρ, it does so at a point to the right of $\rho \cap e_f$. This implies that $e_f \underset{s}{<} e_i$, which means $e_f \underset{s}{<} e_i$ because $\underset{v^*/x}{<}$ extends $\underset{s}{<}$. □

Hence the original problem is reduced to the following restricted problem:

> *Given a sequence $\mathcal{E}$ of m segments sorted according to $\underset{v^*/x}{<}$, preprocess $\mathcal{E}$ so that for any (rightward-directed) query ray ρ emanating from a point p and lying on a line that passes below the left endpoints of all segments in $\mathcal{E}$, we can quickly determine $e_f(\rho)$, the first segment of $\mathcal{E}$ hit by the ray ρ.*

A possible approach to solving this problem is to do a binary search on $\mathcal{E}$, where each step of the search tests whether ρ intersects a segment in some contiguous subsequence of t segments of $\mathcal{E}$. If ρ were a full line ℓ, then such an intersection could be easily detected in $O(\log t)$ time after $O(t \log t)$ preprocessing (in which we construct the convex hull of the right endpoints of these t segments). However, no equally fast procedure is known to detect an intersection between a ray and such a set of segments. To overcome this problem, we next show how to reduce the intersection detection problem to one involving the line containing ρ rather than ρ itself.

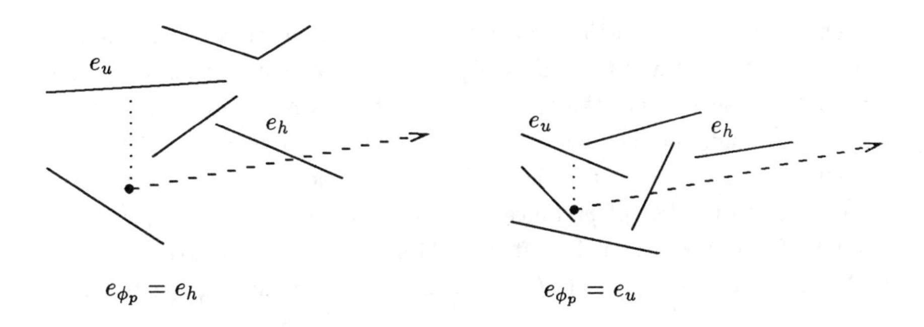

Figure 6.7: Segments e_h, e_u and e_{ϕ_p}

For any point q in the plane, let $e_h = e_{h(q)}$ denote the first segment of $\mathcal{E}$ whose left endpoint lies to the right of (or above) q (see figure 6.7), and let $e_u = e_{u(q)}$ denote the segment in $\mathcal{E}$ lying immediately above q, that is, the vertical ray emanating from q in the upward direction hits e_u before any other segment. If e_h (resp. e_u) is not defined, we put $h = m + 1$ (resp. $u = m + 1$). Finally, put $\phi_q = \min\{h, u\}$.

To compute e_h, construct a balanced binary tree L whose leaves store the segments of $\mathcal{E}$ in their order in $\mathcal{E}$. For each interior node z of L we store the rightmost left endpoint of the segments stored at the subtree rooted at z. L can be constructed in $O(m \log m)$ time, and it is easily seen that e_h can be determined, by searching for q through L, in $O(\log m)$ time. As for e_u, we can easily calculate it in time $O(\log m)$ after $O(m \log m)$ preprocessing, as in [111].

Lemma 6.9 *The query ray ρ emanating from a point p cannot intersect any segment $e_i \in \mathcal{E}$, for $i < \phi_p$. Moreover, ρ intersects e_i for $i \geq \phi_p$ if and only if its right endpoint lies below the line containing ρ.*

Proof: If the first part of the lemma were not true, then there would exist a segment e_i, for $i < \phi_p$, intersecting the ray ρ. In this case the left endpoint of e_i must lie to the left of p, so the vertical ray η from p in the upward direction must intersect e_i. But then the first segment e_u hit by η must satisfy $i \geq u \geq \phi_p$ (because $e_u \underset{v^*}{<} e_i$ and by definition of ϕ_p), a contradiction that proves the first half of the lemma.

The "only if" part of the second half of the lemma follows from the fact that if both the left and the right endpoints of a segment e lie above ℓ, then e cannot intersect ℓ. For the "if" part let e_i be a segment of $\mathcal{E}$, for $i \geq \phi_p$, whose right endpoint lies below ℓ, but e_i does not intersect ρ. If the left endpoint l_{e_i} of e_i lies to the right of p, then obviously e_i intersects ρ, so l_{e_i} must lie to the left of p. Since e_i does not intersect ρ, the intersection point ξ of e_i and ℓ lies to the left of p. Moreover if $x(r_{e_i}) < x(l_{e_{u(p)}})$, then by Lemma 6.7 $e_i \underset{v^*/x}{<} e_{u(p)}$. If $x(r_{e_i}) \geq x(l_{e_{u(p)}})$, then e_i and $e_{u(p)}$ must have x-projections that overlap at some point between ξ and ρ; since e_i lies below $e_{u(p)}$ at this point, we again have $e_i \underset{v^*}{<} e_{u(p)}$. Similarly we can show that $e_i \underset{v^*/x}{<} e_{h(p)}$. Hence $i < \min\{u, h\}$, contradicting the assumption that $i \geq \phi_p$. □

Lemma 6.9 implies that the binary search technique proposed above will work, provided we can detect quickly whether any right endpoint of a segment in some contiguous subsequence of $\mathcal{E}$ lies below ℓ. In other words, the problem now has been reduced to that of detecting an intersection between a set of points and a query half plane. Clearly, this is equivalent to detecting

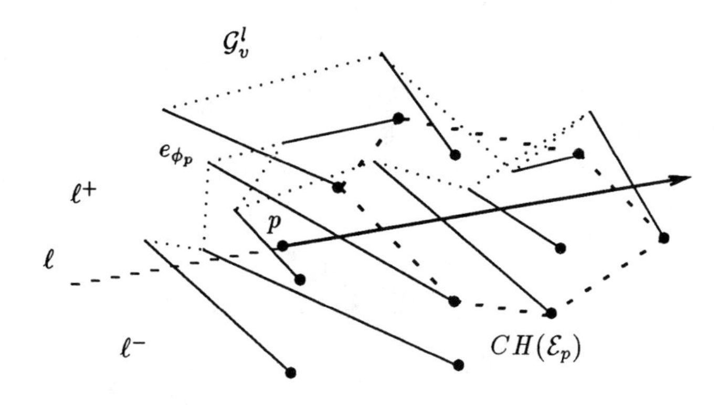

Figure 6.8: Convex hull $\text{CH}(\mathcal{E}_p)$ intersecting ℓ^-

an intersection between the convex hull of these points and the half plane (see figure 6.8).

We are now ready to describe how to preprocess $\mathcal{E}$ so that $e_f(\rho)$ can be computed quickly, for any ray ρ with the above properties. Let r_i denote the right endpoint of $e_i \in \mathcal{E}$, and let $R = \{r_1, \ldots, r_m\}$. We construct a binary tree $\mathcal{T}$ on R in the same way as we constructed $\mathcal{B}$, i.e. the points r_i are stored in the leaves of $\mathcal{T}$ in order, and each node w of $\mathcal{T}$ is associated with the subsequence R_w of R containing all points stored at the leaves of the subtree rooted at w.

At every node w of $\mathcal{T}$, we store the convex hull of R_w. Using $\mathcal{T}$ we can determine $e_f(\rho)$ in time $O(\log^2 m)$: We first find ϕ_p, as described above, in $O(\log m)$ time. Then we treat the suffix $\{r_{\phi_p}, \ldots, r_m\}$ of R as the union of $\log m$ subsets R_w, $w \in \mathcal{T}$, which we can compute in $O(\log m)$ time. We test each R_w in increasing, left-to-right order, to find the first w for which the line ℓ containing ρ intersects the hull of R_w. Then we do a binary search within R_w until we find $e_f(\rho)$. All this takes $O(\log^2 m)$ time. However, we can easily reduce the time to $O(\log m)$, using fractional cascading. This is possible since, as in [29], detecting intersection between ℓ and a convex polygon amounts to searching for the slope of ℓ in the sequence of slopes of the edges of the polygon (see [28], [29] for more details). Therefore we have

Lemma 6.10 *Given a set $\mathcal{E}$ of m non-intersecting line segments in the plane, we can preprocess them, in time $O(m \log m)$, into a data structure of size $O(m \log m)$ so that, given a (rightward-directed) query ray ρ whose containing line lies below the left endpoints of all the segments in $\mathcal{E}$, we can compute the first segment of $\mathcal{E}$ hit by ρ in time $O(\log m)$.*

□

Returning to the original problem, Lemma 6.10 and the preceding discussion imply that we can compute $\Phi(\mathcal{G}_v^l, \rho)$ (and symmetrically $\Phi(\mathcal{G}_v^r, \rho)$), for each $v \in V_{\mathcal{B}}(\ell)$, in time $O(\log n)$. Equation (6.1) and Lemma 6.1 then imply

Theorem 6.11 *Given a set $\mathcal{G}$ of n non-intersecting line segments, we can preprocess it in time $O(n^{3/2} \log^{\omega} n)$, for some $\omega < 4.33$, into a data structure of size $O(n \log^3 n)$, using $O(n^{3/2})$ working storage, so that, given a query ray ρ, its first intersection $\Phi(\mathcal{G}, \rho)$ with $\mathcal{G}$ can be computed in time $O(\sqrt{n} \log^2 n)$.*

□

Remark 6.12:

(i) The space used can be reduced to $O(n \log^2 n)$, without affecting the query time, if we use a single tree structure instead of a family of $O(\log n)$ trees. But then the preprocessing time increases to $O(n^3 \log n)$ (see [55]).

(ii) If we allow randomization, the (expected) preprocessing time can be reduced to $O(n^{4/3} \log^2 n)$ using Matoušek's algorithm [89], but then the query time bound increases by a factor of $\log n$.

6.4 Tradeoff between Space and Query Time

In this section we show that the query time for the ray shooting problem in arrangements of non-intersecting segments can be improved if we allow ourselves more storage. Similar tradeoffs have been obtained for several related problems, such as computing a face in an arrangement of lines (see

[55]) and simplex range searching (see Section 5.10). The main result of this section is an algorithm for computing $\Phi(\mathcal{G}, \rho)$ with $O(\frac{n}{\sqrt{m}} \log^{7/2} \frac{n}{\sqrt{m}} + \log n)$ query time, using $O(m)$ space, where $n \log^3 n \leq m \leq n^2$.

6.4.1 The case of quadratic storage

First, we show that if we allow $O(n^2)$ space, the query time can be reduced to $O(\log n)$.

Let $\mathcal{G} = \{e_1, \ldots, e_n\}$ be a collection of n non-intersecting segments. Dualize all segments $e \in \mathcal{G}$, obtaining a set $\mathcal{G}^*$ of n double wedges (see Section 4.2). Let $\mathcal{L}^*$ denote the set of lines bounding the double wedges of $\mathcal{G}^*$ (i.e. the duals of the endpoints of segments in $\mathcal{G}$). Let $\mathcal{A}(\mathcal{L}^*)$ denote the arrangement of $\mathcal{L}^*$, and let w_f be the set of double wedges of $\mathcal{G}^*$ containing the face $f \in \mathcal{A}(\mathcal{L}^*)$. Standard duality arguments yield:

Lemma 6.13 *Let p be a point lying in the interior of a face f of $\mathcal{A}(\mathcal{L}^*)$. Then p^* intersects each segment $e \in w_f$ transversally at an interior point, and is disjoint from any other segment of $\mathcal{G}$.*

□

Lemma 6.14 *If the segments of $\mathcal{G}$ are non-intersecting, then for all points p in the face f of $\mathcal{A}(\mathcal{L}^*)$, the line p^* intersects the segments of w_f in the same order.*

Proof: Suppose there are two points x and y in a face f such that the lines x^* and y^* intersect the segments of w_f in two different orders. Since the segments in $\mathcal{G}$ are non-intersecting, rotating x^* towards y^* (in the direction that avoids a vertical orientation) we must reach a line p^* that either contains a segment of w_f, or passes through an endpoint of a segment of w_f. (Note that this claim does not hold if the segments can intersect.) The dual p of p^* is a point that lies on the segment $\overline{xy}$, hence in f. This, however, contradicts Lemma 6.13, thus showing that the duals of all points in f intersect the segments of w_f in the same order.

□

We sort the segments in w_f in the order provided by Lemma 6.14. For a ray ρ, let the *image* of ρ be the dual of the line containing ρ. If the image of a ray ρ lies in the face $f \in \mathcal{A}(\mathcal{L}^*)$, then $\Phi(\mathcal{G}, \rho)$ can be computed in $O(\log n)$ time by a binary search on w_f. Therefore, it suffices to show how to store all the lists w_f using only $O(n^2)$ space, so that binary search in each of them can still be done in $O(\log n)$ time.

Let $\mathcal{D}$ denote the dual graph of $\mathcal{A}(\mathcal{L}^*)$, i.e. the graph whose nodes represent faces of $\mathcal{A}(\mathcal{L}^*)$ and whose edges connect pairs of nodes representing adjacent faces. Let $\mathcal{T}$ denote a spanning tree of $\mathcal{D}$. We can convert $\mathcal{T}$ into a path Π by tracing an Eulerian tour around the tree. Observe that if two vertices v_1, v_2 in Π represent faces f_1, f_2 sharing an edge γ, which is a portion of a line ℓ, then $w_{f_1} \oplus w_{f_2}$ is the set of segments having the dual of ℓ as an endpoint. Let δ_γ denote this set of segments. The set w_{f_2} can be obtained from w_{f_1} by deleting the segments of $\delta_\gamma \cap w_{f_1}$ and inserting the segments of $\delta_\gamma - w_{f_1}$. Therefore we can maintain all lists w_f using a persistent data structure (see [42], [111], [51]). Since at each edge γ of Π, only the segments of δ_γ are inserted or deleted, the total space required to store all w_f is $O(n+\sum_{\gamma\in\Pi} |\delta_\gamma|)$ and the total preprocessing time is $O((n+\sum_{\gamma\in\Pi} |\delta_\gamma|) \log n)$. Moreover, using this persistent data structure, $\Phi(\mathcal{G}, \rho)$ can be computed in $O(\log n)$ time (see [111]). Thus, it suffices to prove that $\sum_\gamma |\delta_\gamma| = O(n^2)$. Suppose the segments of $\mathcal{G}$ have $t \leq 2n$ distinct endpoints and ν_i segments are incident to the ith endpoint. It is easy to check that if γ is a portion of the line dual to the ith endpoint, then $|\delta_\gamma| \leq \nu_i$. Obviously $\sum_{i=1}^t \nu_i = 2n$ and each line of $\mathcal{L}^*$ is split into $\leq t+1$ edges, which implies that

$$\sum_{\gamma\in\mathcal{D}} |\delta_\gamma| \quad\leq\quad \sum_{i=1}^{t} (t+1)\nu_i \quad\leq\quad 2n(2n+1).$$

Hence, we have

Theorem 6.15 *Given a collection $\mathcal{G}$ of n non-intersecting segments, we can preprocess $\mathcal{G}$, in $O(n^2 \log n)$ time, into a data structure of size $O(n^2)$ so that, for any query ray ρ, $\Phi(\mathcal{G}, \rho)$ can be computed in $O(\log n)$ time.*

□

6.4.2 The general case

Theorem 6.11 and Theorem 6.15 represent roughly two extremes of the spectrum, because we need at least $O(n)$ space, and we cannot hope to answer a query in $o(\log n)$ time. The general case where the allowed storage m assumes an intermediate value between $n \log^3 n$ and n^2 is handled as follows. For technical reasons we assume for the time being that no endpoint of a segment in $\mathcal{G}$ has degree > 3 (that is incident to more than three segments of $\mathcal{G}$). In Section 6.4.3 we show how to handle degenerate cases (i.e. when there are endpoints of degree > 3).

Using the partitioning algorithm of Chapter 4, partition the dual plane, in time $O(nr \log n \cdot \log^{\tau-1} r)$, into $M = O(r^2)$ triangles $\triangle_1, \ldots, \triangle_M$, each meeting at most $\frac{n}{r}$ lines dual to the endpoints of the given segments, where r is a parameter to be chosen later. Let $\mathcal{L}_i^*$ denote the set of dual lines that intersect the triangle $\triangle_i$, for $i = 1, \ldots, M$; then $|\mathcal{L}_i^*| \le \frac{n}{r}$. For each $\triangle_i$, define the subset $\mathcal{G}_i$ of $\mathcal{G}$ to consist of all segments e having at least one endpoint whose dual is in $\mathcal{L}_i^*$. Obviously $|\mathcal{G}_i| \le \frac{3n}{r}$. We also define $W_i \subset \mathcal{G}$ as

$$W_i = \{e \mid (e \in \mathcal{G}) \wedge (\triangle_i \subset e^*)\}.$$

It is easily checked that

Lemma 6.16 *For each point p lying inside $\triangle_i$, the line p^* does not intersect any segment of $\mathcal{G} - (\mathcal{G}_i \cup W_i)$.*

□

Lemma 6.16 implies that

$$\Phi(\mathcal{G}, \rho) = \min\{\Phi(W_i, \rho), \Phi(\mathcal{G}_i, \rho)\} \tag{6.2}$$

where $\triangle_i$ is the triangle containing the image of ρ. Using the same argument as in Lemma 6.14, we can prove

Lemma 6.17 *All lines whose dual points lie inside $\triangle_i$ intersect the segments of W_i in the same order.*

□

We can thus order the segments of W_i in the order provided by Lemma 6.17, and compute $\Phi(W_i, \rho)$, for any ray ρ whose image lies in $\triangle_i$, in $O(\log n)$ time, using binary search. Let $\triangle_1$ and $\triangle_2$ be two adjacent triangles and let

$$\mathcal{G}_{12} = \{e \mid (e \in \mathcal{G}_1) \wedge (\triangle_2 \subset e^*)\}. \tag{6.3}$$

It is easily seen that $W_2 = (W_1 \cup \mathcal{G}_{12}) - \mathcal{G}_2$. Since $|\mathcal{G}_1|, |\mathcal{G}_2| = O(\frac{n}{r})$, we have $|W_2 - W_1| = O(\frac{n}{r})$. As earlier, we define a graph $\mathcal{D}$, whose vertices are the triangles Δ_i and whose edges connect pairs of vertices representing adjacent triangles. Now an edge between v_1 and v_2 has the set $\mathcal{G}_1 \cup \mathcal{G}_2$ associated with it. Again, we construct a path Π on a spanning tree of $\mathcal{D}$, and obtain a persistent data structure $\Upsilon_1(\mathcal{G})$ to store W_i for all triangles. It can be easily shown that $\Upsilon_1(\mathcal{G})$ requires $O(nr)$ space, and can be constructed in $O(nr \log n)$ time. Now, for any ray ρ, $\Phi(W_i, \rho)$ can still be computed in $O(\log n)$ time, where Δ_i is the triangle containing the image of ρ.

We preprocess each $\mathcal{G}_i$ into a data structure $\Upsilon_2(\mathcal{G}_i)$ of size $O(|\mathcal{G}_i| \log^3 |\mathcal{G}_i|)$ for ray shooting queries, as described in Section 6.3, so that for any ray having its image in Δ_i, we can find $\Phi(\mathcal{G}_i, \rho)$ in $O(\sqrt{\frac{n}{r}} \log^2 \frac{n}{r})$ time.

To compute $\Phi(\mathcal{G}, \rho)$, for a given ray ρ, we first find the triangle Δ_i that contains its image; this can be done in $O(\log n)$ time, using an efficient point location algorithm [60], [111]. It follows from (6.2) that $\Phi(\mathcal{G}, \rho)$ can be computed by calculating each of $\Phi(W_i, \rho)$ and $\Phi(\mathcal{G}_i, \rho)$, as described above; therefore the query time is

$$Q(n) \;=\; O\left(\sqrt{\frac{n}{r}} \log^2 \frac{n}{r} + \log n\right).$$

As for the space complexity $S(n)$, we need $O(r^2)$ space to store the triangle $\Delta_1, \ldots, \Delta_M$, $O(nr)$ space to store Υ_1 and $O(\frac{n}{r} \log^3 \frac{n}{r})$ to store each $\mathcal{G}_i$ (cf. Theorem 6.11). Thus,

$$S(n) \;=\; O(nr) + O\left(r^2 \cdot \frac{n}{r} \log^3 \frac{n}{r}\right) \;=\; O\left(nr \log^3 \frac{n}{r}\right).$$

If we choose $r = \dfrac{m}{n \log^3 \frac{n}{\sqrt{m}}}$, then $S(n) = O(m)$ and the query time becomes

$$\begin{aligned} Q(n) \;&=\; O\left(\sqrt{\frac{n}{m/(n \log^3 \frac{n}{\sqrt{m}})}} \cdot \log^2 \frac{n}{\sqrt{m}} + \log n\right) \\ &=\; O\left(\frac{n}{\sqrt{m}} \log^{7/2} \frac{n}{\sqrt{m}} + \log n\right). \end{aligned}$$

Next, we bound the preprocessing time $P(n)$. $\Delta_1, \ldots, \Delta_M$ can be computed in time $O(nr \log n \log^{\tau-1} r)$ (see Theorem 4.36). Since Υ_1 can be constructed in $O(nr\log n)$ time and each $\mathcal{G}_i$ can be preprocessed in $O((\frac{n}{r})^{3/2}\log^{\tau}\frac{n}{r})$ time (cf. Theorem 6.11), we have

$$\begin{aligned} P(n) &= O(nr\log n\log^{\tau-1} r) + O\left(r^2\cdot\frac{n^{3/2}}{r^{3/2}}\log^{\tau}\frac{n}{r}\right) \\ &= O\left(nr\log n\log^{\tau-1} r + n^{3/2}\sqrt{r}\log^{\tau}\frac{n}{r}\right) \\ &= O\left(n\frac{m}{n\log^3\frac{n}{\sqrt{m}}}\log n\log^{\tau-1}\frac{m}{n} + n^{3/2}\sqrt{\frac{m}{n\log^3\frac{n}{\sqrt{m}}}}\log^{\tau}\frac{n}{\sqrt{m}}\right) \\ &= O\left(m\log^{\tau} n + n\sqrt{m}\log^{\tau-3/2}\frac{n}{\sqrt{m}}\right). \end{aligned}$$

Since we need $O(nr)$ space to compute $\triangle_1,\ldots,\triangle_M$, and $O(\frac{n^{3/2}}{r^{3/2}})$ to preprocess each $\mathcal{G}_i$, the total space required for preprocessing is

$$\begin{aligned} O\left(nr + \frac{n^{3/2}}{r^{3/2}}\right) &= O\left(\frac{m}{\log^3\frac{n}{\sqrt{m}}} + \frac{n^{3/2}}{m^{3/2}/(n\log^3\frac{n}{\sqrt{m}})^{3/2}}\right) \\ &= O\left(m + \frac{n^3}{m^{3/2}}\log^{3/2}\frac{n}{\sqrt{m}}\right). \end{aligned}$$

Hence, we can conclude

Theorem 6.18 *Given a collection $\mathcal{G}$ of n non-intersecting segments in the plane with the property that no endpoint has degree > 3, and a parameter $n\log^3 n < m < n^2$, we can preprocess $\mathcal{G}$, in $O(m\log^{\tau} n + n\sqrt{m}\log^{\tau-3/2}\frac{n}{\sqrt{m}})$ time, into a data structure of size $O(m)$ so that, for query ray ρ, we can compute $\Phi(\mathcal{G},\rho)$ in $O\left(\frac{n}{\sqrt{m}}\log^{7/2}\frac{n}{\sqrt{m}} + \log n\right)$ time. The working storage required for preprocessing is $O\left(m + \frac{n^3}{m^{3/2}}\log^{3/2}\frac{n}{\sqrt{m}}\right)$.* □

Remark 6.19:

(i) If we allow randomization, then using Matoušek's algorithm $\mathcal{G}_i$ can be preprocessed in $O\left((\frac{n}{r})^{4/3}\log^2 n\right)$ time, but the query time increases by a factor of $O(\log n)$. Therefore, following the same analysis we obtain

$$P(n) = O\left(m^{2/3}n^{2/3} + m\log^{\tau} n\right)$$

$$Q(n) = O\left(\frac{n}{\sqrt{m}} \log^{9/2} \frac{n}{\sqrt{m}} + \log n\right).$$

(ii) If we maintain a single tree data structure for each $\mathcal{G}_i$, the query time can be reduced to $O(\frac{n}{\sqrt{m}} \log^3 n)$, but the preprocessing time increases considerably.

6.4.3 Coping with degenerate cases

The analysis of the algorithm described in the previous subsection breaks down if the segments of $\mathcal{G}$ have endpoints of arbitrarily large degree, because then we cannot guarantee that $|\mathcal{G}_i| = O(\frac{n}{r})$, and the analysis of the bound for the total space required to store Υ_1 relies heavily on this bound for $|\mathcal{G}_i|$. In this subsection we overcome this difficulty by showing that, given a set $\mathcal{G}$ of n non-intersecting segments, we can transform it into another set $\mathcal{G}'$ of at most $3n$ (non-intersecting) segments such that no endpoint of a segment in $\mathcal{G}'$ has degree > 3, and $\Phi(\mathcal{G}, \rho)$ can be determined from $\Phi(\mathcal{G}', \rho)$ in $O(1)$ time.

Let $\mathcal{G}_p = \{e_1, \ldots, e_t\}$ be a subset of segments of $\mathcal{G}$ all having a common endpoint p. Let δ be the minimum distance from p to its closest neighbor in $\mathcal{G} - \mathcal{G}_p$, and let c be the circle of radius $\frac{\delta}{2}$ with p as center. For a segment $e_i \in \mathcal{G}_p$, let q_i denote the intersection point of c and e_i. Assume that the segments of $\mathcal{G}_p$ are ordered in counter-clockwise direction along p. There are two cases to consider:

(i) There exist two consecutive segments in $\mathcal{G}_p$, say e_t and e_1, such that the angle between e_t and e_1 is $> 180°$. For $1 < i < t$, we remove the portion of e_i that lies in the interior of c (i.e. $\overline{pq_i}$) and add the segments $\overline{q_1q_2}, \ldots, \overline{q_{t-1}q_t}$ to $\mathcal{G}$ (see figure 6.9a).

(ii) The angle between every two consecutive segments of $\mathcal{G}_p$ is $< 180°$. For each $e \in \mathcal{G}_p$, we remove the portion of e that lies in the interior of c and add the segments $\overline{q_1q_2}, \ldots, \overline{q_{t-1}q_t}, \overline{q_tq_1}$ to $\mathcal{G}$ (see figure 6.9b).

We repeat this process for each endpoint of the segments of $\mathcal{G}$ whose degree is > 3. Let $\mathcal{G}'$ be the new set of segments; obviously $|\mathcal{G}'| \leq 3n$, and each endpoint has degree ≤ 3. If the source point, s, of ρ lies inside one of

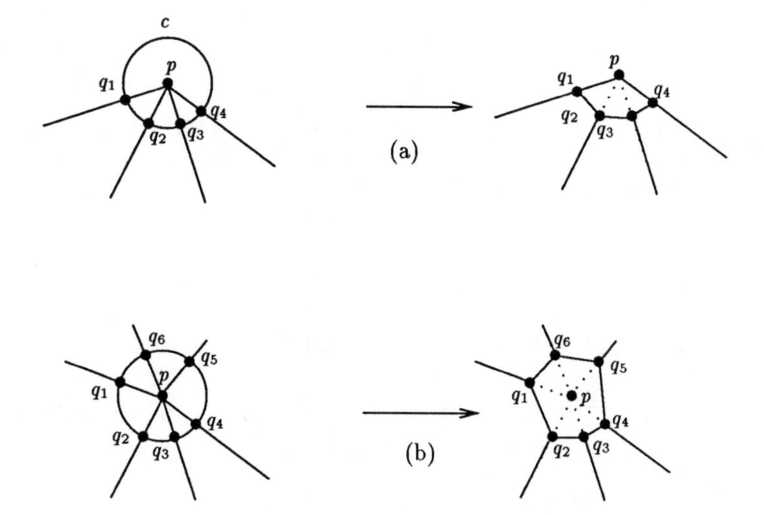

Figure 6.9: Modifying segments having a common endpoint of degree > 3.

of the newly created polygons, say in $\triangle pq_{i-1}q_i$, then $\Phi(\mathcal{G},\rho)$ lies on one of the segments incident to q_{i-1}, q_i. On the other hand, if s lies outside all such polygons and $\Phi(\mathcal{G}',\rho)$ lies on a segment of $\mathcal{G}$, then $\Phi(\mathcal{G},\rho) = \Phi(\mathcal{G}',\rho)$. Finally, if s lies outside all little polygons but $\Phi(\mathcal{G}',\rho)$ lies on a segment $\overline{q_{i-1}q_i}$ and e_{i-1} (resp. e_i) is the segment of $\mathcal{G}$ incident to q_{i-1} (resp. q_i), then $\Phi(\mathcal{G},\rho)$ lies on either e_{i-1} or e_i. Since s can be located in $O(\log n)$ time by preprocessing $\mathcal{A}(\mathcal{G}')$ for planar point location, $\Phi(\mathcal{G},\rho)$ can be computed from $\Phi(\mathcal{G}',\rho)$ in $O(1)$ time. Moreover for each endpoint p, the minimum distance δ_p can be computed in $O(n\log n)$ time by constructing the closest point Voronoi diagram of $\mathcal{G}$ [135].

Hence by Theorem 6.18, we have

Theorem 6.20 *Given a set $\mathcal{G}$ of n segments and a parameter $n\log^3 n < m < n^2$, we can preprocess it, in time $O(m\log^\tau n + n\sqrt{m}\log^{\tau-3/2}\frac{n}{\sqrt{m}})$, into a data structure of size $O(m)$ so that, for a query ray ρ, we can compute $\Phi(\mathcal{G},\rho)$ in time $O(\frac{n}{\sqrt{m}}\log^{7/2}\frac{n}{\sqrt{m}} + \log n)$.*

□

6.5 Reporting All Intersections

In the last two sections we gave algorithms to compute $\Phi(\mathcal{G},\rho)$ for a collection of non-intersecting segments. We now extend these algorithms to solve the following problem:

> *Given a set $\mathcal{G}$ of n non-intersecting segments, preprocess it so that, for a query ray ρ, we can quickly compute all intersections $\mathcal{I}_\rho$ between $\mathcal{G}$ and ρ in their order along ρ.*

Dobkin and Edelsbrunner [50] have given an algorithm that preprocesses $\mathcal{G}$ into a data structure of linear size so that, for a query ray ρ, $\mathcal{I}_\rho$ can be computed in $O(n^{0.695} + |\mathcal{I}_\rho|)$ time. (In fact their algorithm works for an arbitrary collection of segments.) We first present an algorithm that uses roughly linear space, by generalizing the algorithm described in Section 6.3.

We preprocess $\mathcal{G}$, as in Section 6.3, in $O(n^{3/2}\log^\tau n)$ time using $O(n\log^3 n)$ space. For a given ray ρ, we compute $\mathcal{I}_\rho$ as follows. Let ℓ denote the line containing the ray ρ, and let C be a spanning path in $\mathbf{C}$ that intersects ℓ in $O(\sqrt{n})$ edges. As described in Section 5.9, compute $V_{\mathcal{B}}(\ell)$ in $O(\sqrt{n}\log n)$ time. Now we report all intersection points in $\mathcal{I}_p$ by walking along the ray ρ and stopping at each point of $\mathcal{I}_\rho$. For a point $q \in \rho$, let ρ_q be the ray emanating from q and contained in ρ.

The algorithm maintains the following invariant: When we are at a point $q \in \rho$, we maintain a list of all points $\Phi(\mathcal{G}_v, \rho_q)$, for all $v \in V_{\mathcal{B}}(\ell)$, as a priority queue $\mathcal{Q}$ (with respect to their order along ρ). Observe that $\mathcal{Q}$ remains the same between two consecutive points of $\mathcal{I}_\rho$, and that the root of $\mathcal{Q}$ stores the point of $\mathcal{I}_\rho$ that we are going to encounter next. Therefore, it suffices to show how to update $\mathcal{Q}$ after visiting a point of $\mathcal{I}_\rho$. Suppose, when we are at a point q, the root of $\mathcal{Q}$ stores $\sigma = \Phi(\mathcal{G}_u, \rho_q)$, for some $u \in V_{\mathcal{B}}(\ell)$. It is easily seen that when we cross σ, the next intersection point of ρ and $\mathcal{G}_v$, for all $v \in V_{\mathcal{B}}(\ell) - \{u\}$, does not change. Thus $\mathcal{Q}$ can be updated by deleting σ from $\mathcal{Q}$, and inserting $\Phi(\mathcal{G}_u, \rho_\sigma)$ in $\mathcal{Q}$ provided $\Phi(\mathcal{G}_u, \rho_\sigma) \neq \infty$. Continue this process until $\mathcal{Q}$ becomes empty. It is easily seen that this procedure reports all intersection points of ρ and the segments of $\mathcal{G}$ in their order along ρ.

To bound the running time of the algorithm, observe that initially we spend $O(\sqrt{n}\log^2 n)$ time to construct the queue $\mathcal{Q}$ for $q = p$, and then spend $O(\log n)$ time in updating $\mathcal{Q}$ after each intersection. Hence, we have

Theorem 6.21 *Given a collection $\mathcal{G}$ of n non-intersecting segments, we can preprocess it, in time $O(n^{3/2}\log^\tau n)$, into a data structure of size $O(n\log^3 n)$ so that, given a query ray ρ, $\mathcal{I}_\rho$ can be computed in $O(\sqrt{n}\log^2 n + |\mathcal{I}_\rho|\log n)$ time.*

□

An immediate corollary of Theorem 6.21 is

Corollary 6.22 *Given a collection $\mathcal{G}$ of n non-intersecting segments, we can preprocess it, in time $O(n^{3/2}\log^\tau n)$, into a data structure of size $O(n\log^3 n)$ so that, given a query segment e, we can compute all K intersections between e and $\mathcal{G}$ in time $O(\sqrt{n}\log^2 n + K\log n)$.*

Next we show that, as in Section 6.4, the query time can be improved if we allow more space. Now preprocess $\mathcal{G}$ as described in Section 6.4 (if the segments of $\mathcal{G}$ have endpoints with degree > 3, we modify the set $\mathcal{G}$, as described in Section 6.4.3). Recall that, in Section 6.4, we maintain two data structures — (i) the persistent data structure Υ_1 to store W_i for each triangle Δ_i, and (ii) $\Upsilon_2(\mathcal{G}_i)$ for ray shooting queries. For a query ray ρ, we compute $\mathcal{I}_\rho$ as follows.

Suppose the ray origin p lies in the triangle Δ_i, let the sorted W_i be $(e_1, e_2, \ldots, e_m)$, and suppose $\Phi(W_i, \rho) \in e_k$. Then by Lemma 6.14, $e_k, \ldots, e_m$ intersect ρ in that order along ρ, and we thus obtain all intersections between W_i and ρ. The intersections between $\mathcal{G}_i$ and ρ are obtained by the procedure described above, except that the the size of $\mathcal{Q}$ is now only $O(\sqrt{\frac{n}{r}} \log \frac{n}{r})$ because $|\mathcal{G}_i| \leq \frac{n}{r}$. $\mathcal{I}_\rho$ is then obtained by merging the two output lists of intersections with W_i and $\mathcal{G}_i$. Hence, following the same analysis as in Section 6.4, we can conclude

Theorem 6.23 *Given a collection $\mathcal{G}$ of non-intersecting segments and a parameter $n \log^3 n < m < n^2$, we can preprocess it, in time $O(m \log^\tau n + n\sqrt{m} \log^{\tau-3/2} \frac{n}{\sqrt{m}})$, into a data structure of size $O(m)$ so that, given a query ray ρ, $\mathcal{I}_\rho$ can be computed in $O(\frac{n}{\sqrt{m}} \log^{7/2} \frac{n}{\sqrt{m}} + \log n + |\mathcal{I}_\rho| \log \frac{n}{\sqrt{m}})$ time.* □

Corollary 6.24 *Given a collection $\mathcal{G}$ of n non-intersecting segments and a parameter $n \log^3 n < m < n^2$, we can preprocess it, in time $O(m \log^\tau n + n\sqrt{m} \log^{\tau-3/2} \frac{n}{\sqrt{m}})$, into a data structure of size $O(m)$ so that, given a query segment e, we can compute all K intersections between e and $\mathcal{G}$ in time $O(\frac{n}{\sqrt{m}} \log^{7/2} \frac{n}{\sqrt{m}} + \log n + K \log \frac{n}{\sqrt{m}})$.* □

6.6 Ray Shooting in General Arrangements of Segments

In this section we extend our algorithm to general arrangements of possibly intersecting segments. The section is organized as follows. In Section 6.6.1 we describe how to preprocess $\mathcal{G}$ for ray shooting queries, and in Section 6.6.2 we show how to answer a query. We analyze the time and space complexity of our algorithm in Section 6.6.3 and finally derive a tradeoff between space and query time, similar to that of Section 6.4, in Section 6.6.4.

6.6.1 Preprocessing the segments

In this section $\mathcal{G}$ denotes an arbitrary collection of n segments in the plane. To simplify the exposition, we assume that the segments of $\mathcal{G}$ are bounded. The preprocessing of $\mathcal{G}$ is done as follows. We construct a partition tree $\mathcal{T}$, and associate with each node $v \in \mathcal{T}$ a collection $\mathcal{G}_v \subseteq \mathcal{G}$ of n_v segments, a triangle Δ_v, and another auxiliary set $\mathcal{G}'_v$ of n'_v segments. If $n_v \leq c$, for some fixed constant c, then v is a leaf of $\mathcal{T}$. Otherwise it is an internal node of $\mathcal{T}$, which is further processed as follows. For some fixed constant $r \geq 2$, partition Δ_v into $M = O(r^2)$ triangles $\Delta_1, \ldots, \Delta_M$, using the algorithm described in Chapter 4 (see also [89]), so that each triangle Δ_i meets at most $\frac{n_v}{r}$ lines containing the segments of $\mathcal{G}_v$. Create M children $w_1, \ldots, w_M$ of v, and associate with each child w_i the corresponding triangle $\Delta_{w_i} = \Delta_i$. We put a segment e of $\mathcal{G}_v$ in $\mathcal{G}_{w_i}$ if at least one of the endpoints of e lies in Δ_i. We also associate with w_i an auxiliary set $\mathcal{G}'_{w_i}$ of all segments of $\mathcal{G}_v$ that cross Δ_i but do not have any endpoint inside Δ_i. Let $\mathcal{M}_v$ be the planar map formed by the triangles $\Delta_1, \ldots, \Delta_M$. The root u of $\mathcal{T}$ is associated with $\mathcal{G}$ itself, and Δ_u is a triangle that contains all the segments of $\mathcal{G}$. Moreover, $\mathcal{G}'_u = \emptyset$, by definition.

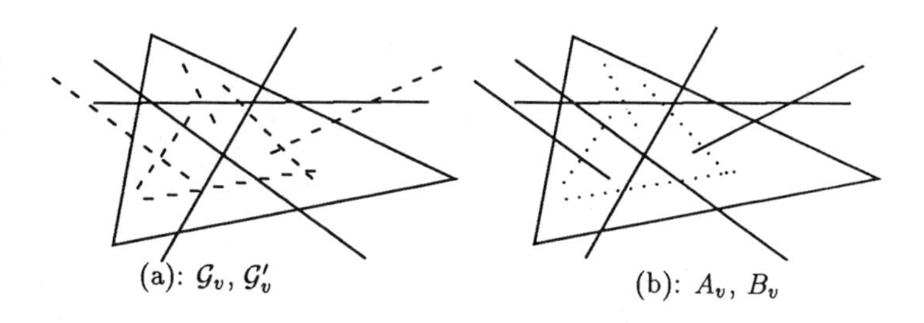

Figure 6.10: (a): solid lines denote $\mathcal{G}'_v$; (b): solid lines denote A_v

We preprocess each node $v \in \mathcal{T}$ as follows. We preprocess the planar map $\mathcal{M}_v$ for point location queries (see [60], [111]) and store the resulting data structure at v. Let $\mathcal{L}'_v$ denote the set of lines containing the segments of $\mathcal{G}'_v$; $|\mathcal{L}'_v| \leq n'_v$. Then we preprocess $\mathcal{L}'_v$ into a data structure $\Upsilon_1(\mathcal{L}'_v)$ for computing $\Phi(\mathcal{L}'_v, \rho)$ using the algorithm of Edelsbrunner et al. [55] for ray shooting in arrangements of lines (see also Section 5.9). If $\Phi(\mathcal{L}'_v, \rho)$ lies outside Δ_v, then we reset it to $+\infty$.

Next, we partition the segments of $\mathcal{G}_v \cup \mathcal{G}'_v$ into two parts:

(i) A_v: "long" segments having at most one of their endpoints inside Δ_v; $|A_v| \leq n_v + n'_v$.

(ii) B_v: "short" segments having both of their endpoints inside Δ_v, that is, segments lying fully inside Δ_v; obviously $|B_v| \leq n_v$.

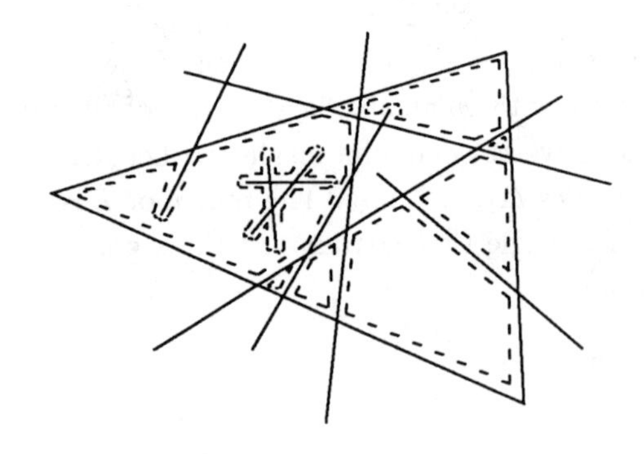

Figure 6.11: Zone of a triangle Δ_i; some faces are non-simple polygons

Definition 6.25: The *zone* of a triangle Δ in an arrangement $\mathcal{A}(\mathcal{G})$ of a set $\mathcal{G}$ of segments is the collection of the face portions $f \cap \Delta$, for all faces $f \in \mathcal{A}(\mathcal{G})$ that intersect $\partial\Delta$ (see figure 6.11).

Using the same argument as in [56], it can be proved that the total number of edges in the zone of a triangle in an arrangement of n segments is $O(n\alpha(n))$. Let $\mathcal{H}_v$ denote the zone of Δ_v in $\mathcal{A}(A_v)$. Since $|A_v| \leq n_v + n'_v$, there are $O((n_v + n'_v)\alpha(n_v + n'_v))$ edges in $\mathcal{H}_v$. Observe that every segment of A_v intersects $\partial\Delta_v$, therefore $\mathcal{H}_v \cup \partial\Delta_v$ is a collection of $k \leq n_v$ simple polygons $\mathbf{P} = \{\mathcal{P}_1, \ldots, \mathcal{P}_k\}$, each of which touches $\partial\Delta_v$ (see figure 6.12). In other words, each $\mathcal{P}_i$ has at least one edge that is a portion of $\partial\Delta_v$.

Moreover, a point $p \in \partial\Delta_v - \bigcup A_v$ lies in exactly one polygon $\mathcal{P} \in \mathbf{P}$. Sort the intersection points between segments of A_v and $\partial\Delta_v$ so that, for a point $p \in \partial\Delta_v - \bigcup A_v$, the polygon containing p can be computed quickly, using binary search. Preprocess each simple polygon $\mathcal{P} \in \mathbf{P}$ into a data structure for ray shooting queries using the algorithm of Chazelle and Guibas [27] (see also [72]). If $\Phi(\mathcal{P}, \rho)$ lies on $\partial\Delta_v$ (that is, ρ does not intersect $\mathcal{P}$ in the interior of Δ_v), then $\Phi(\mathcal{P}, \rho)$ is reset to $+\infty$. Let $\Upsilon_2(A_v)$ be the union of all these structures.

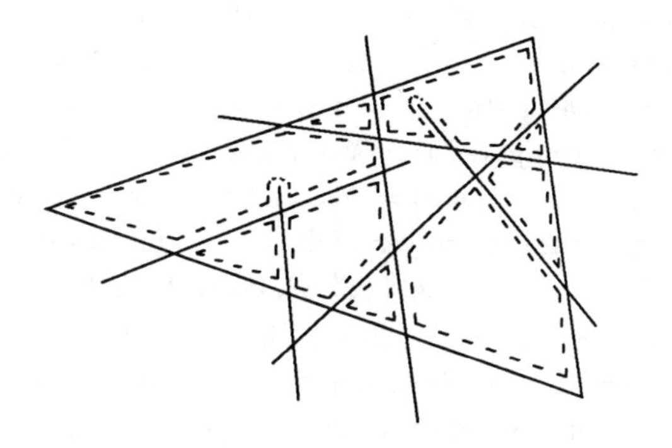

Figure 6.12: Zone of Δ_v for A_v; each face is a simple polygon (possibly with some overlapping edges)

Finally, we preprocess the short segments in B_v for ray shooting in non-simple polygons. Let $\mathcal{F}_v$ denote the boundary of the unbounded face in $\mathcal{A}(\bigcup_{j \le M} B_{w_j})$, where $w_1, \ldots, w_M$ are the children of v. By the result of Pollack et al. [105] (see also [75]), $\mathcal{F}_v$ has $O(n_v \alpha(n_v))$ segments, since $|\bigcup_{j \le M} B_{w_j}| \le |B_v| \le n_v$. Since the segments of $\mathcal{F}_v$ are non-intersecting, we preprocess them into a data structure $\Upsilon_3(B_v)$ for computing $\Phi(\mathcal{F}_v, \rho)$, using the algorithm described in Section 6.3.

We repeat this preprocessing for every node v of $\mathcal{T}$. The resulting collection of data structures is the output of the preprocessing stage.

6.6.2 Answering a query

Let ρ be a query ray emanating from a point p in direction d. The query is answered by traversing a path Π_ρ of $\mathcal{T}$ and computing $\sigma_v = \Phi(\mathcal{G}_v, \rho)$ at each node $v \in \Pi_\rho$ in a bottom-up fashion. At the end of this process we obtain at the root u, $\sigma_u = \Phi(\mathcal{G}_u, \rho) = \Phi(\mathcal{G}, \rho)$.

The path Π_ρ is defined so that for each node v along Π_ρ the ray origin p lies in Δ_v. The processing of a node v of Π_ρ is done as follows. If v is a leaf, then we compute σ_v directly using a brute-force method; otherwise we locate the triangle $\Delta_t \in \mathcal{M}_v$ containing the ray origin p and obtain the corresponding child t of v along Π_ρ. We next test whether $\Phi(\mathcal{G}_v, \rho)$ lies inside Δ_t. This is done by calculating each of $\sigma_t = \Phi(\mathcal{G}_t, \rho)$ and $\sigma'_t = \Phi(\mathcal{G}'_t, \rho)$. The former of the two is computed recursively and the latter is computed using $\Upsilon_1(\mathcal{L}'_t)$. If $\Phi(\mathcal{L}'_t, \rho) \in \Delta_t$, then $\sigma'_t = \Phi(\mathcal{L}'_t, \rho)$, because segments of $\mathcal{G}'_t$ do not have endpoints inside Δ_t; moreover $\Phi(\mathcal{L}'_t, \rho) \notin \Delta_t$ implies $\sigma'_t \notin \Delta_t$. Thus σ'_t can be determined by computing $\Phi(\mathcal{L}'_t, \rho)$, using $\Upsilon_1(\mathcal{L}'_t)$.

If either σ_t or σ'_t lies inside Δ_t, then we have already found $\sigma_v = \Phi(\mathcal{G}_v, \rho)$. Otherwise we know that σ_v lies outside Δ_t, in which case we proceed as follows. Let W_ρ denote the set of children of v, other than Δ_t, that intersect ρ. W_ρ can be computed by traversing the ray ρ in the planar map $\mathcal{M}_v$. It is easily seen that

$$\begin{aligned}
\Phi(\mathcal{G}_v, \rho) &= \min_{w \in W_\rho} \{\Phi(\mathcal{G}_w \cup \mathcal{G}'_w, \rho)\} \\
&= \min_{w \in W_\rho} \{\min\{\Phi(A_w, \rho), \Phi(B_w, \rho)\}\} \\
&= \min \left\{ \min_{w \in W_\rho} \{\Phi(A_w, \rho)\}, \min_{w \in W_\rho} \{\Phi(B_w, \rho)\} \right\} \\
&= \min \left\{ \min_{w \in W_\rho} \{\Phi(A_w, \rho)\}, \{\Phi(\bigcup_{w \in W_\rho} B_w, \rho)\} \right\}. \qquad (6.4)
\end{aligned}$$

For $w \in W_\rho$, p does not lie in Δ_w, therefore either ρ does not intersect A_w, or if it does, then $\Phi(A_w, \rho)$ lies on an edge of $\mathcal{H}_w$. Let ξ be the first intersection point of ρ and $\partial\Delta_w$, and let $\mathcal{P}$ be the polygon of $\mathcal{H}_w$ containing ξ (see figure 6.13). (Assume that $\xi \notin \cup A_w$, because then we already know $\Phi(A_w, \rho)$.) If ρ' is the ray emanating from ξ in the same direction as ρ, then

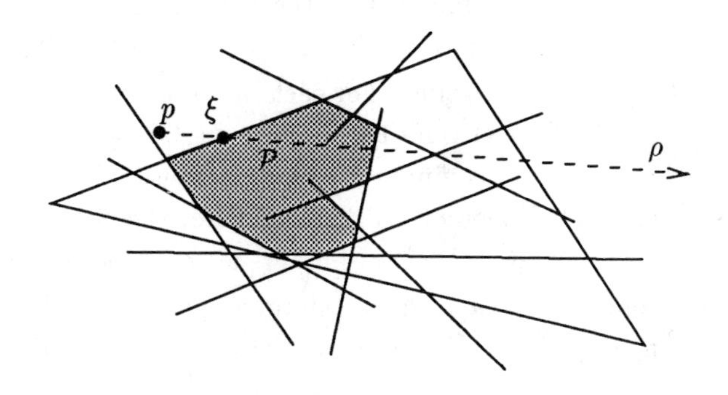

Figure 6.13: $\mathcal{A}(A_{w_j})$ and $\mathcal{P}$

it is obvious that

$$\Phi(A_w,\rho) = \Phi(A_w,\rho') = \Phi(\mathcal{P},\rho').$$

But $\Phi(\mathcal{P},\rho')$, and thus also $\Phi(A_w,\rho)$, can be computed from $\Upsilon_2(A_w)$ in logarithmic time.

Since ρ does not intersect B_{w_t}, the ray origin p lies in the unbounded face of $\mathcal{A}(B_{w_t})$. Moreover, observe that the segments of B_{w_j} are contained in Δ_{w_j}, which implies that p lies in the unbounded face of $\mathcal{A}(B_{w_j})$, for any child of v. Therefore p lies in the unbounded face of $\mathcal{A}(\bigcup_{j\leq M} B_{w_j})$ (where M is the number of children of v). Hence,

$$\Phi(\bigcup_{w\in W_\rho} B_w,\rho) = \Phi(\mathcal{F}_v,\rho).$$

The desired $\Phi(\mathcal{G}_v,\rho)$ is now obtained using (6.4) and $\Upsilon_3(\mathcal{F}_v)$.

6.6.3 Analysis of the algorithm

The correctness of the algorithm follows from the above discussion, so we only have to analyze the time and space complexity of the algorithm. First consider the query time $Q(n)$. Let Π_ρ be the path followed by the algorithm as it computes $\Phi(\mathcal{G},\rho)$. We bound the time spent at each node

$v \in \Pi_\rho$. We spend $O(\log r)$ time to find the triangle Δ_t containing the ray origin p. It follows from [55] (see also Section 5.9) that $\Phi(\mathcal{L}'_{w_t}, \rho)$ can be computed in $O(\sqrt{n'_{w_t}} \log n'_{w_t})$ time. For each of these triangles Δ_w intersected by ρ, we spend $O(\log n)$ time to compute $\Phi(A_w, \rho)$ [27]; we spend $O(\sqrt{n_v \alpha(n_v)} \log^2 n_v)$ time to compute $\Phi(\mathcal{F}_v, \rho)$ because $\mathcal{F}_v$ has at most $O(n_v \alpha(n_v))$ segments (cf. Theorem 6.11). Therefore, the time spent at a node v is at most $O(\sqrt{n_v \alpha(n_v)} \log^2 n_v + \sqrt{n'_{w_t}} \log n_v + s \log n_v)$. Since $s \leq r^2 = O(1)$ and $n'_{w_t} \leq n'_v$, the total time spent at the node v is $O(\sqrt{n_v \alpha(n_v)} \log^2 n + \sqrt{n'_v} \log n)$. Summing over all nodes of Π_ρ, we obtain

$$Q(n) = \sum_{v \in \Pi_\rho} O\left(\sqrt{n_v \alpha(n_v)} \log^2 n + \sqrt{n'_v} \log n\right). \tag{6.5}$$

It is easily checked that for a node v at level i of $\mathcal{T}$, we have $n_v \leq \frac{n}{r^i}$. Moreover if z is the father of v, then by construction, Δ_v meets at most $\frac{n_z}{r}$ segments of $\mathcal{G}_z$, which implies $n'_v \leq \frac{n_z}{r}$. Therefore

$$n_v \leq \frac{n}{r^i} \quad \text{and} \quad n'_v \leq \frac{n}{r^i} \tag{6.6}$$

Thus, (6.5) becomes

$$\begin{aligned} Q(n) &= \sum_{i=0}^{\log n} O\left(\sqrt{\frac{n\alpha(n)}{r^i}} \log^2 n\right) \\ &= O(\sqrt{n\alpha(n)} \log^2 n) \qquad \text{(because } r \geq 2\text{)}. \end{aligned}$$

Next, let us analyze the space complexity $S(n)$ and the preprocessing time $P(n)$ of our algorithm. At each node $v \in \mathcal{T}$ we store the following data structures:

(i) $\mathcal{M}_v$: The node v is partitioned into $O(r^2)$ triangles in time $O(n_v r \log n_v \cdot \log^{\tau-1} r)$ (cf. Theorem 4.36), therefore by [60], $\mathcal{M}_v$ can be preprocessed, in time $O(r^2 \log r)$, into a data structure of size $O(r^2)$ for point location queries. Since r is chosen to be constant, the time bound is just $O(n_v \log n_v)$ and the space required is $O(n_v)$.

(ii) $\Upsilon_1(\mathcal{L}'_v)$: It follows from the result of Edelsbrunner et al. [55] (see also Corollary 5.25) that $\Upsilon_1(\mathcal{L}'_v)$ requires $O(n'_v \log^2 n'_v)$ space, and can be constructed in $O(n'^{3/2}_v \cdot \log^\tau n'_v)$ time.

(iii) $\Upsilon_2(A_v)$: Since $\mathcal{H}_v$ has at most $O((n_v + n'_v)\alpha(n_v + n'_v))$ edges, $\Upsilon_2(A_v)$ requires only $O((n_v + n'_v)\alpha(n_v + n'_v)))$ space (cf. [27]), and can be constructed in $O((n_v + n'_v)\alpha(n_v + n'_v)\log(n_v + n'_v))$ time.

(iv) $\Upsilon_3(\mathcal{F}_v)$: $\mathcal{F}_v$ has $O(n_v\alpha(n_v))$ edges, therefore by Theorem 6.11, $\Upsilon_3(\mathcal{F}_v)$ requires only $O(n_v\alpha(n_v)\log^3 n_v)$ space, and can be constructed in $O(n_v^{3/2}\alpha^{3/2}(n_v)\log^\tau n_v)$ time (which subsumes the $O(n_v\alpha(n_v)\log^2 n_v)$ time needed to compute $\mathcal{F}_v$, as in [59]).

Thus, the space used at v is $O(n_v\alpha(n_v)\log^3 n_v + n'_v \log^2 n_v)$. Summing over all nodes of $\mathcal{T}$, we get

$$S(n) = \sum_{v\in\mathcal{T}} (n_v\alpha(n_v)\log^3 n_v + n'_v \log^2 n'_v).$$

Observe that each triangle of $\mathcal{M}_v$ meets $O(\frac{n_v}{r})$ segments of $\mathcal{G}_v$, therefore if W_v is the set of children of v, then

$$\sum_{w\in W_v} n'_w = O(n_v r), \tag{6.7}$$

which implies that

$$\begin{aligned} S(n) &= \sum_{v\in\mathcal{T}} O(n_v\alpha(n_v)\log^3 n_v + n_v r \log^2 n_v) \\ &= \sum_{v\in\mathcal{T}} O(n_v\alpha(n_v)\log^3 n_v). \end{aligned}$$

Since each endpoint of a segment $e \in \mathcal{G}$ falls inside one triangle Δ_v for each level of $\mathcal{T}$, e is associated with at most two nodes at each level. Let $l(v)$ denote the level of the node v in $\mathcal{T}$. Then for every $i \le \log n$, we have

$$\sum_{l(v)=i} n_v \le 2n. \tag{6.8}$$

Hence,

$$S(n) = \sum_{i=1}^{\log n} O(n\alpha(n)\log^3 n) = O(n\alpha(n)\log^4 n).$$

Finally, we bound the preprocessing time $P(n)$ of our algorithm The above discussion implies that the total time spent in preprocessing is at most

$$\begin{aligned} P(n) &= \sum_{v \in T} O(n_v^{3/2} \alpha^{3/2}(n_v) \log^\tau n + n_v'^{3/2} \alpha^{3/2}(n_v') \log^\tau n) \\ &= \sum_{v \in T} O(n_v^{3/2} \alpha^{3/2}(n_v) \log^\tau n). \end{aligned} \tag{6.9}$$

Since τ is a constant < 4.33, we can replace the term $\alpha^{3/2}(n_v) \log^\tau n$ in (6.9) with $\log^\tau n$, where τ is a different constant but still < 4.33. Therefore

$$\begin{aligned} P(n) &= \sum_{i=1}^{\log n} \sum_{l(v)=i} O(n_v^{3/2} \log^\tau n) \\ &= \sum_{i=1}^{\log n} O\left(r^i \left(\frac{2n}{r^i} \right)^{3/2} \log^\tau n \right) \\ &= O\left(n^{3/2} \log^\tau n \sum_{i=1}^{\log n} \frac{1}{(\sqrt{r})^i} \right) \\ &\leq O(n^{3/2} \log^\tau n) \qquad \text{(because } r \geq 2\text{)}. \end{aligned}$$

Hence, we can conclude that

Theorem 6.26 *Given a collection $\mathcal{G}$ of n (possibly intersecting) segments, we can preprocess $\mathcal{G}$, in time $O(n^{3/2} \log^\tau n)$, into a data structure of size $O(n\alpha(n) \log^4 n)$ so that, for any query ray ρ, we can compute $\Phi(\mathcal{G}, \rho)$ in $O(\sqrt{n\alpha(n)} \log^2 n)$ time.*

□

Remark 6.27: If $\mathcal{G}$ contains unbounded segments, then the triangle Δ_u associated with the root u of $\mathcal{T}$ should be a triangle that contains all intersection points and all bounded segments of $\mathcal{G}$. Such a Δ_u can be easily computed in $O(n \log n)$ time. Now for each segment $e \in \mathcal{G}$, we compute e' $= e \cap \Delta_u$ and apply our algorithm to the new set of segments. The portions of the segments lying in the exterior of Δ_u do not intersect each other, and are ordered in the non-decreasing order of their slopes along $\partial \Delta_u$ in counter-clockwise direction. Therefore if a query ray does not hit a segment of $\mathcal{G}$ inside Δ_u, we can determine, in additional $O(\log n)$ time, the first segment hit by the ray outside Δ_u, which shows that our algorithm works for unbounded segments as well.

6.6.4 Tradeoff between space and query time

In this subsection we establish a tradeoff between space and query time for ray shooting in general arrangements of segments. As in Section 6.4 we first give a very simple algorithm that preprocesses $\mathcal{G}$, in time $O(n^2\alpha^2(n)\log n)$, into a data structure of size $O(n^2\alpha^2(n))$ so that, given a query ray ρ, $\Phi(\mathcal{G},\rho)$ can be computed in $O(\log n)$ time.

We compute the arrangement $\mathcal{A}(\mathcal{G})$ in time $O(n^2 \log n)$ using the line sweep method [107] (or in time $O(n^2)$ using a more involved algorithm [62]), and preprocess $\mathcal{A}(\mathcal{G})$ for point location queries [60], [111]. Since the edges of $\mathcal{A}(\mathcal{G})$ are non-intersecting, we can preprocess each face of $\mathcal{A}(\mathcal{G})$ into a data structure Υ_f for logarithmic-time ray shooting queries, using $O(|n_f|^2)$ space, where n_f is the number of edges bounding f, as described in Section 6.4.

To compute $\Phi(\mathcal{G},\rho)$, for a query ray ρ, we first locate the face f of $\mathcal{A}(\mathcal{G})$ containing the ray origin p. Obviously $\Phi(\mathcal{G},\rho)$ lies on the boundary of f and therefore $\Phi(\mathcal{G},\rho) = \Phi(\partial f,\rho)$ can be computed in $O(\log n)$ time, using Υ_f. Thus, the overall query time is $O(\log n)$.

As for the storage, $\mathcal{A}(\mathcal{G})$ can be preprocessed for point location queries using $O(n^2)$ space (cf. [60], [111]). The total space required to store all Υ_f is $O(\sum_f n_f^2)$. Theorem 6.15 implies that the preprocessing time is $O(\sum_f n_f^2 \log n)$, and it has been shown in [56] that

$$\sum_{f\in\mathcal{A}(\mathcal{G})} n_f^2 = O(n^2\alpha^2(n)).$$

Hence, we have

Theorem 6.28 *Given a collection $\mathcal{G}$ of n segments in the plane, we can preprocess $\mathcal{G}$, in time $O(n^2\alpha^2(n)\log n)$, into a data structure of size $O(n^2\alpha^2(n))$ so that, given a ray ρ, $\Phi(\mathcal{G},\rho)$ can be computed in $O(\log n)$ time.*

□

Next we give an algorithm for the general case, where $n^{1+\epsilon_0} \le m \le n^{2-\epsilon_1}$, for some constants $\epsilon_0, \epsilon_1 > 0$. We thus write $m = n^\gamma$, for some $1+\epsilon_0 \le \gamma \le$

$2 - \epsilon_1$. To preprocess $\mathcal{G}$ into a data structure of size $O(m)$, we proceed in the same way as in Section 6.6.1 except that at each node v we are allowed more space, so we construct larger-size data structures that facilitate faster ray shooting in $\mathcal{L}'_v$, $\mathcal{F}_v$, etc.

Edelsbrunner et al. [55] (see also Corollary 5.25) have shown that, given a set $\mathcal{L}$ of n lines and a parameter $1 \leq \beta_1 < n$, $\mathcal{L}$ can be preprocessed, in $O(n^{3/2}\sqrt{\beta_1}\log^\tau n)$ time[1], into a data structure $\Upsilon_1(\mathcal{L})$ of size $O(n\beta_1 \log^2 n)$ so that, for any query ray ρ we can compute $\Phi(\mathcal{L}, \rho)$ in $O(\sqrt{\frac{n}{\beta_1}}\log n)$ time. At each node v of level i, we store $\Upsilon_1(\mathcal{L}'_v)$ with an appropriate value of $\beta_1 = \beta_1^i$ (to be specified later).

Similarly, we have shown in Section 6.4 that, given a set $\mathcal{E}$ of n non-intersecting segments and a parameter β_2, we can preprocess $\mathcal{E}$, in time $O(n^{3/2}\sqrt{\beta_2}\log^\tau n)$, into a data structure $\Upsilon_3(\mathcal{E})$ of size $O(n\beta_2 \log^3 n)$ so that, for a query ray ρ, we can compute $\Phi(\mathcal{E}, \rho)$ in time $O(\sqrt{\frac{n}{\beta_2}}\log^2 n)$. For a node v at level i, we store $\Upsilon_3(\mathcal{F}_v)$ with an appropriate value of $\beta_2 = \beta_2^i$. (Recall that if $\mathcal{F}_v$ has a vertex with degree > 3, then we need to modify the segments of $\mathcal{F}_v$, as described in Section 6.4.3.)

Besides these two types of data structures, we also store $\Upsilon_2(A_v)$ of Section 6.6.1 at each node v of $\mathcal{T}$. For answering a query, we use the same procedure, as described in Section 6.6.2.

Next, we analyze the complexity of this algorithm. First, consider the space used by our algorithm. Since $|\mathcal{L}'_v| = n'_v$, $|\mathcal{F}_v| = O(n_v\alpha(n_v))$ and $|A_v| \leq n'_v + n_v$, the space used by a node v of $\mathcal{T}$ at level i is

$$O(n'_v\beta_1^i \log^2 n'_v + n_v\alpha(n_v)\beta_2^i \log^3 n_v + (n_v + n'_v)\alpha(n_v + n'_v)) = O(n'_v\beta_1^i \log^2 n'_v + n_v\beta_2^i \log^3 n_v).$$

Therefore the total space used is

$$S(n) = \sum_{i=0}^{\log n} \sum_{l(v)=i} O(n'_v\beta_1^i \log^2 n'_v + n_v\alpha(n_v)\beta_2^i \log^3 n_v)$$

[1] Actually the preprocessing time is $O((n\beta_1 + n^{3/2}\sqrt{\beta_1})\log^\tau n)$, but it can be verified that for our choice of β_1 the first term never dominates, so for simplicity we only write the second term.

$$= \sum_{i=0}^{\log n} O\Big(\sum_{l(v)=i} n'_v \beta_1^i \log^2 n'_v\Big) + \sum_{i=0}^{\log n} O\Big(\sum_{l(v)=i} n_v \beta_2^i \alpha(n) \log^3 n_v\Big).$$

By (6.7) and (6.8), we obtain

$$S(n) = \sum_{i=0}^{\log n} O(\beta_1^i n \log^2 n'_v) + \sum_{i=0}^{\log n} O(\beta_2^i n \alpha(n) \log^3 n_v).$$

Let n_i (resp. n'_i) denote the maximum value of $|\mathcal{G}_v|$ (resp. $|\mathcal{G}'_v|$). By (6.6), we have $n_i, n'_i \leq \frac{n}{r^i}$. If we choose $\beta_1^i = \dfrac{n'^{\gamma-1}_i}{\log^3 n'_i}$ and $\beta_2^i = \dfrac{n_i^{\gamma-1}}{\alpha(n)\log^3 n_i}$, which is easily seen to satisfy $\beta_1^i, \beta_2^i > 1$, then we obtain

$$\begin{aligned} S(n) &= \sum_{i=0}^{\log n} O\Big(\frac{n'^{\gamma-1}_i}{\log^3 n'_i} n \log^2 n'_i\Big) + \\ &\quad \sum_{i=0}^{\log n} O\Big(\frac{n_i^{\gamma-1}}{\alpha(n)\log^3 n_i}\alpha(n)\log^3 n_i\Big) \\ &= O\Big(\sum_{i=0}^{\log n} \frac{n^\gamma}{r^{i(\gamma-1)}\log n'_i}\Big) + O\Big(\sum_{i=0}^{\log n} \frac{n^\gamma}{r^{i(\gamma-1)}}\Big) \\ &= O(n^\gamma) \qquad \text{(because } \gamma > 1) \\ &= O(m). \end{aligned}$$

As for the query time,

$$Q(n) = \sum_{i=0}^{\log n} O\left(\sqrt{\frac{n'_i}{\beta_1^i}} \log n'_i\right) + \sum_{i=0}^{\log n} O\left(\sqrt{\frac{n_i\alpha(n_i)}{\beta_2^i}} \log^2 n_i\right).$$

Substituting the values of β_1^i and β_2^i, we obtain

$$\begin{aligned} Q(n) &= \sum_{i=0}^{\log n} O\left(\sqrt{\frac{n'_i}{n'^{\gamma-1}_i/\log^3 n'_i}} \log n'_i\right) + \\ &\quad \sum_{i=0}^{\log n} O\left(\sqrt{\frac{n_i\alpha(n_i)}{n_i^{\gamma-1}/(\alpha(n)\log^3 n_i)}} \log^2 n_i\right) \\ &= \sum_{i=0}^{\log n} O\left(\left(\frac{n}{r^i}\right)^{1-\gamma/2} \alpha(n) \log^{7/2} n\right) \end{aligned}$$

$$= O(n^{1-\gamma/2}\alpha(n)\log^{7/2} n)$$
$$= O\left(\frac{n\alpha(n)}{\sqrt{m}}\log^{7/2} n\right).$$

Finally, the time spent in preprocessing a node v of $\mathcal{T}$ at level i is

$$O({n'_v}^{3/2}\sqrt{\beta_1^i}\log^\tau n'_v) + O(n_v^{3/2}\sqrt{\beta_2^i}\alpha^{3/2}(n_v)\log^\tau n_v).$$

Using the same argument as for (6.9), we can ignore the term $\alpha^{3/2}(n_v)$ in the above equality. Therefore summing over all nodes, we get

$$\begin{aligned}
P(n) &= \sum_{i=0}^{\log n}\sum_{l(v)=i} O({n'_v}^{3/2}\sqrt{\beta_1^i}\log^\tau n'_v + n_v^{3/2}\sqrt{\beta_2^i}\log^\tau n_v) \\
&= O\left(\sum_{i=0}^{\log n}\left(n^{3/2}\sqrt{\frac{n^{\gamma-1}}{r^{i(\gamma-1)}\log^3 n}}\right)\log^\tau n\right) \\
&\quad \text{(Substituting the values of } \beta_1^i \text{ and } \beta_2^i\text{, and using (6.6))} \\
&= O(n\sqrt{n^\gamma}\log^{\tau-3/2} n) \\
&= O(n\sqrt{m}\log^{\tau-3/2} n).
\end{aligned}$$

Hence, we can conclude that

Theorem 6.29 *Given a set $\mathcal{G}$ of n segments and a parameter $n^{1+\epsilon_0} \le m \le n^{2-\epsilon_1}$, for some constants $\epsilon_0, \epsilon_1 > 0$, we can preprocess $\mathcal{G}$, in time $O(n\sqrt{m}\cdot\log^{\tau-3/2} n)$, into a data structure of size $O(m)$ so that, for any query ray ρ, we can compute $\Phi(\mathcal{G},\rho)$ in time $O\left(\frac{n\alpha(n)}{\sqrt{m}}\log^{7/2} n\right)$.*

□

Remark 6.30: The algorithm of [EGH*] actually constructs $\Upsilon_1(\mathcal{L})$, in time

$$O(n\beta_1\log n\cdot\log^{\tau-1}\beta_1 + n^{3/2}\sqrt{\beta_1}\log^\tau\frac{n}{\beta_1}),$$

using $O(n\beta_1\log\frac{n}{\beta_1})$ space, and answers a query in time $O(\sqrt{\frac{n}{\beta_1}}\log\frac{n}{\beta_1})$. Similarly the algorithm described in Section 6.4 constructs $\Upsilon_3(\mathcal{G})$, in time

$$O(n\beta_2\log n\cdot\log^{\tau-1}\beta_2 + n^{3/2}\sqrt{\beta_2}\log^\tau\frac{n}{\beta_2}),$$

using $O(n\beta_2 \log \frac{n}{\beta_2})$ space, and answers a query in $O(\sqrt{\frac{n}{\beta_2}} \log \frac{n}{\beta_2})$ time. Using these bounds in the above analysis, we can improve the query time $Q(n)$ to

$$O\left(\frac{n\alpha(n)}{\sqrt{m}} \log^{7/2}\left(\frac{n\alpha(n)}{\sqrt{m}}\right) + \log n\right).$$

The preprocessing time is now $O\left(m \log^\tau n + n\alpha(n)\sqrt{m} \log^{\tau-3/2}(\frac{n\alpha(n)}{\sqrt{m}})\right)$.

6.7 Implicit Point Location — Preprocessing Version

In this section we describe an efficient algorithm for the preprocessing version of the implicit point location problem, which can be formally defined as follows.

> *We are given a collection $\mathcal{G} = \{e_1, \ldots, e_n\}$ of n segments. With each segment e we associate a function ψ_e defined on the entire plane, which assumes values in some associative and commutative semigroup S (denote its operation by $+$). We define $\Psi(x) = \sum_{e \in \mathcal{G}} \psi_e(x)$. We want to preprocess $\mathcal{G}$ so that, for any query point p, we can quickly compute $\Psi(p)$.*

We assume that ψ_e and Ψ satisfy the same conditions as in Section 5.7.

It is shown in [73] that many natural problems including the problem of determining whether p lies in the union of the given objects, or of counting how many objects contain p, fall into this scheme. See also the following section for details.

The goal is to come up with an algorithm that uses $O(n \log^{O(1)} n)$ space and computes $\Psi(p)$, for any query point p, in sublinear time. Guibas et al. [73] gave a randomized algorithm, with $O(n \log^{k+1} n)$ expected running time, to construct a data structure of $O(n)$ size so that, for a query point p, $\Psi(p)$ can be computed in $O(n^{2/3+\delta})$ time, for any $\delta > 0$. In this section we present an algorithm that improves the query time to $O(\sqrt{n} \log^2 n)$, and makes the preprocessing deterministic (albeit no longer close to linear).

Let $\mathcal{L}$ denote the set of lines containing the segments of $\mathcal{G}$. Dualize the lines of $\mathcal{L}$ to obtain a set $\mathcal{L}^*$ of n points. Let $\mathbf{C} = \{\mathcal{C}_1, \ldots, \mathcal{C}_k\}$ denote a family of $k = O(\log n)$ spanning paths on $\mathcal{L}^*$, with $\sigma(\mathbf{C}) = O(\sqrt{n})$. We show how to preprocess a single path $\mathcal{C} \in \mathbf{C}$.

First, construct a binary tree $\mathcal{B} = \mathcal{B}(\mathcal{C})$ as in Section 6.2. With each node v of $\mathcal{B}$ we associate a set $\mathcal{G}_v$ of segments $e \in \mathcal{G}$ such that the dual of the line containing e belongs to S_v (as defined in Section 6.2). At each node v, we store $\mathcal{D}(\mathcal{G}_v)$ so that, for any query point p lying either above all the lines containing the segments of $\mathcal{G}_v$ or below all of them, $\Psi_v(p) = \sum_{e \in \mathcal{G}_v} \psi_e(p)$ can be computed in $O(\log n)$ time.

For a given query point p, we compute $\Psi(p)$ as follows. Let p^* denote the dual of p. Obviously

$$\sum_{i=1}^{n} \psi_{e_i}(p) = \sum_{v \in V_{\mathcal{B}}(p^*)} \Psi_v(p).$$

Therefore, it suffices to show how to compute $\Psi_v(p)$, for a node $v \in V_{\mathcal{B}}(p^*)$. Observe that for any $v \in V_{\mathcal{B}}(p^*)$, p^* lies either above all the points of S_v, or below all of them, say below. Since duality preserves the above-below relationship, p lies below all the lines containing the segments of $\mathcal{G}_v$. Therefore, $\Psi_v(p)$ can be easily computed in $O(\log n)$ time using $\mathcal{D}(\mathcal{G}_v)$.

Next, let us analyze the complexity of our algorithm. First consider the time spent in answering a query. By Theorem 3.5, we can determine, in $O(\log n)$ time, a path $\mathcal{C} \in \mathbf{C}$ that intersects p^* in at most $O(\sqrt{n})$ edges, and it follows from the discussion in Section 6.2 that, for a given line p^*, $V_{\mathcal{B}}(p^*)$ can be computed in $O(\sqrt{n} \log n)$ time. By property (iii), for each $v \in V_{\mathcal{B}}(p^*)$, $\Psi_v(p)$ can be calculated in $O(\log n)$ time. Thus the total time spent is $O(\sqrt{n} \log^2 n)$. As for the space complexity, $\mathcal{D}(\mathcal{G}_v)$ requires $O(|\mathcal{G}_v|)$ space. Since the segments associated with the nodes of $\mathcal{B}$ at the same level are pairwise disjoint, the total space required to store $\mathcal{B}$ is $O(n \log n)$. Finally, the preprocessing time is bounded by the time spent in computing $\mathbf{C}$ plus the time spent in preprocessing $\mathcal{G}_v$ for all $v \in \mathcal{B}$. Hence, the total preprocessing time is $O(n^{3/2} \log^\tau n + n \log^{k+2} n) = O(n^{3/2} \log^\tau n)$.

Therefore, we can conclude

Theorem 6.31 *Given a collection $\mathcal{G}$ of n segments, and function ψ_e associated with each segment satisfying properties (i)–(iii), we can preprocess $\mathcal{G}$,* *in* $O(n^{3/2}\log^{\tau} n)$ *time, into a data structure of size* $O(n\log^2 n)$ *so that, for any query point* p, $\Psi(p)$ *can be computed in* $O(\sqrt{n}\log^2 n)$ *time.*

□

Remark 6.32:

(i) As in Section 6.3, we can reduce the space complexity to $O(n\log n)$ by maintaining a single tree structure instead of a family of $O(\log n)$ trees. Also, if we allow randomization, then the (expected) preprocessing time is $O(n^{4/3}\log^2 n)$, but the query time increases by a factor of $\log n$.

(ii) In some applications, where calculation of $\Psi(x)$ in (iii) above is accomplished by a binary search, it is possible to reduce the query time to $O(\sqrt{n}\log n)$, using fractional cascading.

(iii) As in the case of the ray shooting problem, if we allow more space, the query time can be improved. Instead of describing the tradeoff for the general case, we will describe it in the next section for a specific example.

(iv) In Section 5.7 we solved the *batched* version of this problem, where all the query points p are given in advance. We present there a solution that runs in time

$$O(m^{2/3}n^{2/3}\log^{2/3} n\log^{\tau/3}\frac{n}{\sqrt{m}} + n\log^k n\log\frac{n}{\sqrt{m}} + m\log n),$$

where m is the number of given query points.

6.8 Other Applications

In this section we consider other applications of our technique. All these problems were studied in [73]; the authors obtained algorithms with $O(n^{2/3+\delta})$ query time, for any $\delta > 0$. We show that using our approach the query time can be reduced to roughly $\sqrt{n}$.

6.8.1 Polygon containment problem — preprocessing version

First consider the following problem:

> *Given a set* **T** *of n (possibly intersecting) triangles, we want to preprocess* **T** *so that, given a query point p, we can quickly count the number of triangles in* **T** *containing p (or just determine whether p lies in the union of these triangles).*

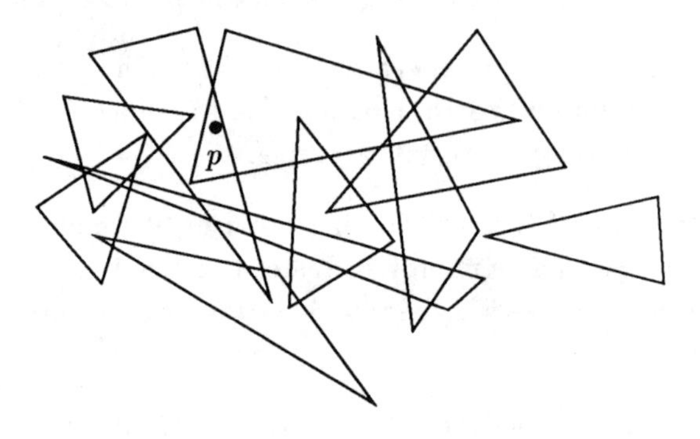

Figure 6.14: Polygon containment problem

We first present an algorithm that uses roughly linear space, and then show that the query time can be improved by using more space. Our algorithm is based on the following observation of [73]. Let $\mathcal{G}$ denote the set of edges bounding the triangles in **T** and, for each $e \in \mathcal{G}$, let $B(e)$ denote the semi-infinite trapezoidal strip lying below e. Define a function ψ_e in the plane so that $\psi_e(p) = 0$ for a point p outside B_e, and for $p \in B_e$, $\psi_e(p) = 1$, if the triangle corresponding to e lies below the line containing the segment e, otherwise $\psi_e(p) = -1$. It can be checked that $\Psi(p)$, for a point p, gives the number of triangles of **T** containing p. Moreover, ψ_e obviously satisfies properties (i)–(ii). As to property (iii), if a point p lies above all lines containing the given edges then $\Psi(p) = 0$, by definition. On the other hand if p lies below all these lines, we do the following. Let $\hat{e}$ denote the x-projection of an edge e of some triangle. It is easily checked that

$$\Psi(p) = \sum_{p_x \in \hat{e}_j} \epsilon_j,$$

where ϵ_j is the non-zero value of ψ_{e_j} at p. Note that the sum of the right hand side remains the same between two consecutive endpoints of the projected segments, and the constant values of Ψ over these intervals can be computed, in overall time $O(n \log n)$, by scanning the projected segments from left to right. Hence, we can preprocess $\mathbf{T}$, in time $O(n \log n)$, into a data structure $\mathcal{D}$ so that, for a point p lying below all lines of $\mathcal{L}$, $\Psi(p)$ can be computed in $O(\log n)$ time.

Thus, the observation of [73] and Theorem 6.31 imply that by preprocessing $\mathcal{G}_v$ into the above data structure $\mathcal{D}_v$, for each node v of $\mathcal{B}$, the number of triangles in $\mathbf{T}$ containing a query point p can be counted in $O(\sqrt{n} \log^2 n)$ time. But observe that each of the data structures $\mathcal{D}_v$ is a sorted list, and at each node v we do a binary search in $\mathcal{D}_v$ to compute Ψ_v. We can therefore apply fractional cascading technique of [28] to the collection of lists $\mathcal{D}_v$ attached to the nodes v of $\mathcal{B}$. This will allow us to search through the lists $\mathcal{D}_v$ of all nodes $v \in V_{\mathcal{B}}(\ell)$ in overall time $O(\log n + |V_{\mathcal{B}}(\ell)|) = O(\sqrt{n} \log n)$. Hence, we have

Theorem 6.33 *Given a set* $\mathbf{T}$ *of* n *triangles in the plane, we can preprocess* $\mathbf{T}$*, in time* $O(n^{3/2} \log^{\tau} n)$*, into a data structure of size* $O(n \log^2 n)$ *so that, given a query point* p*, we can determine, in time* $O(\sqrt{n} \log n)$*, the number of triangles in* $\mathbf{T}$ *containing the point* p*.*

□

We next establish a tradeoff between space and query time for the polygon containment problem. If we allow $O(n^2)$ space, then we can construct the entire arrangement $\mathcal{H}$ of $\bigcup_{e \in \mathcal{G}} B_e$. It is easily seen that the value of Ψ does not change within a face of $\mathcal{H}$, and while constructing $\mathcal{H}$ we can compute Ψ for each of its face. Now given a point p, we can compute $\Psi(p)$ in $O(\log n)$ time by locating p in $\mathcal{H}$. Thus if we allow quadratic storage, the query time can be reduced to $O(\log n)$. Next we give an algorithm for the general case when $n \log^2 n < m < n^2$.

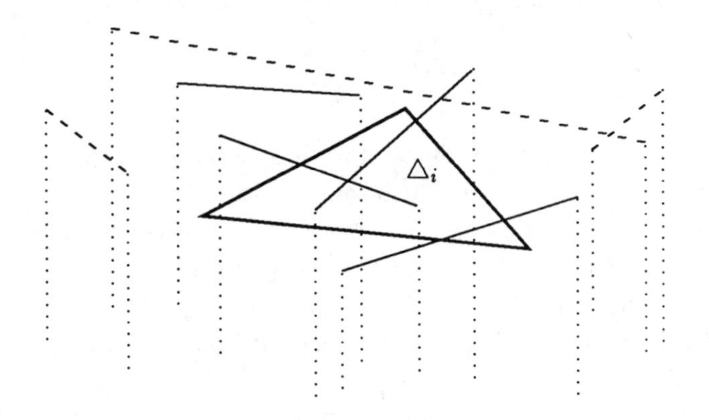

Figure 6.15: Triangle Δ_i and segments of $\mathcal{G}$: Solid lines are $\mathcal{G}_i$; dashed lines are $\mathcal{G}'_i$

Let Γ denote the set of lines bounding the trapezoidal strips B_e, that is, the lines containing the segments of $\mathcal{G}$ and the vertical lines passing through the endpoints of segments in $\mathcal{G}$. Partition the plane into $M = O(r^2)$ triangles $\Delta_1, \ldots, \Delta_M$, each meeting at most $\frac{n}{r}$ lines of Γ (cf. Theorem 4.36). With each Δ_i we associate a set $\mathcal{G}_i$ consisting of all segments $e \in \mathcal{G}$ such that either e or one of the two downward-directed vertical rays emanating from its endpoints intersects Δ_i (see figure 6.15). Let $\mathcal{G}'_i = \mathcal{G} - \mathcal{G}_i$. We can compute $\mathcal{G}_i$, for each i, in total time $O(nr \log n)$. Since Δ_i does not intersect the boundary of B_e for $e \in \mathcal{G}'_i$, ψ_e remains constant over Δ_i. Moreover, $\sum_{e \in \mathcal{G}'_i} \psi_e$, for every Δ_i, can be computed in $O(nr)$ time, as described in Section 5.10. We preprocess $\mathcal{G}_i$ into a data structure of size $O(\frac{n}{r} \log^2 \frac{n}{r})$, using the method just mentioned. For answering a query, we first locate the triangle Δ_k containing the query point p. Once we know Δ_k, $\sum_{e \in \mathcal{G}'_k} \psi_e(p)$ can be determined in $O(1)$ time, and $\sum_{e \in \mathcal{G}_k} \psi_e(p)$ can be computed as described above. Since $|\mathcal{G}_i| = O(\frac{n}{r})$, the query time is

$$Q(n) = O\left(\sqrt{\frac{n}{r}} \log \frac{n}{r} + \log n\right).$$

We need $O(r^2)$ space to store the planar map formed by $\Delta_1, \ldots, \Delta_M$ and $O(\frac{n}{r}\log^2\frac{n}{r})$ to store the data structure constructed for each $\mathcal{G}_i$. Therefore, the total space used is

$$\begin{aligned} S(n) &= O(r^2) + O\left(r^2 \cdot \frac{n}{r}\log^2\frac{n}{r}\right) \\ &= O(nr\log^2\frac{n}{r}). \end{aligned}$$

If we choose $r = \dfrac{m}{n\log^2\frac{n}{\sqrt{m}}}$, which is easily seen to satisfy $1 \le r < n$, then $S(n) = O(m)$, and the query time is

$$\begin{aligned} Q(n) &= O\left(\sqrt{\frac{n}{m/(n\log^2\frac{n}{\sqrt{m}})}}\log\frac{n}{\sqrt{m}} + \log n\right) \\ &= O\left(\frac{n}{\sqrt{m}}\log^2\frac{n}{\sqrt{m}} + \log n\right). \end{aligned}$$

Finally, the preprocessing time is

$$\begin{aligned} P(n) &= O(nr\log n\log^{\tau-1} r) + O\left(r^2 \cdot \left(\frac{n}{r}\right)^{3/2}\log^{\tau}\frac{n}{r}\right) \\ &= O\left(m\log^{\tau} n + n^{3/2}\sqrt{\frac{m}{n\log^2\frac{n}{\sqrt{m}}}}\log^{\tau}\frac{n}{\sqrt{m}}\right) \\ &= O(m\log^{\tau} n + n\sqrt{m}\log^{\tau-1}\frac{n}{\sqrt{m}}). \end{aligned}$$

Hence, we can conclude

Theorem 6.34 *Given a collection* $\mathbf{T}$ *of* n *(possibly intersecting) triangles in the plane, we can preprocess* $\mathbf{T}$*, in time* $O(m\log^{\tau} n + n\sqrt{m}\log^{\tau-1}\frac{n}{\sqrt{m}})$*, into a data structure of size* $O(m)$ *so that, for a query point* p*, we can count the number of triangles of* $\mathbf{T}$ *containing* p *in time* $O(\frac{n}{\sqrt{m}}\log^2\frac{n}{\sqrt{m}} + \log n)$*.*

□

Remark 6.35: In Section 5.7, we have described a deterministic algorithm to solve the batched version of this problem, when all points are given in advance, using a different technique. If m is the number of query points, then it runs in time $O(m^{2/3}n^{2/3}\log^{2/3} n\log^{\tau/3}\frac{n}{\sqrt{m}} + (m+n)\log n)$.

6.8.2 Implicit hidden surface removal — preprocessing version

Next we consider the preprocessing version of the implicit hidden surface removal, defined in Section 5.7. Formally, it can be stated as follows:

> *Let* $\mathbf{T} = \{\triangle_1, \ldots, \triangle_n\}$ *be a collection of* n *non-intersecting triangles such that* $\triangle_i$ *lies in the plane* $z = c_i$*, where* $0 < c_1 \le c_2 \le \cdots \le c_n$ *are some fixed heights. Preprocess* $\mathbf{T}$ *so that given a query point* p *on the* xy*-plane, we can determine the lowest triangle* $\triangle_i$ *hit by the upward-directed vertical ray emanating from* p.

Guibas et al. [73] have given an algorithm for this problem that uses randomized processing and has $O(n^{2/3+\delta})$ query time, for any $\delta > 0$. Their algorithm first projects all triangles on the xy-plane, and then performs a binary search through the sequence $(\triangle_1^*, \ldots, \triangle_n^*)$ of projected triangles to find the first index j such that $\triangle_j^*$ contains the query point p. Each step of the binary search tests whether p lies in the union of some contiguous block of projected triangles, using the polygon containment algorithm. Therefore the preprocessing step consists of constructing a binary tree $\mathcal{Z}$ on $\mathbf{T}$ whose leaves store the triangles of $\mathbf{T}$ in increasing height, and each internal node w is associated with a set of triangles $\mathbf{T}_w$, stored at the leaves of the subtree rooted at w. For each node w of $\mathcal{Z}$, preprocess $\mathbf{T}_w$ for the polygon containment problem, using the algorithm described in Section 6.8.1. It now follows from the above discussion that a query can be answered by following a path π in $\mathcal{Z}$ and solving the polygon containment problem at each node of π. Hence using Theorem 6.33, we can conclude

Theorem 6.36 *The implicit hidden surface removal problem for an ordered collection of* n *triangles in three-dimensional space can be solved in* $O(\sqrt{n}\log^2 n)$ *query time,* $O(n\log^3 n)$ *space and* $O(n^{3/2}\log^\tau n)$ *preprocessing.*

□

Remark 6.37:

(i) Recently several algorithms for other variants of the implicit hidden surface removal problem have been developed; see [112], [12].

(ii) As in the case of the polygon containment problem, the query time can be improved if we allow more space. In particular, if we allow $O(m)$ space, where $n < m < n^2$, then $Q(n) = O(\frac{n}{\sqrt{m}} \log^2 \frac{n}{\sqrt{m}} + \log^2 n)$ and $P(n) = O(m \log^\tau n + n\sqrt{m} \log^{\tau-1} \frac{n}{\sqrt{m}})$.

(iii) We can easily modify this algorithm without affecting its complexity so that the query point p can lie anywhere in $\mathbb{R}^3$, instead of the xy–plane. We leave it for the reader to verify the details.

6.8.3 Polygon placement problem

Finally consider the following problem:

> *Let P be a k-gon (not necessarily simple) and let $\Delta = \{\triangle_1, \ldots, \triangle_n\}$ be a set of n (possibly intersecting) triangles. Preprocess them so that, given a (translated) placement of P, we can quickly determine whether P intersects any of the obstacles at that placement.*

Such a situation arises in several applications [17]. A special case, in which P is convex and the triangles are non-intersecting, has been widely studied (see e.g. [13], [33], [68], [84]). But the best known solution for the general case is by Guibas et al. [73], who have given an algorithm with randomized preprocessing and $O((kn)^{2/3+\delta})$ query time, for any $\delta > 0$, by reducing this problem to the polygon containment problem. Using their technique, and applying Theorem 6.33, we can easily obtain

Theorem 6.38 *We can preprocess Δ and P, in $O((kn)^{3/2} \log^\tau kn)$ time, into a data structure of size $O(kn \log^2 kn)$ so that, given a translated placement of P, we can determine in time $O(\sqrt{kn} \log kn)$, whether P collides with the obstacles at that placement.*

□

Remark 6.39: The tradeoff between space and query time described in Section 6.8.1 works here as well. Therefore, if we allow $O(m)$ space, where $n < m < n^2$, then $Q(n) = O(\frac{kn}{\sqrt{m}} \log^2 \frac{kn}{\sqrt{m}} + \log kn)$ and $P(n) = O(m \log^\tau kn + kn\sqrt{m} \log^{\tau-1} kn)$.

6.9 Discussion and Open Problems

In this chapter we presented efficient algorithms for various problems involving collections of segments in the plane, using spanning trees with low stabbing number. Recently Matoušek [92] gave an $O(n^{3/2} \log^2 n)$ algorithm to compute a single spanning tree of n points with $O(\sqrt{n})$ stabbing number, which improves the space complexity of these algorithms by a logarithmic factor and the preprocessing time by a polylog factor. Although these algorithms are significantly faster than previously known algorithms, several interesting questions still remain unanswered:

(i) The most challenging open problem is to give non-trivial lower bounds for the ray shooting and the implicit point location problems. Recently Chazelle [20] showed that if we allow only $O(n)$ space, then a simplex range query (i.e. counting the number of points of a given set contained in a query triangle) requires $\Omega(\sqrt{n})$ time. We conjecture that similar lower bounds hold for these problems as well.

(ii) A weakness of our algorithms is that, unlike the algorithms of [73], the preprocessing time is not close to linear. The most expensive step in the preprocessing of our algorithms is the construction of spanning trees with low stabbing number, which runs in $O(n^{3/2} \log^\tau n)$ time. Thus an intriguing open problem is to come up with a close-to-linear algorithm for constructing such spanning trees. In fact, a deterministic algorithm with $O(n^{4/3})$ time complexity will also be quite interesting (recall that Matoušek's randomized algorithm runs in roughly $n^{4/3}$ (expected) time).

(iii) We have shown in Section 5.7 that in the batched version of the implicit point location problem (i.e. when all query points are given in advance) the average cost of computing $\Psi(p)$, over all given query points P, can be made roughly $n^{1/3}$ per point while still using close-to-linear space. It is not known whether we can obtain similar bounds for the batched version of the ray shooting problem. A more challenging problem is: Can we answer a ray shooting query in $o(\sqrt{n})$ time without increasing storage significantly, if we know the lines containing the query rays (but not their origins) in advance?

(iv) Mark Overmars has asked the following question, which is a generalization of the polygon containment problem: *Given a set* T *of triangles, preprocess them so that, for a query segment* e, *one can quickly determine if* e *is contained in the union of triangles of* T. It will be interesting to come up with an efficient algorithm using spanning trees of low stabbing number.

(v) Finally, there remains the task of looking for other interesting problems that can be solved efficiently using the spanning trees of low stabbing number.

Bibliography

[1] W. Ackermann, Zum Hilbertschen Aufbau der reellen Zahlen, *Mathematical Annals* 99 (1928), 118–133.

[2] M. Ajtai, J. Komlos and E. Szemerédi, Sorting in $c \log n$ parallel steps, *Combinatorica* 3 (1983), 1–19.

[3] P. Alevizos, J. Boissonnat and F. Preparata, On the boundary of a union of rays, *Proceedings 6th Annual Symposium on Theoretical Aspects of Computer Science*, Lecture Notes in Computer Science, Vol. 349, Springer Verlag, 1989, pp. 72–83.

[4] B. Aronov, H. Edelsbrunner, L. Guibas and M. Sharir, Improved bounds on the complexity of many faces in arrangements of segments, Technical Report 459, Dept. Computer Science, New York University, July 1989.

[5] B. Aronov and M. Sharir, Triangles in space or building (and analyzing) castles in the air, *Proceedings 4th Annual Symposium on Computational Geometry*, 1988, pp. 381–391. (Also to appear in *Combinatorica.*)

[6] M. Atallah, Some dynamic computational geometry problems, *Computers and Mathematics with Applications* 11 (1985), 1171–1181.

[7] F. Aurenhammer, Voronoi diagrams — A survey, manuscript, 1988.

[8] C. Bajaj and M. Kim, Compliant motion planning with geometric models, *Proceedings 3rd Annual Symposium on Computational Geometry*, 1987, pp. 171–180.

[9] A. Balstan and M. Sharir, On shortest paths between two convex polyhedra, *J. ACM* 35 (1988), 267–287.

[10] K. E. Batcher, Sorting networks and their applications, *Proceedings AFIPS Spring Joint Summer Computer Conference* 32 (1968), pp. 307–314.

[11] J. Bentley and M. Ottmann, Algorithms for reporting and counting intersections, *IEEE Transactions on Computers* C-28 (1979), 643–647.

[12] M. Bern, Hidden surface removal for rectangles, *J. Computer and Systems Sciences* 40 (1990), 49–69.

[13] B. Bhattacharya and J. Zorbas, Solving the two-dimensional findpath problem using a line-triangle representation of the robot, *J. Algorithms* 9 (1988), 449–469.

[14] K. Q. Brown, Geometric transforms for fast geometric algorithms, PhD Thesis, Carnegie–Mellon University, Pittsburgh, Pennsylvania, 1980.

[15] K. Brown, Algorithms for reporting and counting geometric intersections, *IEEE Transactions on Computers* 30 (1981), 147–148.

[16] R. Canham, A theorem on arrangements of lines in the plane, *Israel J. Math.* 7 (1969), 393–397.

[17] B. Chazelle, The polygon containment problem, in *Advances in Computing Research, Vol. I: Computational Geometry*, (F. P. Preparata, Ed.), JAI Press, Greenwich, Connecticut, 1983, 1–33.

[18] B. Chazelle, Fast searching in a real algebraic manifold with applications to geometric complexity, *Proceedings CAAP'85*, Lecture Notes in Computer Science, Springer-Verlag, Berlin 1985, pp. 145–156.

[19] B. Chazelle, Reporting and counting segment intersections, *J. Computer and Systems Sciences* 32 (1986), 156–182.

[20] B. Chazelle, Lower bounds on the complexity of polytope range searching, *J. American Mathematical Society* 2 (1989), 637–666.

[21] B. Chazelle, Tight bounds on the stabbing number of trees in Euclidean plane, Technical Report CS-TR-155-58, Dept. Computer Science, Princeton University, May 1988.

[22] B. Chazelle, Private Communication.

[23] B. Chazelle and H. Edelsbrunner, An optimal algorithm for intersecting line segments in the plane, *Proceedings* 29th *Annual IEEE Symposium on Foundations of Computer Science*, 1988, pp. 590–600.

[24] B. Chazelle, H. Edelsbrunner, L. Guibas and M. Sharir, Algorithms for bichromatic line segment problems and polyhedral terrains, manuscript, 1989.

[25] B. Chazelle, H. Edelsbrunner, L. Guibas and M. Sharir, Lines in space: Combinatorics, algorithms and applications, *Proceedings* 21st *Annual ACM Symposium on Theory of Computing*, 1989.

[26] B. Chazelle and J. Friedman, A deterministic view of random sampling and its use in geometry, *Proceedings* 29th *Annual IEEE Symposium on Foundations of Computer Science*, 1988, pp. 539–549.

[27] B. Chazelle and L. Guibas, Visibility and intersection problems in plane geometry, *Discrete and Computational Geometry* 4 (1989), 551–581.

[28] B. Chazelle and L. Guibas, Fractional Cascading: I. A data structuring technique, *Algorithmica* 1 (1986), 133–162.

[29] B. Chazelle and L. Guibas. Fractional cascading: II. Applications, *Algorithmica* 1 (1986), 163–191.

[30] B. Chazelle, L. Guibas and D. T. Lee, The power of geometric duality, *BIT* 25 (1985), 76–90.

[31] B. Chazelle and M. Sharir, An algorithm for generalized point location and its application, to appear in *J. Symbolic Computation*, 1989.

[32] B. Chazelle and E. Welzl, Quasi optimal range searching in spaces with finite VC-dimension, *Discrete and Computational Geometry*4 (1989), 467–489.

[33] P. Chew and L. Drysdale, Voronoi diagrams based on convex distance functions, *Proceedings* 1st *Annual Symposium on Computational Geometry*, 1985, pp. 235–244.

[34] K. Clarkson, A probabilistic algorithm for the post office problem, *Proceedings* 17th *Annual ACM Symposium on Theory of Computing*, 1985, pp. 75–84.

[35] K. Clarkson, Approximate algorithms for shortest path motion planning, *Proceedings* 19th *Annual ACM Symposium on Theory of Computing*, 1987, pp. 56–65.

[36] K. Clarkson, New applications of random sampling in computational geometry, *Discrete and Computational Geometry* 2 (1987), 195–222.

[37] K. Clarkson, New applications of random sampling in computational geometry II, *Proceedings* 4th *Annual Symposium on Computational Geometry*, 1988, pp. 1–11.

[38] K. Clarkson, H. Edelsbrunner, L. Guibas, M. Sharir and E. Welzl, Combinatorial complexity bounds for arrangements of curves and surfaces, *Discrete and Computational Geometry* 5 (1990), 99–162.

[39] K. Clarkson and P. Shor, Algorithms for diametral pairs and convex hulls that are optimal, randomized and incremental, *Proceedings* 4th *Annual Symposium on Computational Geometry*, 1988, 12–17.

[40] K. Clarkson, R. E. Tarjan and C. J. Van Wyk, A fast Las Vegas algorithm for triangulating a simple polygon, *Discrete and Computational Geometry* 4 (1989), 423–432.

[41] R. Cole, Slowing down sorting networks to obtain faster sorting algorithms, *J. ACM* 31 (1984), 200–208.

[42] R. Cole, Searching and storing similar lists, *J. Algorithms* 7 (1986), 202–220.

[43] R. Cole, J. Salowe, W. Steiger and E. Szemerédi, Optimal slope selection, *SIAM J. Computing*, 18 (1989), 792–810.

[44] R. Cole and M. Sharir, Visibility problems for polyhedral terrains, *J. Symbolic Computation* 7 (1989), 11–30.

[45] R. Cole, M. Sharir and C. K. Yap, On k-hulls and related problems, *SIAM J. Computing* 16 (1987), 61–77.

[46] R. Courant, *Differential and integral calculus vol. II*, Interscience Publishers, Inc. New York, 1936.

[47] H. Davenport, A combinatorial problem connected with differential equations II, *Acta Arithmetica* 17 (1971), 363–372.

[48] H. Davenport and A. Schinzel, A combinatorial problem connected with differential equations, *Amer. J. Math.* 87 (1965), 684–689.

[49] F. Dévai, Quadratic bounds for hidden line elimination, *Proceedings* 2nd *Annual Symposium on Computational Geometry*, 1986, pp. 269–275.

[50] D. Dobkin and H. Edelsbrunner, Space searching for intersecting objects, *J. Algorithms* 8 (1987), 348–361.

[51] J. Driscoll, N. Sarnak, D. Sleator and R. Tarjan, Making data structures persistent, *J. Computer and Systems Sciences* 38 (1989), 86–124.

[52] H. Edelsbrunner, *Algorithms in Combinatorial Geometry*, Springer–Verlag, Heidelberg, 1987.

[53] H. Edelsbrunner, The upper envelope of piecewise linear functions; tight complexity bounds in higher dimensions, *Discrete and Computational Geometry*4 (1989), 337–343.

[54] H. Edelsbrunner and L. Guibas, Topologically sweeping an arrangement, *J. Computer and Systems Sciences* 38 (1989), 165–194.

[55] H. Edelsbrunner, L. Guibas, J. Hershberger, R. Seidel, M. Sharir, J. Snoeyink and E. Welzl, Implicitly representing arrangements of lines or segments, *Discrete and Computational Geometry* 4 (1989), 433–466.

[56] H. Edelsbrunner, L. Guibas, J. Pach, R. Pollack, R. Seidel and M. Sharir, Arrangements of curves in the plane — topology, combinatorics, and algorithms, *Proceedings* 15th *International Colloquium on Automata, Languages and Programming*, Lecture Notes in Computer Science, Vol. 318, Springer-Verlag, New York, 1988, pp. 214–229.

[57] H. Edelsbrunner, L. Guibas and M. Sharir, The upper envelope of piecewise linear functions: Algorithms and applications, *Discrete and Computational Geometry* 4 (1989), 311–336.

[58] H. Edelsbrunner, L. Guibas and M. Sharir, The complexity of many faces in arrangement of lines and of segments, *Discrete and Computational Geometry* 5 (1990), 161–196.

[59] H. Edelsbrunner, L. Guibas and M. Sharir, The complexity of many cells in arrangements of planes and related problems, *Discrete and Computational Geometry* 5 (1990), 197–216.

[60] H. Edelsbrunner, L. Guibas and G. Stolfi, Optimal point location in monotone subdivisions, *SIAM J. Computing* 15 (1986), 317–340.

[61] H. Edelsbrunner, H. Maurer, F. Preparata, A. Rosenberg, E. Welzl and D. Wood, Stabbing line segments, *BIT* 22 (1982), 274–281.

[62] H. Edelsbrunner, J. O'Rourke and R. Seidel, Constructing arrangements of lines and hyperplanes with applications, *SIAM J. Computing* 15 (1986), 341–363.

[63] H. Edelsbrunner and R. Seidel, Voronoi diagrams and arrangements, *Discrete and Computational Geometry* 1 (1986), 25–44.

[64] H. Edelsbrunner and M. Sharir, The maximum number of ways to stab n convex non-intersecting objects in the plane is $2n-2$, *Discrete and Computational Geometry* 5 (1990), 35–42.

[65] H. Edelsbrunner and E. Welzl, Constructing belts in two-dimensional arrangements with applications, *SIAM J. Computing* 15 (1986), 271–284.

[66] H. Edelsbrunner and E. Welzl, On the maximal number of edges of many faces in an arrangement, *Journal of Combinatorial Theory, Series A* 41 (1986), 159–166.

[67] H. Edelsbrunner and E. Welzl, Halfplanar range search in linear space and $O(n^{0.695})$ query time, *Information Processing Letters* 23 (1986), 289–293.

[68] S. Fortune, Fast algorithms for polygon containment, *Proceedings* 12th *International Colloquium on Automata, Languages and Programming*, Lecture Notes in Computer Science, 194, Springer Verlag, New York, 1985, pp. 189–198.

[69] B. Grünbaum, *Convex polytopes*, Interscience, London, 1967.

[70] B. Grünbaum, Arrangements and hyperplanes, *Congressum Numerantium III, Louisiana Conference on Combinatorics Graph Theory and Computing*, 1971, pp. 41–106.

[71] B. Grünbaum, *Arrangements and spreads*, CBMS Regional Conference Series in Applied Mathematics 10, American Mathematical Society, Providence, 1972.

[72] L. Guibas, J. Hershberger, D. Leven, M. Sharir and R. Tarjan, Linear time algorithms for shortest path and visibility problems, *Algorithmica* 2 (1987) 209–233.

[73] L. Guibas, M. Overmars and M. Sharir, Ray shooting, implicit point location, and related queries in arrangements of segments, Technical Report 433, Dept. Computer Science, New York University, March 1989.

[74] L. Guibas, M. Overmars and M. Sharir, Counting and reporting intersections in arrangements of line segments, Technical Report 434, Dept. Computer Science, New York University, March 1989.

[75] L. Guibas, M. Sharir and S. Sifrony, On the general motion planning problem with two degrees of freedom, *Discrete and Computational Geometry* 4 (1989), 491–521.

[76] L. Guibas and F. Yao, On translating a set of rectangles, in *Advances in Computer Research, Vol. I: Computational Geometry* (F. P. Preparata, Ed.), JAI Press, 1983, 61–77.

[77] S. Hart and M. Sharir, Nonlinearity of Davenport-Schinzel sequences and of generalized path compression schemes, *Combinatorica* 6 (1986), 151–177.

[78] D. Haussler and E. Welzl, ϵ-nets and simplex range queries, *Discrete and Computational Geometry* 2 (1987), 127–151.

[79] J. Jaromczyk and M. Kowaluk, Skewed projections with an application to line stabbing in $\mathbb{R}^3$, *Proceedings* 4th *Annual Symposium on Computational Geometry*, 1988, pp. 362–370.

[80] K. Kedem, R. Livne, J. Pach and M. Sharir, On the union of Jordan regions and collision-free translational motion amidst polygonal obstacles, *Discrete and Computational Geometry* 1 (1986), 59–71.

[81] K. Kedem and M. Sharir, An efficient motion planning algorithm for a convex polygonal object in two-dimensional polygonal space, *Discrete and Computational Geometry* 5 (1990), 43–75.

[82] D. Kirkpatrick, Optimal search in planar subdivisions, *SIAM J. Computing* 12 (1983), 28–35.

[83] D. Leven and M. Sharir, Planning a purely translational motion for a convex object in two-dimensional space using generalized Voronoi Diagrams, *Discrete and Computational Geometry* 2 (1987), 9–31.

[84] D. Leven and M. Sharir, On the number of critical free contacts of a convex polygonal objects moving in 2-D polygonal space, *Discrete and Computational Geometry* 2 (1987), 255–270.

[85] F. Levi, Die Teilung der projektiven Ebene durch Gerade oder Pseudogerade, *Ber. Math. Phys. Kl. Sächs. Akad. Wiss. Leipzig* 78 (1926), 256–267.

[86] E. Lockwood, *A Book of Curves*, Cambridge University Press, Cambridge, 1961.

[87] S. Maddila and C. Yap, Moving a polygon around the corner in a corridor, *Proceedings* 2nd *Annual Symposium on Computational Geometry*, 1986, pp. 187–192.

[88] H. Mairson and J. Stolfi, Reporting and counting intersections between two sets of line segments, *Theoretical Foundations of Computer Graphics and CAD*, (R. Earnshaw, Ed.), NATO ASI Series, F40, Springer Verlag, Berlin, 1988, 307–325.

[89] J. Matoušek, Construction of ϵ-nets, *Proceedings* 5th *Annual Symposium on Computational Geometry*, 1989, pp. 1–10. (Also to appear in *Discrete and Computational Geometry.*)

[90] J. Matoušek, Constructing spanning trees with low crossing numbers, to appear in *Informatique Theoretique et Applications.*

[91] J. Matoušek, Cutting hyperplane arrangements, *Proceedings* 6th *Annual Symposium on Computational Geometry*, 1990, pp. 1–9.

[92] J. Matoušek, More on cutting arrangements and spanning trees with low stabbing number, Technical Report B-90-2, Fachbereich Mathematik, Freie Universitaät, February 1990.

[93] J. Matoušek and E. Welzl, Good splitters for counting points in triangles, *Proceedings* 5th *Annual Symposium on Computational Geometry*, 1989, pp. 124–130.

[94] M. McKenna, Worst case optimal hidden surface removal, *ACM Transactions on Graphics* 6 (1987), 9–31.

[95] M. McKenna and J. O'Rourke, Arrangements of lines in 3-space: A data structure with applications, *Proceedings* 5th *Annual Symposium on Computational Geometry*, 1988, pp. 371–380.

[96] N. Megiddo, Applying parallel computation algorithms in design of serial algorithms, *J. ACM* 30 (1983), 852–865.

[97] E. Melchior, Uber Vielseite der projektiven Ebene, *Deutsche Mathematik* 5 (1940), 461–475.

[98] K. Mulmuley, A fast planar partition algorithm, I, *Proceedings* 29th *Annual IEEE Symposium on Foundations of Computer Science*, 1988, pp. 580–589.

[99] K. Mulmuley, A fast planar partition algorithm, II, *Proceedings* 5th *Annual Symposium on Computational Geometry*, 1989, pp. 33–43.

[100] C. Ó'Dúnlaing and C. Yap, A "retraction" method for planning the motion of a disk, *J. Algorithms* 6 (1985), 104–111.

[101] C. Ó'Dúnlaing , M. Sharir and C. Yap, Generalized Voronoi diagrams for a ladder: II. Efficient construction of the diagram, *Algorithmica* 2 (1987), 27–59.

[102] M. Overmars and M. Sharir, Output sensitive Hidden surface removal algorithms, *Proceedings* 30th *Annual IEEE Symposium on Foundations of Computer Science*, 1990, pp. 598–603.

[103] J. Pach and M. Sharir, The upper envelope of piecewise linear functions and the boundary of a region enclosed by convex planes, to appear in *Discrete and Computational Geometry* 4 (1989), 291–309.

[104] J. Pach and G. Woeginger, Some new bounds for epsilon-nets, *Proceedings* 6th *Annual Symposium on Computational Geometry*, 1990, pp. 10–15.

[105] R. Pollack, M. Sharir and S. Sifrony, Separating two simple polygons by a sequence of translations, *Discrete and Computational Geometry* 3 (1988), 123–136.

[106] F. Preparata, A note on locating a set of points in a planar subdivision, *SIAM J. Computing* 8 (1979), 542–545.

[107] F. Preparata and M. Shamos, *Computational Geometry: An Introduction*, Springer Verlag, Heidelberg, 1985.

[108] J. Reif and S. Sen, Optimal randomized parallel algorithms for computational geometry, *Proceedings* 16th *International Conference on Parallel Processing*, 1987.

[109] J. Reif and S. Sen, Polling: A new randomized sampling technique for computational geometry, *Proceedings* 21st *Annual ACM Symposium on Theory of Computing*, 1989.

[110] D. P. Roselle and R. G. Stanton, Some properties of Davenport-Schinzel sequences, *Acta Arithmetica* 17 (1971), 355–362.

[111] N. Sarnak and R. Tarjan, Planar point location using persistent search trees, *Communications ACM* 29 (1986), 669–679.

[112] A. Schmitt, H. Muller and W. Leister, Ray tracing algorithm — Theory and practice, in *Theoretical Foundations of Computer Graphics and CAD*, (R. Earnshaw, Ed.) NATO ASI Series, Vol. F-40, Springer-Verlarg, New York, 1988, 997–1030.

[113] J. Schwartz and M. Sharir, On the two dimensional Davenport-Schinzel problem, to appear in *J. Symbolic Computation.*

[114] M. Shamos, Computational Geometry, PhD Thesis, Yale University, New Haven, 1978.

[115] M. Shamos and D. Hoey, Geometric intersection problems, *Proceedings* 17th *Annual IEEE Symposium on Foundations of Computer Science*, 1976, pp. 208–215.

[116] M. Sharir, Intersection and closest-pair problems for a set of planar discs, *SIAM J. Computing* 14 (2), 1985, 448–468.

[117] M. Sharir, Almost linear upper bounds on the length of general Davenport-Schinzel sequences, *Combinatorica* 7 (1987), 131–143.

[118] M. Sharir, Improved lower bounds on the length of Davenport-Schinzel sequences, *Combinatorica* 8 (1988), 117–124.

[119] M. Sharir, Davenport-Schinzel sequences and their geometric applications, *Theoretical Foundations of Computer Graphics and CAD*, (R. Earnshaw, Ed.), NATO ASI Series, F40, Springer Verlag, Berlin, 1988, 253–278.

[120] M. Sharir, R. Cole, K. Kedem, D. Leven, R. Pollack and S. Sifrony, Geometric applications of Davenport-Schinzel sequences, *Proceedings* 27th *Annual IEEE Symposium on Foundations of Computer Science*, 1986, 77–86.

[121] P. Shor, Geometric realization of superlinear Davenport-Schinzel sequences, manuscript, 1988.

[122] G. K. von Staudt, *Geometrie der Lage*, Nürenberg, 1847.

[123] J. Steiner, Einige Gesetze über die Theilung der Ebene und des Raumes, *J. Reine Angew. Math.* 1 (1826), 349–364.

[124] S. Suri, A linear algorithm for minimum link paths inside a simple polygon, *Computer Vision, Graphics and Image Processing* 35 (1986), 99–110.

[125] S. Suri and J. O'Rourke, Worst case optimal algorithms for constructing visibility polygons with holes, *Proceedings* 2nd *Annual Symposium on Computational Geometry*, 1986, pp. 14–23.

[126] J. Sylvester, Problem 2473, *Mathematical Questions and Solutions from the Euclidean Times* 8 (1867), 104–107.

[127] E. Szemerédi, On a problem by Davenport and Schinzel, *Acta Arithmetica* 25 (1974) 213–224.

[128] E. Szemerédi and W. Trotter Jr., Extremal problems in discrete geometry, *Combinatorica*, 3 (1983), 381-392.

[129] R. E. Tarjan, *Data Structure and Network Algorithms*, SIAM Publications, Pennsylvania, 1983.

[130] B. van der Waerden, *Algebra, vol. 2,* Fredrick Unger Publishing co., New York, 1970.

[131] E. Welzl, More on k-sets of finite sets in the plane, *Discrete and Computational Geometry* 1 (1986), 83–94.

[132] E. Welzl, Partition trees for triangle counting and other range searching problems, *Proceedings* 4th *Annual Symposium on Computational Geometry*, 1988, pp. 23–33.

[133] A. Wiernik and M. Sharir, Planar realizations of nonlinear Davenport-Schinzel sequences by segments, *Discrete and Computational Geometry*, 3 (1988), 15–47.

[134] G. Woeginger, Epsilon-nets for half planes, Technical Report B-88-02, Dept. of Mathematics, Free University, Berlin, March 1988.

[135] C. K. Yap, An $O(n \log n)$ algorithm for the Voronoi diagram of a set of simple curve segments, *Discrete and Computational Geometry* 2 (1987), 365–393.

Index of Symbols

Index of Keywords